KK

Texts and Monographs in
Symbolic Computation

A Series of the Research Institute for Symbolic Computation,
Johannes-Kepler-University, Linz, Austria
Edited by B. Buchberger and G. E. Collins

Bernd Sturmfels

Algorithms in Invariant Theory

Springer-Verlag Wien New York

Dr. Bernd Sturmfels
Department of Mathematics
Cornell University, Ithaca, New York, U.S.A.

This work is subject to copyright.
All rights are reserved, whether the whole or part of the material is concerned, specifically those of translation, reprinting, re-use of illustrations, broadcasting, reproduction by photo-copying machines or similar means, and storage in data banks.
© 1993 Springer-Verlag/Wien
Printed in Austria

Data conversion by H.-D. Ecker, Büro für Textverarbeitung, Bonn
Printed by Novographic, Ing. Wolfgang Schmid, A-1230 Wien
Printed on acid-free paper

With 5 Figures

ISSN 0943-853X
ISBN 3-211-82445-6 Springer-Verlag Wien New York
ISBN 0-387-82445-6 Springer-Verlag New York Wien

Preface

The aim of this monograph is to provide an introduction to some fundamental problems, results and algorithms of invariant theory. The focus will be on the three following aspects:

(i)   *Algebraic algorithms* in invariant theory, in particular algorithms arising from the theory of Gröbner bases;
(ii) *Combinatorial algorithms* in invariant theory, such as the straightening algorithm, which relate to representation theory of the general linear group;
(iii) *Applications* to projective geometry.

Part of this material was covered in a graduate course which I taught at RISC-Linz in the spring of 1989 and at Cornell University in the fall of 1989. The specific selection of topics has been determined by my personal taste and my belief that many interesting connections between invariant theory and symbolic computation are yet to be explored.

In order to get started with her/his own explorations, the reader will find exercises at the end of each section. The exercises vary in difficulty. Some of them are easy and straightforward, while others are more difficult, and might in fact lead to research projects. Exercises which I consider "more difficult" are marked with a star.

This book is intended for a diverse audience: graduate students who wish to learn the subject from scratch, researchers in the various fields of application who want to concentrate on certain aspects of the theory, specialists who need a reference on the algorithmic side of their field, and all others between these extremes. The overwhelming majority of the results in this book are well known, with many theorems dating back to the 19th century. Some of the algorithms, however, are new and not published elsewhere.

I am grateful to B. Buchberger, D. Eisenbud, L. Grove, D. Kapur, Y. Lakshman, A. Logar, B. Mourrain, V. Reiner, S. Sundaram, R. Stanley, A. Zelevinsky, G. Ziegler and numerous others who supplied comments on various versions of the manuscript. Special thanks go to N. White for introducing me to the beautiful subject of invariant theory, and for collaborating with me on the topics in Chapters 2 and 3. I am grateful to the following institutions for their support: the Austrian Science Foundation (FWF), the U.S. Army Research Office (through MSI Cornell), the National Science Foundation, the Alfred P. Sloan Foundation, and the Mittag-Leffler Institute (Stockholm).

Ithaca, June 1993   Bernd Sturmfels

Contents

1 Introduction 1
1.1 Symmetric polynomials 2
1.2 Gröbner bases 7
1.3 What is invariant theory? 14
1.4 Torus invariants and integer programming 19

2 Invariant theory of finite groups 25
2.1 Finiteness and degree bounds 25
2.2 Counting the number of invariants 29
2.3 The Cohen–Macaulay property 37
2.4 Reflection groups 44
2.5 Algorithms for computing fundamental invariants 50
2.6 Gröbner bases under finite group action 58
2.7 Abelian groups and permutation groups 64

3 Bracket algebra and projective geometry 77
3.1 The straightening algorithm 77
3.2 The first fundamental theorem 84
3.3 The Grassmann–Cayley algebra 94
3.4 Applications to projective geometry 100
3.5 Cayley factorization 110
3.6 Invariants and covariants of binary forms 117
3.7 Gordan's finiteness theorem 129

4 Invariants of the general linear group 137
4.1 Representation theory of the general linear group 137
4.2 Binary forms revisited 147
4.3 Cayley's Ω-process and Hilbert finiteness theorem 155
4.4 Invariants and covariants of forms 161
4.5 Lie algebra action and the symbolic method 169
4.6 Hilbert's algorithm 177
4.7 Degree bounds 185

References 191
Subject index 196

1 Introduction

Invariant theory is both a classical and a new area of mathematics. It played a central role in 19th century algebra and geometry, yet many of its techniques and algorithms were practically forgotten by the middle of the 20th century.

With the fields of combinatorics and computer science reviving old-fashioned algorithmic mathematics during the past twenty years, also classical invariant theory has come to a renaissance. We quote from the expository article of Kung and Rota (1984):

"Like the Arabian phoenix rising out of its ashes, the theory of invariants, pronounced dead at the turn of the century, is once again at the forefront of mathematics. During its long eclipse, the language of modern algebra was developed, a sharp tool now at last being applied to the very purpose for which it was invented."

This quote refers to the fact that three of Hilbert's fundamental contributions to modern algebra, namely the *Nullstellensatz*, the *Basis Theorem* and the *Syzygy Theorem* were first proved as lemmas in his invariant theory papers (Hilbert 1890, 1893). It is also noteworthy that, contrary to a common belief, Hilbert's main results in invariant theory yield an explicit finite algorithm for computing a fundamental set of invariants for all classical groups. We will discuss Hilbert's algorithm in Chap. 4.

Throughout this text we will take the complex numbers $\mathbf{C}$ to be our ground field. The ring of polynomials $f(x_1, x_2, \ldots, x_n)$ in n variables with complex coefficients is denoted $\mathbf{C}[x_1, x_2, \ldots, x_n]$. All algorithms in this book will be based upon arithmetic operations in the ground field only. This means that if the scalars in our input data are contained in some subfield $K \subset \mathbf{C}$, then all scalars in the output also lie in K. Suppose, for instance, we specify an algorithm whose input is a finite set of $n \times n$-matrices over $\mathbf{C}$, and whose output is a finite subset of $\mathbf{C}[x_1, x_2, \ldots, x_n]$. We will usually apply such an algorithm to a set of input matrices which have entries lying in the field $\mathbf{Q}$ of rational numbers. We can then be sure that all output polynomials will lie in $\mathbf{Q}[x_1, x_2, \ldots, x_n]$.

Chapter 1 starts out with a discussion of the ring of symmetric polynomials, which is the simplest instance of a ring of invariants. In Sect. 1.2 we recall some basics from the theory of Gröbner bases, and in Sect. 1.3 we give an elementary exposition of the fundamental problems in invariant theory. Section 1.4 is independent and can be skipped upon first reading. It deals with invariants of algebraic tori and their relation to integer programming. The results of Sect. 1.4 will be needed in Sect. 2.7 and in Chap. 4.

1.1. Symmetric polynomials

Our starting point is the fundamental theorem on symmetric polynomials. This is a basic result in algebra, and studying its proof will be useful to us in three ways. First, we illustrate some fundamental questions in invariant theory with their solution in the easiest case of the symmetric group. Secondly, the main theorem on symmetric polynomials is a crucial lemma for several theorems to follow, and finally, the algorithm underlying its proof teaches us some basic computer algebra techniques.

A polynomial $f \in \mathbf{C}[x_1, \ldots, x_n]$ is said to be *symmetric* if it is invariant under every permutation of the variables $x_1, x_2, \ldots, x_n$. For example, the polynomial $f_1 := x_1 x_2 + x_1 x_3$ is not symmetric because $f_1(x_1, x_2, x_3) \neq f_1(x_2, x_1, x_3) = x_1 x_2 + x_2 x_3$. On the other hand, $f_2 := x_1 x_2 + x_1 x_3 + x_2 x_3$ is symmetric.

Let z be a new variable, and consider the polynomial

$$g(z) = (z - x_1)(z - x_2) \ldots (z - x_n)$$
$$= z^n - \sigma_1 z^{n-1} + \sigma_2 z^{n-2} - \ldots + (-1)^n \sigma_n.$$

We observe that the coefficients of g with respect to the new variable z,

$$\sigma_1 = x_1 + x_2 + \ldots + x_n,$$
$$\sigma_2 = x_1 x_2 + x_1 x_3 + \ldots + x_2 x_3 + \ldots + x_{n-1} x_n,$$
$$\sigma_3 = x_1 x_2 x_3 + x_1 x_2 x_4 + \ldots + x_{n-2} x_{n-1} x_n,$$
$$\ldots \ldots \ldots \ldots \ldots \ldots \ldots$$
$$\sigma_n = x_1 x_2 x_3 \cdots x_n,$$

are symmetric in the old variables $x_1, x_2, \ldots, x_n$. The polynomials $\sigma_1, \sigma_2, \ldots, \sigma_n \in \mathbf{C}[x_1, x_2, \ldots, x_n]$ are called the *elementary symmetric polynomials*.

Since the property to be symmetric is preserved under addition and multiplication of polynomials, the symmetric polynomials form a subring of $\mathbf{C}[x_1, \ldots, x_n]$. This implies that every polynomial expression $p(\sigma_1, \sigma_2, \ldots, \sigma_n)$ in the elementary symmetric polynomials is symmetric in $\mathbf{C}[x_1, \ldots, x_n]$. For instance, the monomial $c \cdot \sigma_1^{\mu_1} \sigma_2^{\mu_2} \ldots \sigma_1^{\mu_1}$ in the elementary symmetric polynomials is symmetric and homogeneous of degree $\mu_1 + 2\mu_2 + \ldots + n\mu_n$ in the original variables $x_1, x_2, \ldots, x_n$.

Theorem 1.1.1 (Main theorem on symmetric polynomials). *Every symmetric polynomial f in $\mathbf{C}[x_1, \ldots, x_n]$ can be written uniquely as a polynomial*

$$f(x_1, x_2, \ldots, x_n) = p\big(\sigma_1(x_1, \ldots, x_n), \ldots, \sigma_n(x_1, \ldots, x_n)\big)$$

in the elementary symmetric polynomials.

Proof. The proof to be presented here follows the one in van der Waerden

1.1. Symmetric polynomials

(1971). Let $f \in \mathbf{C}[x_1, \ldots, x_n]$ be any symmetric polynomial. Then the following algorithm rewrites f uniquely as a polynomial in $\sigma_1, \ldots, \sigma_n$.

We sort the monomials in f using the *degree lexicographic order*, here denoted "$\prec$". In this order a monomial $x_1^{\alpha_1} \ldots x_n^{\alpha_n}$ is smaller than another monomial $x_1^{\beta_1} \ldots x_n^{\beta_n}$ if it has lower total degree (i.e., $\sum \alpha_i < \sum \beta_i$), or if they have the same total degree and the first non-vanishing difference $\alpha_i - \beta_i$ is negative.

For any monomial $x_1^{\alpha_1} \ldots x_n^{\alpha_n}$ occurring in the symmetric polynomial f also all its images $x_1^{\alpha_{\sigma 1}} \ldots x_n^{\alpha_{\sigma n}}$ under any permutation σ of the variables occur in f. This implies that the initial monomial $\text{init}(f) = c \cdot x_1^{\gamma_1} x_2^{\gamma_2} \ldots x_n^{\gamma_n}$ of f satisfies $\gamma_1 \geq \gamma_2 \geq \ldots \geq \gamma_n$. By definition, the *initial monomial* is the largest monomial with respect to the total order "$\prec$" which appears with a nonzero coefficient in f.

In our algorithm we now replace f by the new symmetric polynomial $\tilde{f} := f - c \cdot \sigma_1^{\gamma_1 - \gamma_2} \sigma_2^{\gamma_2 - \gamma_3} \cdots \sigma_{n-1}^{\gamma_{n-1} - \gamma_n} \sigma_n^{\gamma_n}$, we store the summand $c \cdot \sigma_1^{\gamma_1 - \gamma_2} \sigma_2^{\gamma_2 - \gamma_3} \cdots \sigma_{n-1}^{\gamma_{n-1} - \gamma_n} \sigma_n^{\gamma_n}$, and, if $\tilde{f}$ is non-zero, then we return to the beginning of the previous paragraph.

Why does this process terminate? By construction, the initial monomial of $c \cdot \sigma_1^{\gamma_1 - \gamma_2} \sigma_2^{\gamma_2 - \gamma_3} \cdots \sigma_{n-1}^{\gamma_{n-1} - \gamma_n} \sigma_n^{\gamma_n}$ equals $\text{init}(f)$. Hence in the difference defining $\tilde{f}$ the two initial monomials cancel, and we get $\text{init}(\tilde{f}) \prec \text{init}(f)$. The set of monomials m with $m \prec \text{init}(f)$ is finite because their degree is bounded. Hence the above rewriting algorithm must terminate because otherwise it would generate an infinite decreasing chain of monomials.

It remains to be seen that the representation of symmetric polynomials in terms of elementary symmetric polynomials is unique. In other words, we need to show that the elementary symmetric polynomials $\sigma_1, \ldots, \sigma_n$ are algebraically independent over $\mathbf{C}$.

Suppose on the contrary that there is a non-zero polynomial $p(y_1, \ldots, y_n)$ such that $p(\sigma_1, \ldots, \sigma_n) = 0$ in $\mathbf{C}[x_1, \ldots, x_n]$. Given any monomial $y_1^{\alpha_1} \cdots y_n^{\alpha_n}$ of p, we find that $x_1^{\alpha_1 + \alpha_2 + \ldots + \alpha_n} x_2^{\alpha_2 + \ldots + \alpha_n} \cdots x_n^{\alpha_n}$ is the initial monomial of $\sigma_1^{\alpha_1} \cdots \sigma_n^{\alpha_n}$. Since the linear map

$$(\alpha_1, \alpha_2, \ldots, \alpha_n) \mapsto (\alpha_1 + \alpha_2 + \ldots + \alpha_n, \alpha_2 + \ldots + \alpha_n, \ldots, \alpha_n)$$

is injective, all other monomials $\sigma_1^{\beta_1} \ldots \sigma_n^{\beta_n}$ in the expansion of $p(\sigma_1, \ldots, \sigma_n)$ have a different initial monomial. The lexicographically largest monomial $x_1^{\alpha_1 + \alpha_2 + \ldots + \alpha_n} x_2^{\alpha_2 + \ldots + \alpha_n} \cdots x_n^{\alpha_n}$ is not cancelled by any other monomial, and therefore $p(\sigma_1, \ldots, \sigma_n) \neq 0$. This contradiction completes the proof of Theorem 1.1.1. $\triangleleft$

As an example for the above rewriting procedure, we write the bivariate symmetric polynomial $x_1^3 + x_2^3$ as a polynomial in the elementary symmetric polynomials:

$$\underline{x_1^3 + x_2^3} \longrightarrow \underline{\sigma_1^3 - 3x_1^2 x_2 - 3x_1 x_2^2} \longrightarrow \sigma_1^3 - 3\sigma_1 \sigma_2.$$

The subring $\mathbf{C}[\mathbf{x}]^{S_n}$ of symmetric polynomials in $\mathbf{C}[\mathbf{x}] := \mathbf{C}[x_1, \ldots, x_n]$ is the prototype of an invariant ring. The elementary symmetric polynomials $\sigma_1, \ldots, \sigma_n$ are said to form a *fundamental system of invariants*. Such fundamental systems are generally far from being unique. Let us describe another generating set for the symmetric polynomials which will be useful later in Sect. 2.1. The polynomial $p_k(\mathbf{x}) := x_1^k + x_2^k + \ldots + x_n^k$ is called the *k-th power sum*.

Proposition 1.1.2. The ring of symmetric polynomials is generated by the first n power sums, i.e.,

$$\mathbf{C}[\mathbf{x}]^{S_n} = \mathbf{C}[\sigma_1, \sigma_2, \ldots, \sigma_n] = \mathbf{C}[p_1, p_2, \ldots, p_n].$$

Proof. A *partition* of an integer d is an integer vector $\lambda = (\lambda_1, \lambda_2, \ldots, \lambda_n)$ such that $\lambda_1 \geq \lambda_2 \geq \ldots \geq \lambda_n \geq 0$ and $\lambda_1 + \lambda_2 + \ldots + \lambda_n = d$. We assign to a monomial $x_1^{i_1} \ldots x_n^{i_n}$ of degree d the partition $\lambda(i_1, \ldots, i_n)$ which is the decreasingly sorted string of its exponents.

This gives rise to the following total order on the set of degree d monomials in $\mathbf{C}[\mathbf{x}]$. We set $x_1^{i_1} \ldots x_n^{i_n} \prec x_1^{j_1} \ldots x_n^{j_n}$ if the partition $\lambda(i_1, \ldots, i_n)$ is lexicographically larger than $\lambda(j_1, \ldots, j_n)$, or if the partitions are equal and $(i_1, \ldots, i_n)$ is lexicographically smaller than $(j_1, \ldots, j_n)$. We note that this total order on the set of monomials in $\mathbf{C}[\mathbf{x}]$ is *not* a monomial order in the sense of Gröbner bases theory (cf. Sect. 1.2). As an example, for $n = 3$, $d = 4$ we have
$x_3^4 \prec x_2^4 \prec x_1^4 \prec x_2 x_3^3 \prec x_2^3 x_3 \prec x_1 x_3^3 \prec x_1 x_2^3 \prec x_1^3 x_3 \prec x_1^3 x_2 \prec x_2^2 x_3^2 \prec x_1^2 x_3^2 \prec x_1^2 x_2^2 \prec x_1 x_2 x_3^2 \prec x_1 x_2^2 x_3 \prec x_1^2 x_2 x_3$.

We find that the initial monomial of a product of power sums equals

$$\mathrm{init}(p_{i_1} p_{i_2} \ldots p_{i_n}) = c_{i_1 i_2 \ldots i_n} \cdot x_1^{i_1} x_2^{i_2} \ldots x_n^{i_n} \quad \text{whenever } i_1 \geq i_2 \geq \ldots \geq i_n$$

where $c_{i_1 i_2 \ldots i_n}$ is a positive integer.

Now we are prepared to describe an algorithm which proves Proposition 1.1.2. It rewrites a given symmetric polynomial $f \in \mathbf{C}[\mathbf{x}]$ as a polynomial function in $p_1, p_2, \ldots, p_n$. By Theorem 1.1.1 we may assume that f is one of the elementary symmetric polynomials. In particular, the degree d of f is less or equal to n. Its initial monomial $\mathrm{init}(f) = c \cdot x_1^{i_1} \ldots x_n^{i_n}$ satisfies $n \geq i_1 \geq \ldots \geq i_n$. Now replace f by $\tilde{f} := f - \frac{c}{c_{i_1 \ldots i_n}} p_{i_1} \ldots p_{i_n}$. By the above observation the initial monomials in this difference cancel, and we get $\mathrm{init}(\tilde{f}) \prec \mathrm{init}(f)$. Since both f and $\tilde{f}$ have the same degree d, this process terminates with the desired result. ◁

Here is an example for the rewriting process in the proof of Proposition 1.1.2. We express the three-variate symmetric polynomial $f := x_1 x_2 x_3$ as a polynomial function in p_1, p_2 and p_3. Using the above method, we get

1.1. Symmetric polynomials

$$x_1 x_2 x_3 \longrightarrow \tfrac{1}{6} p_1^3 - \tfrac{1}{2} \sum_{i \neq j} x_i^2 x_j - \tfrac{1}{6} \sum_k x_k^3$$
$$\longrightarrow \tfrac{1}{6} p_1^3 - \tfrac{1}{2}\bigl(p_1 p_2 - \sum_k x_k^3\bigr) - \tfrac{1}{6} \sum_k x_k^3$$
$$\longrightarrow \tfrac{1}{6} p_1^3 - \tfrac{1}{2} p_1 p_2 + \tfrac{1}{3} p_3.$$

Theorem 1.1.1 and Proposition 1.1.2 show that the monomials in the elementary symmetric polynomials and the monomials in the power sums are both **C**-vector space bases for the ring of symmetric polynomials $\mathbf{C}[\mathbf{x}]^{S_n}$. There are a number of other important such bases, including the *complete symmetric polynomials*, the *monomial symmetric polynomials* and the *Schur polynomials*. The relations between these bases is of great importance in algebraic combinatorics and representation theory. A basic reference for the theory of symmetric polynomials is Macdonald (1979).

We close this section with the definition of the Schur polynomials. Let A_n denote the *alternating group*, which is the subgroup of S_n consisting of all even permutations. Let $\mathbf{C}[\mathbf{x}]^{A_n}$ denote the subring of polynomials which are fixed by all even permutations. We have the inclusion $\mathbf{C}[\mathbf{x}]^{S_n} \subseteq \mathbf{C}[\mathbf{x}]^{A_n}$. This inclusion is proper, because the polynomial

$$D(x_1, \ldots, x_n) := \prod_{1 \leq i < j \leq n} (x_i - x_j)$$

is fixed by all even permutations but not by any odd permutation.

Proposition 1.1.3. *Every polynomial* $f \in \mathbf{C}[\mathbf{x}]^{A_n}$ *can be written uniquely in the form* $f = g + h \cdot D$ *where* g *and* h *are symmetric polynomials.*

Proof. We set

$$g(x_1, \ldots, x_n) := \tfrac{1}{2}\bigl[f(x_1, x_2, x_3, \ldots, x_n) + f(x_2, x_1, x_3, \ldots, x_n)\bigr] \quad \text{and}$$
$$\tilde{h}(x_1, \ldots, x_n) := \tfrac{1}{2}\bigl[f(x_1, x_2, x_3, \ldots, x_n) - f(x_2, x_1, x_3, \ldots, x_n)\bigr].$$

Thus f is the sum of the symmetric polynomial g and the antisymmetric polynomial $\tilde{h}$. Here $\tilde{h}$ being *antisymmetric* means that

$$\tilde{h}(x_{\sigma_1}, \ldots, x_{\sigma_n}) = \mathrm{sign}(\sigma) \cdot \tilde{h}(x_1, \ldots, x_n)$$

for all permutations $\sigma \in S_n$. Hence $\tilde{h}$ vanishes identically if we replace one of the variables x_i by some other variable x_j. This implies that $x_i - x_j$ divides $\tilde{h}$, for all $1 \leq i < j \leq n$, and therefore D divides $\tilde{h}$. To show uniqueness, we suppose that $f = g + hD = g' + h'D$. Applying an odd permutation π, we get $f \circ \pi = g - hD = g' - h'D$. Now add both equations to conclude $g = g'$ and therefore $h = h'$. ◁

With any partition $\lambda = (\lambda_1 \geq \lambda_2 \geq \ldots \geq \lambda_n)$ of an integer d we associate the homogeneous polynomial

$$a_\lambda(x_1, \ldots, x_n) = \det \begin{pmatrix} x_1^{\lambda_1+n-1} & x_2^{\lambda_1+n-1} & \cdots & x_n^{\lambda_1+n-1} \\ x_1^{\lambda_2+n-2} & x_2^{\lambda_2+n-2} & \cdots & x_n^{\lambda_2+n-2} \\ \vdots & \vdots & \ddots & \vdots \\ x_1^{\lambda_n} & x_2^{\lambda_n} & \cdots & x_n^{\lambda_n} \end{pmatrix}.$$

Note that the total degree of $a_\lambda(x_1, \ldots, x_n)$ equals $d + \binom{n}{2}$.

The polynomials a_λ are precisely the nonzero images of monomials under *antisymmetrization*. Here by *antisymmetrization* of a polynomial we mean its canonical projection into the subspace of antisymmetric polynomials. Therefore the a_λ form a basis for the **C**-vector space of all antisymmetric polynomials. We may proceed as in the proof of Proposition 1.1.3 and divide a_λ by the discriminant. The resulting expression $s_\lambda := a_\lambda/D$ is a symmetric polynomial which is homogeneous of degree $d = |\lambda|$. We call $s_\lambda(x_1, \ldots, x_n)$ the *Schur polynomial* associated with the partition λ.

Corollary 1.1.4. *The set of Schur polynomials s_λ, where $\lambda = (\lambda_1 \geq \lambda_2 \geq \ldots \geq \lambda_n)$ ranges over all partitions of d into at most n parts, forms a basis for the **C**-vector space $\mathbf{C}[\mathbf{x}]_d^{S_n}$ of all symmetric polynomials homogeneous of degree d.*

Proof. It follows from Proposition 1.1.3 that multiplication with D is an isomorphism from the vector space of symmetric polynomials to the space of antisymmetric polynomials. The images of the Schur polynomials s_λ under this isomorphism are the antisymmetrized monomials a_λ. Since the latter are a basis, also the former are a basis. ◁

Exercises

(1) Write the symmetric polynomials $f := x_1^3 + x_2^3 + x_3^3$ and $g := (x_1 - x_2)^2(x_1 - x_3)^2(x_2 - x_3)^2$ as polynomials in the elementary symmetric polynomials $\sigma_1 = x_1 + x_2 + x_3$, $\sigma_2 = x_1x_2 + x_1x_3 + x_2x_3$, and $\sigma_3 = x_1x_2x_3$.

(2) Study the complexity of the algorithm in the proof of Theorem 1.1.1. More precisely, find an upper bound in terms of $\deg(f)$ for the number of steps needed to express a symmetric $f \in \mathbf{C}[x_1, \ldots, x_n]$ as a polynomial in the elementary symmetric polynomials.

(3) Write the symmetric polynomials $\sigma_4 := x_1x_2x_3x_4$ and $p_5 := x_1^5 + x_2^5 + x_3^5 + x_4^5$ as polynomials in the first four power sums $p = x_1 + x_2 + x_3 + x_4$, $p_2 = x_1^2 + x_2^2 + x_3^2 + x_4^2$, $p_3 = x_1^3 + x_2^3 + x_3^3 + x_4^3$, $p_4 = x_1^4 + x_2^4 + x_3^4 + x_4^4$.

(4) Consider the vector space $V = \mathbf{C}[x_1, x_2, x_3]_6^{S_3}$ of all symmetric polynomials in three variables which are homogeneous of degree 6. What is the dimension of V? We get three different bases for V by considering Schur

polynomials $s_{(\lambda_1,\lambda_2,\lambda_3)}$, monomials $\sigma_1^{i_1}\sigma_2^{i_2}\sigma_3^{i_3}$ in the elementary symmetric polynomials, and monomials $p_1^{i_1}p_2^{i_2}p_3^{i_3}$ in the power sum symmetric polynomials. Express each element in one of these bases as a linear combination with respect to the other two bases.

(5) Prove the following explicit formula for the elementary symmetric polynomials in terms of the power sums (Macdonald 1979, p. 20):

$$\sigma_k = \frac{1}{k!}\det\begin{pmatrix} p_1 & 1 & 0 & \cdots & 0 \\ p_2 & p_1 & 2 & \cdots & 0 \\ \vdots & \vdots & \ddots & \ddots & \vdots \\ p_{k-1} & p_{k-2} & \cdots & p_1 & k-1 \\ p_k & p_{k-1} & \cdots & \cdots & p_1 \end{pmatrix}.$$

1.2. Gröbner bases

In this section we review background material from computational algebra. More specifically, we give a brief introduction to the theory of Gröbner bases. Our emphasis is on how to use Gröbner bases as a basic building block in designing more advanced algebraic algorithms. Readers who are interested in "how this black box works" may wish to consult either of the text books Cox et al. (1992) or Becker et al. (1993). See also Buchberger (1985, 1988) and Robbiano (1988) for additional references and details on the computation of Gröbner bases.

Gröbner bases are a general purpose method for multivariate polynomial computations. They were introduced by Bruno Buchberger in his 1965 dissertation, written at the University of Innsbruck (Tyrol, Austria) under the supervision of Wolfgang Gröbner. Buchberger's main contribution is a finite algorithm for transforming an arbitrary generating set of an ideal into a Gröbner basis for that ideal.

The basic principles underlying the concept of Gröbner bases can be traced back to the late 19th century and the early 20th century. One such early reference is P. Gordan's 1900 paper on the invariant theory of binary forms. What is called *"Le système irréductible N"* on page 152 of Gordan (1900) is a Gröbner basis for the ideal under consideration.

Buchberger's Gröbner basis method generalizes three well-known algebraic algorithms:

- the Euclidean algorithm (for univariate polynomials)
- Gaussian elimination (for linear polynomials)
- the Sylvester resultant (for eliminating one variable from two polynomials)

So we can think of Gröbner bases as a version of the Euclidean algorithm which works also for more than one variable, or as a version of Gaussian elimination which works also for higher degree polynomials. The basic algorithms are implemented in many computer algebra systems, e.g., MAPLE, REDUCE, AX-

IOM, MATHEMATICA, MACSYMA, MACAULAY, COCOA, and playing with one of these systems is an excellent way of familiarizing oneself with Gröbner bases. In MAPLE, for instance, the command "gbasis" is used to compute a Gröbner basis for a given set of polynomials, while the command "normalf" reduces any other polynomial to normal form with respect to a given Gröbner basis.

The mathematical set-up is as follows. A total order "$\prec$" on the monomials $x_1^{\lambda_1} \ldots x_n^{\lambda_n}$ in $\mathbf{C}[x_1, \ldots, x_n]$ is said to be a *monomial order* if $1 \preceq m_1$ and $(m_1 \prec m_2 \Rightarrow m_1 \cdot m_3 \prec m_2 \cdot m_3)$ for all monomials $m_1, m_2, m_3 \in \mathbf{C}[x_1, \ldots, x_n]$. Both the degree lexicographic order discussed in Sect. 1.1 and the *(purely) lexicographic order* are important examples of monomial orders. Every linear order on the variables $x_1, x_2, \ldots, x_n$ can be extended to a lexicographic order on the monomials. For example, the order $x_1 \prec x_3 \prec x_2$ on three variables induces the (purely) lexicographic order $1 \prec x_1 \prec x_1^2 \prec x_1^3 \prec x_1^4 \prec \ldots \prec x_3 \prec x_3 x_1 \prec x_3 x_1^2 \prec \ldots \prec x_2 \prec x_2 x_1 \prec x_2 x_1^2 \prec \ldots$ on $\mathbf{C}[x_1, x_2, x_3]$.

We now fix any monomial order "$\prec$" on $\mathbf{C}[x_1, \ldots, x_n]$. The largest monomial of a polynomial $f \in \mathbf{C}[x_1, \ldots, x_n]$ with respect to "$\prec$" is denoted by init(f) and called the *initial monomial* of f. For an ideal $I \subset \mathbf{C}[x_1, \ldots, x_n]$, we define its *initial ideal* as init$(I) := \langle \{\text{init}(f) : f \in I\} \rangle$. In other words, init$(I)$ is the ideal generated by the initial monomials of all polynomials in I. An ideal which is generated by monomials, such as init(I), is said to be a *monomial ideal*. The monomials $m \notin \text{init}(I)$ are called *standard*, and the monomials $m \in \text{init}(I)$ are *non-standard*.

A *finite* subset $\mathcal{G} := \{g_1, g_2, \ldots, g_s\}$ of an ideal I is called a *Gröbner basis* for I provided the initial ideal init(I) is generated by $\{\text{init}(g_1), \ldots, \text{init}(g_s)\}$. One last definition: the Gröbner basis $\mathcal{G}$ is called *reduced* if init(g_i) does not divide any monomial occurring in g_j, for all distinct $i, j \in \{1, 2, \ldots, s\}$. Gröbner bases programs (such as "gbasis" in MAPLE) take a finite set $\mathcal{F} \subset \mathbf{C}[\mathbf{x}]$ and they output a reduced Gröbner basis $\mathcal{G}$ for the ideal $\langle \mathcal{F} \rangle$ generated by $\mathcal{F}$. They are based on the Buchberger algorithm.

The previous paragraph is perhaps the most compact way of defining Gröbner bases, but it is not at all informative on what Gröbner bases theory is all about. Before proceeding with our theoretical crash course, we present six concrete examples $(\mathcal{F}, \mathcal{G})$ where $\mathcal{G}$ is a reduced Gröbner basis for the ideal $\langle \mathcal{F} \rangle$.

Example 1.2.1 (Easy examples of Gröbner bases). In (1), (2), (5), (6) we also give examples for the *normal form reduction* versus a Gröbner bases $\mathcal{G}$ which rewrites every polynomial modulo $\langle \mathcal{F} \rangle$ as a $\mathbf{C}$-linear combination of standard monomials (cf. Theorem 1.2.6). In all examples the used monomial order is specified and the initial monomials are underlined.

(1) For any set of univariate polynomials $\mathcal{F}$, the reduced Gröbner basis $\mathcal{G}$ is always a singleton, consisting of the greatest common divisor of $\mathcal{F}$. Note that $1 \prec x \prec x^2 \prec x^3 \prec x^4 \prec \ldots$ is the only monomial order on $\mathbf{C}[x]$.
$\mathcal{F} = \{\underline{12x^3} - x^2 - 23x - 11, \underline{x^4} - x^2 - 2x - 1\}$
$\mathcal{G} = \{\underline{x^2} - x - 1\}$
Normal form: $x^3 + x^2 \to_{\mathcal{G}} 3x + 2$

1.2. Gröbner bases

$x_n]$. Its ideal is finitely generated by Lemma 1.2.2. Hence there exists an integer j such that $m_j \in \langle m_1, m_2, \ldots, m_{j-1}\rangle$. This means that m_i divides m_j for some $i < j$. Since "$\prec$" is a monomial order, this implies $m_i \prec m_j$ with $i < j$. This proves Corollary 1.2.3. ◁

Theorem 1.2.4.
(1) Any ideal $I \subset \mathbf{C}[x_1, \ldots, x_n]$ has a Gröbner basis $\mathcal{G}$ with respect to any monomial order "$\prec$".
(2) Every Gröbner basis $\mathcal{G}$ generates its ideal I.

Proof. Statement (1) follows directly from Lemma 1.2.2 and the definition of Gröbner bases. We prove statement (2) by contradiction. Suppose the Gröbner basis $\mathcal{G}$ does not generate its ideal, that is, the set $I \setminus \langle \mathcal{G} \rangle$ is non-empty. By Corollary 1.2.3, the set of initial monomials $\{\mathrm{init}(f) : f \in I \setminus \langle \mathcal{G}\rangle\}$ has a minimal element $\mathrm{init}(f_0)$ with respect to "$\prec$". The monomial $\mathrm{init}(f_0)$ is contained in $\mathrm{init}(I) = \langle \mathrm{init}(\mathcal{G})\rangle$. Let $g \in \mathcal{G}$ such that $\mathrm{init}(g)$ divides $\mathrm{init}(f_0)$, say $\mathrm{init}(f_0) = m \cdot \mathrm{init}(g)$.

Now consider the polynomial $f_1 := f_0 - m \cdot g$. By construction, $f_1 \in I \setminus \langle \mathcal{G}\rangle$. But we also have $\mathrm{init}(f_1) \prec \mathrm{init}(f_0)$. This contradicts the minimality in the choice of f_0. This contradiction shows that $\mathcal{G}$ does generate the ideal I. ◁

From this we obtain as a direct consequence the following basic result.

Corollary 1.2.5 (Hilbert's basis theorem). Every ideal in the polynomial ring $\mathbf{C}[x_1, x_2, \ldots, x_n]$ is finitely generated.

We will next prove the normal form property of Gröbner bases.

Theorem 1.2.6. Let I be any ideal and "$\prec$" any monomial order on $\mathbf{C}[x_1, \ldots, x_n]$. The set of (residue classes of) standard monomials is a $\mathbf{C}$-vector space basis for the residue ring $\mathbf{C}[x_1, \ldots, x_n]/I$.

Proof. Let $\mathcal{G}$ be a Gröbner basis for I, and consider the following algorithm which computes the normal form modulo I.
Input: $p \in \mathbf{C}[x_1, \ldots, x_n]$.
1. Check whether all monomials in p are standard. If so, we are done: p is in normal form and equivalent modulo I to the input polynomial.
2. Otherwise let $\mathrm{hnst}(p)$ be the highest non-standard monomial occurring in p. Find $g \in \mathcal{G}$ such that $\mathrm{init}(g)$ divides $\mathrm{hnst}(p)$, say $m \cdot \mathrm{init}(g) = \mathrm{hnst}(p)$.
3. Replace p by $\tilde{p} := p - m \cdot g$, and go to 1.

We have $\mathrm{init}(\tilde{p}) \prec \mathrm{init}(p)$ in Step 3, and hence Corollary 1.2.3 implies that this algorithm terminates with a representation of $p \in \mathbf{C}[x_1, \ldots, x_n]$ as a $\mathbf{C}$-linear combination of standard monomials modulo I. We conclude the proof of Theorem 1.2.6 by observing that such a representation is necessarily unique because, by definition, every polynomial in I contains at least one non-standard monomial. This means that zero cannot be written as non-trivial linear combination of standard monomials in $\mathbf{C}[x_1, \ldots, x_n]/I$. ◁

Sometimes it is possible to give an a priori proof that an explicitly known "nice" subset of a polynomial ideal I happens to be a Gröbner basis. In such a lucky situation there is no need to apply the Buchberger algorithm. In order to establish the Gröbner basis property, tools from algebraic combinatorics are ~~ularly useful. We illustrate this by generalizing the above Example (5) to~~ ~~ber of variables.~~
~~ideal in $\mathbf{C}[\mathbf{x}, \mathbf{y}] = \mathbf{C}[x_1, x_2, \ldots, x_n, y_1, y_2, \ldots, y_n]$ which~~
~~$\sigma_i(x_1, \ldots, x_n) - y_i$ for $i = 1, 2, \ldots, n$. Here~~
~~ic polynomial. In other words, I is the~~
~~roots and coefficients of a generic~~
~~al h_i is defined to be the sum of all~~
~~given set of variables. In particular, we have~~
$x_k^{v_k} x_{k+1}^{v_{k+1}} \cdots x_n^{v_n}$ where the sum ranges over all $\binom{n-k+i}{i}$ non-
~~ger~~ vectors $(v_k, v_{k+1}, \ldots, v_n)$ whose coordinates sum to i.

Theorem 1.2.7. *The unique reduced Gröbner basis of I with respect to the lexicographic monomial order induced from $x_1 \succ x_2 \succ \ldots \succ x_n \succ y_1 \succ y_2 \succ \ldots \succ y_n$ equals*

$$\mathcal{G} = \left\{ h_k(x_k, \ldots, x_n) + \sum_{i=1}^{k} (-1)^i h_{k-i}(x_k, \ldots, x_n) y_i : k = 1, \ldots, n \right\}.$$

Proof. In the proof we use a few basic facts about symmetric polynomials and Hilbert series of graded algebras. We first note the following symmetric polynomial identity

$$h_k(x_k, \ldots, x_n) + \sum_{i=1}^{k} (-1)^i h_{k-i}(x_k, \ldots, x_n) \sigma_i(x_1, \ldots, x_{k-1}, x_k, \ldots, x_n) = 0.$$

This identity shows that $\mathcal{G}$ is indeed a subset of the ideal I.

We introduce a grading on $\mathbf{C}[\mathbf{x}, \mathbf{y}]$ by setting degree(x_i) $= 1$ and degree(y_j) $= j$. The ideal I is homogeneous with respect to this grading. The quotient ring $R = \mathbf{C}[\mathbf{x}, \mathbf{y}]/I$ is isomorphic as a graded algebra to $\mathbf{C}[x_1, \ldots, x_n]$, and hence the Hilbert series of $R = \bigoplus_{d=0}^{\infty} R_d$ equals $H(R, z) = \sum_{d=0}^{\infty} \dim_{\mathbf{C}}(R_d) z^d = (1-z)^{-n}$. It follows from Theorem 1.2.6 that the quotient $\mathbf{C}[\mathbf{x}, \mathbf{y}]/\text{init}_\prec(I)$ modulo the initial ideal has the same Hilbert series $(1-z)^{-n}$.

Consider the monomial ideal $J = \langle x_1, x_2^2, x_3^3, \ldots, x_n^n \rangle$ which is generated by the initial monomials of the elements in $\mathcal{G}$. Clearly, J is contained in the initial ideal $\text{init}_\prec(I)$. Our assertion states that these two ideals are equal. For the proof it is sufficient to verify that the Hilbert series of $R' := \mathbf{C}[\mathbf{x}, \mathbf{y}]/J$ equals the Hilbert series of R.

1.2. Gröbner bases

A vector space basis for R' is given by the set of all monomials $x_1^{i_1} \cdots x_n^{i_n} y_1^{j_1} \cdots y_n^{j_n}$ whose exponents satisfy the constraints $i_1 < 1, i_2 < 2, \ldots, i_n < n$. This shows that the Hilbert series of R' equals the formal power series

$$H(R', z) = \left(\sum z^{i_1 + i_2 + \ldots + i_n}\right) \left(\sum z^{j_1 + 2j_2 + \ldots + nj_n}\right).$$

The second sum is over all $(j_1, \ldots, j_n) \in \mathbf{N}^n$ and thus equals $[(1-z)(1-z^2) \cdots (1-z^n)]^{-1}$. The first sum is over all $(i_1, \ldots, i_n) \in \mathbf{N}^n$ with $i_\mu < \mu$ and hence equals the polynomial $(1+z)(1+z+z^2) \cdots (1+z+z^2+\ldots+z^{n-1})$. We compute their product as follows:

$$H(R', z) = \left(\frac{1}{1-z}\right)\left(\frac{1+z}{1-z^2}\right)\left(\frac{1+z+z^2}{1-z^3}\right) \cdots \left(\frac{1+z+z^2+\ldots+z^{n-1}}{1-z^n}\right)$$

$$= \left(\frac{1}{1-z}\right)\left(\frac{1}{1-z}\right)\left(\frac{1}{1-z}\right) \cdots \left(\frac{1}{1-z}\right) = H(R, z).$$

This completes the proof of Theorem 1.2.7. ◁

The normal form reduction versus the Gröbner basis $\mathcal{G}$ in Theorem 1.2.7 provides an alternative algorithm for the Main Theorem on Symmetric Polynomials (1.1.1). If we reduce any symmetric polynomial in the variables $x_1, x_2, \ldots, x_n$ modulo $\mathcal{G}$, then we get a linear combination of standard monomials $y_1^{i_1} y_2^{i_2} \cdots y_n^{i_n}$. These can be identified with monomials $\sigma_1^{i_1} \sigma_2^{i_2} \cdots \sigma_n^{i_n}$ in the elementary symmetric polynomial.

Exercises

(1) Let "$\prec$" be a monomial order and let I be any ideal in $\mathbf{C}[x_1, \ldots, x_n]$. A monomial m is called *minimally non-standard* if m is non-standard and all proper divisors of m are standard. Show that the set of minimally non-standard monomials is finite.

(2) Prove that the reduced Gröbner basis $\mathcal{G}_{\text{red}}$ of I with respect to "$\prec$" is unique (up to multiplicative constants from $\mathbf{C}$). Give an algorithm which transforms an arbitrary Gröbner basis into $\mathcal{G}_{\text{red}}$.

(3) Let $I \subset \mathbf{C}[x_1, \ldots, x_n]$ be an ideal, given by a finite set of generators. Using Gröbner bases, describe an algorithm for computing the *elimination ideals* $I \cap \mathbf{C}[x_1, \ldots, x_i], i = 1, \ldots, n-1$, and prove its correctness.

(4) Find a characterization for all monomial orders on the polynomial ring $\mathbf{C}[x_1, x_2]$. (Hint: Each variable receives a certain "weight" which behaves additively under multiplication of variables.) Generalize your result to n variables.

(5) * Fix any ideal $I \subset \mathbf{C}[x_1, \ldots, x_n]$. We say that two monomial orders are *I-equivalent* if they induce the same initial ideal for I. Show that there are only finitely many I-equivalence classes of monomial orders.

(6) Let $\mathcal{F}$ be a set of polynomials whose initial monomials are pairwise relatively prime. Show that $\mathcal{F}$ is a Gröbner basis for its ideal.

1.3. What is invariant theory?

Many problems in applied algebra have symmetries or are invariant under certain natural transformations. In particular, all *geometric* magnitudes and properties are invariant with respect to the underlying transformation group. Properties in *Euclidean geometry* are invariant under the Euclidean group of rotations, reflections and translations, properties in *projective geometry* are invariant under the group of projective transformations, etc. ... This identification of geometry and invariant theory, expressed in Felix Klein's Erlanger Programm (cf. Klein 1872, 1914), is much more than a philosophical remark. The practical significance of invariant-theoretic methods as well as their mathematical elegance is our main theme. We wish to illustrate why invariant theory is likely to play an increasingly important role for computer algebra and computational geometry in the future.

We begin with some basic invariant-theoretic terminology. Let Γ be a subgroup of the group $GL(\mathbf{C}^n)$ of invertible $n \times n$-matrices. This is the group of transformations, which defines the geometry or geometric situation under consideration. Given a polynomial function $f \in \mathbf{C}[x_1, \ldots, x_n]$, then every linear transformation $\pi \in \Gamma$ transforms f into a new polynomial function $f \circ \pi$. For example, if $f = x_1^2 + x_1 x_2 \in \mathbf{C}[x_1, x_2]$ and $\pi = \begin{pmatrix} 3 & 5 \\ 4 & 7 \end{pmatrix}$, then

$$f \circ \pi = (3x_1 + 5x_2)^2 + (3x_1 + 5x_2)(4x_1 + 7x_2) = 21x_1^2 + 71x_1 x_2 + 60x_2^2.$$

In general, we are interested in the set

$$\mathbf{C}[x_1, \ldots, x_n]^\Gamma := \{f \in \mathbf{C}[x_1, \ldots, x_n] : \forall \pi \in \Gamma \ (f = f \circ \pi)\}$$

of all polynomials which are invariant under this action of Γ. This set is a subring of $\mathbf{C}[x_1, \ldots, x_n]$ since it is closed under addition and multiplication. We call $\mathbf{C}[x_1, \ldots, x_n]^\Gamma$ the *invariant subring* of Γ. The following questions are often called the fundamental problems of invariant theory.

(1) Find a set $\{I_1, \ldots, I_m\}$ of generators for the invariant subring $\mathbf{C}[x_1, \ldots, x_n]^\Gamma$. All the groups Γ studied in this text do admit such a *finite* set of *fundamental invariants*. A famous result of Nagata (1959) shows that the invariant subrings of certain "exotic" matrix groups are not finitely generated.
(2) Describe the algebraic relations among the fundamental invariants $I_1, \ldots, I_m$. These are called *syzygies*.
(3) Give an algorithm which rewrites an arbitrary invariant $I \in \mathbf{C}[x_1, \ldots, x_n]^\Gamma$ as a polynomial $I = p(I_1, \ldots, I_m)$ in the fundamental invariants.

1.3. What is invariant theory?

For the classical geometric groups, such as the Euclidean group or the projective group, also the following question is important.

(4) Given a geometric property $\mathcal{P}$, find the corresponding invariants (or covariants) and vice versa. Is there an algorithm for this transition between geometry and algebra?

Example 1.3.1 (Symmetric polynomials). Let S_n be the group of permutation matrices in $GL(\mathbf{C}^n)$. Its invariant ring $\mathbf{C}[x_1, \ldots, x_n]^{S_n}$ equals the subring of symmetric polynomials in $\mathbf{C}[x_1, \ldots, x_n]$. For the symmetric group S_n all three fundamental problems were solved in Sect. 1.1.

(1) The elementary symmetric polynomials form a fundamental set of invariants:
$$\mathbf{C}[x_1, \ldots, x_n]^{S_n} = \mathbf{C}[\sigma_1, \ldots, \sigma_n].$$

(2) These fundamental invariants are algebraically independent: There is no nonzero syzygy.
(3) We have two possible algorithms for rewriting symmetric polynomials in terms of elementary symmetric ones: either the method in the proof in Theorem 1.1.1 or the normal form reduction modulo the Gröbner basis in Theorem 1.2.7.

Example 1.3.2 (The cyclic group of order 4). Let $n = 2$ and consider the group
$$Z_4 = \left\{ \begin{pmatrix} 1 & 0 \\ 0 & 1 \end{pmatrix}, \begin{pmatrix} -1 & 0 \\ 0 & -1 \end{pmatrix}, \begin{pmatrix} 0 & 1 \\ -1 & 0 \end{pmatrix}, \begin{pmatrix} 0 & -1 \\ 1 & 0 \end{pmatrix} \right\}$$
of rotational symmetries of the square. Its invariant ring equals
$$\mathbf{C}[x_1, x_2]^{Z_4} = \{ f \in \mathbf{C}[x_1, x_2] : f(x_1, x_2) = f(-x_2, x_1) \}.$$

(1) Here we have three fundamental invariants
$$I_1 = x_1^2 + x_2^2, \quad I_2 = x_1^2 x_2^2, \quad I_3 = x_1^3 x_2 - x_1 x_2^3.$$

(2) These satisfy the algebraic dependence $I_3^2 - I_2 I_1^2 + 4 I_2^2$. This syzygy can be found with the slack variable Gröbner basis method in Example 1.2.1.(4).
(3) Using Gröbner basis normal form reduction, we can rewrite any invariant as a polynomial in the fundamental invariants. For example, $x_1^7 x_2 - x_2^7 x_1 \to I_1^2 I_3 - I_2 I_3$.

We next give an alternative interpretation of the invariant ring from the point of view of elementary algebraic geometry. Every matrix group Γ acts on the vector space $\mathbf{C}^n$, and it decomposes $\mathbf{C}^n$ into Γ-orbits

$$\Gamma \mathbf{v} = \{\sigma \mathbf{v} : \sigma \in \Gamma\} \quad \text{where } \mathbf{v} \in \mathbf{C}^n.$$

Remark 1.3.3. The invariant ring $\mathbf{C}[\mathbf{x}]^\Gamma$ consists of those polynomial functions f which are constant along all Γ-orbits in $\mathbf{C}^n$.

Proof. A polynomial $f \in \mathbf{C}[\mathbf{x}]$ is constant on all Γ-orbits if and only if

$$\forall \mathbf{v} \in \mathbf{C}^n \, \forall \pi \in \Gamma : f(\pi \mathbf{v}) = f(\mathbf{v})$$
$$\iff \forall \pi \in \Gamma \, \forall \mathbf{v} \in \mathbf{C}^n : (f \circ \pi)(\mathbf{v}) = f(\mathbf{v}).$$

Since $\mathbf{C}$ is an infinite field, the latter condition is equivalent to f being an element of $\mathbf{C}[\mathbf{x}]^\Gamma$. ◁

Remark 1.3.3 suggests that the invariant ring can be interpreted as the ring of polynomial functions on the quotient space $\mathbf{C}^n/\Gamma$ of Γ-orbits on $\mathbf{C}^n$. We are tempted to conclude that $\mathbf{C}^n/\Gamma$ is actually an algebraic variety which has $\mathbf{C}[\mathbf{x}]^\Gamma$ as its coordinate ring. This statement is not quite true for most infinite groups: it can happen that two distinct Γ-orbits in $\mathbf{C}^n$ cannot be distinguished by a polynomial function because one is contained in the closure of the other.

For finite groups Γ, however, the situation is nice because all orbits are finite and hence closed subsets of $\mathbf{C}^n$. Here $\mathbf{C}[\mathbf{x}]^\Gamma$ is truly the coordinate ring of the *orbit variety* $\mathbf{C}^n/\Gamma$. The first fundamental problem (1) can be interpreted as finding an embedding of $\mathbf{C}^n/\Gamma$ as an affine subvariety into $\mathbf{C}^m$, where m is the number of fundamental invariants. For example, the orbit space $\mathbf{C}^2/Z_4$ of the cyclic group in Example 1.3.2 equals the hypersurface in $\mathbf{C}^3$ which is defined by the equation $y_3^2 - y_2 y_1^2 + 4 y_2^2 = 0$. The map $(x_1, x_2) \mapsto (I_1(x_1, x_2), I_2(x_1, x_2), I_3(x_1, x_2))$ defines a bijection (check this!!) from the set of Z_4-orbits in $\mathbf{C}^2$ onto this hypersurface.

Let us now come to the fundamental problem (4). We illustrate this question for the Euclidean group of rotations, translations and reflections in the plane. The Euclidean group acts on the polynomial ring $\mathbf{C}[x_1, y_1, x_2, y_2, \ldots, x_n, y_n]$ by rigid motions

$$\begin{pmatrix} x_i \\ y_i \end{pmatrix} \mapsto \begin{pmatrix} \cos \phi & \sin \phi \\ -\sin \phi & \cos \phi \end{pmatrix} \cdot \begin{pmatrix} x_i \\ y_i \end{pmatrix} + \begin{pmatrix} a \\ b \end{pmatrix},$$

and by reflections, such as $(x_i, y_i) \mapsto (-x_i, y_i)$. The invariant polynomials under this action correspond to geometric properties of a configuration of n points (x_i, y_i) in the Euclidean plane. Naturally, for this interpretation we restrict ourselves to the field $\mathbf{R}$ of real numbers.

Example 1.3.4. Consider the three polynomials $L := x_1^2 + y_1^2 - 7$, $D := (x_1 - x_2)^2 + (y_1 - y_2)^2$, and $R := x_1^2 + y_1^2 - x_1 x_2 - y_1 y_2 - x_1 x_3 - y_1 y_3 + x_2 x_3 + y_2 y_3$. The first polynomial L expresses that point "1" has distance 7 from the origin.

1.3. What is invariant theory?

This property is *not* Euclidean because it is not invariant under translations, and L is *not* a Euclidean invariant. The second polynomial D measures the distance between the two points "1" and "2", and it is a Euclidean invariant. Also R is a Euclidean invariant: it vanishes if and only if the lines "$\overline{12}$" and "$\overline{13}$" are perpendicular.

The following general representation theorem was known classically.

Theorem 1.3.5. The subring of Euclidean invariants is generated by the squared distances
$$D_{ij} := (x_i - x_j)^2 + (y_i - y_j)^2, \quad 1 \leq i < j \leq n.$$

For a new proof of Theorem 1.3.5 we refer to Dalbec (1992). In that article an efficient algorithm is given for expressing any Euclidean invariant in terms of the D_{ij}. It essentially amounts to specifying a Gröbner basis for the *Cayley–Menger ideal* of syzygies among the squared distances D_{ij}. Here are two examples for the resulting rewriting process.

Example 1.3.6 (Heron's formula for the squared area of a triangle).
Let $A_{123} \in \mathbf{C}[x_1, y_1, x_2, y_2, x_3, y_3]$ denote the squared area of the triangle "123". The polynomial A_{123} is a Euclidean invariant, and its representation in terms of squared distances equals

$$A_{123} = \det \begin{pmatrix} 0 & 1 & 1 & 1 \\ 1 & 0 & D_{12} & D_{13} \\ 1 & D_{12} & 0 & D_{23} \\ 1 & D_{13} & D_{23} & 0 \end{pmatrix}.$$

Note that the triangle area $\sqrt{A_{123}}$ is not a polynomial in the vertex coordinates.

Example 1.3.7 (Cocircularity of four points in the plane). Four points (x_1, y_1), (x_2, y_2), (x_3, y_3), (x_4, y_4) in the Euclidean plane lie on a common circle if and only if

$$\exists x_0, y_0 : (x_i - x_0)^2 + (y_i - y_0)^2 = (x_j - x_0)^2 + (y_j - y_0)^2 \quad (1 \leq i < j \leq 4).$$

This in turn is the case if and only if the following invariant polynomial vanishes:

$$D_{12}^2 D_{34}^2 + D_{13}^2 D_{24}^2 + D_{14}^2 D_{23}^2 - 2 D_{12} D_{13} D_{24} D_{34} - 2 D_{12} D_{14} D_{23} D_{34} - \\ - 2 D_{13} D_{14} D_{23} D_{24}.$$

Writing Euclidean properties in terms of squared distances is part of a new method for automated geometry theorem proving due to T. Havel (1991).

We have illustrated the basic idea of geometric invariants for the Euclidean plane. Later in Chap. 3, we will focus our attention on projective geometry.

In projective geometry the underlying algebra is better understood than in Euclidean geometry. There we will be concerned with the action of the group $\Gamma = SL(\mathbf{C}^d)$ by right multiplication on a generic $n \times d$-matrix $\mathbf{X} = (x_{ij})$. Its invariants in $\mathbf{C}[\mathbf{X}] := \mathbf{C}[x_{11}, x_{12} \ldots, x_{nd}]$ correspond to geometric properties of a configuration of n points in projective $(d-1)$-space.

The *first fundamental theorem*, to be proved in Sect. 3.2, states that the corresponding invariant ring $\mathbf{C}[\mathbf{X}]^\Gamma$ is generated by the $d \times d$-subdeterminants

$$[i_1 i_2 \ldots i_d] := \det \begin{pmatrix} x_{i_1,1} & \cdots & x_{i_1,d} \\ \vdots & \ddots & \vdots \\ x_{i_d,1} & \cdots & x_{i_d,d} \end{pmatrix}.$$

Example 1.3.8. The expression in Example 1.2.1 (6) is a polynomial function in the coordinates of four points on the projective line (e.g., the point "3" has homogeneous coordinates (x_{31}, x_{32})). This polynomial is invariant, it does correspond to a geometric property, because it can be rewritten in terms of brackets as $[14][23] + [13][24]$. It vanishes if and only if the projective *cross ratio* $(1, 3; 4, 2) = [13][24]/[14][23]$ of the four points equals -1.

The projective geometry analogue to the above rewriting process for Euclidean geometry will be presented in Sects. 3.1 and 3.2. It is our objective to show that the set of *straightening syzygies* is a Gröbner basis for the *Grassmann ideal* of syzygies among the *brackets* $[i_1 i_2 \ldots i_d]$. The resulting Gröbner basis normal form algorithm equals the classical straightening law for *Young tableaux*. Its direct applications are numerous and fascinating, and several of them will be discussed in Sects. 3.4–3.6.

The bracket algebra and the straightening algorithm will furnish us with the crucial technical tools for studying invariants of forms (= homogeneous polynomials) in Chap. 4. This subject is the cornerstone of classical invariant theory.

Exercises

(1) Show that every finite group $\Gamma \subset GL(\mathbf{C}^n)$ does have nonconstant polynomial invariants. Give an example of an infinite matrix group Γ with $\mathbf{C}[\mathbf{x}]^\Gamma = \mathbf{C}$.

(2) Write the Euclidean invariant R in Example 1.3.4 as a polynomial function in the squared distances D_{12}, D_{13}, D_{23}, and interpret the result geometrically.

(3) Fix a set of positive and negative integers $\{a_1, a_2, \ldots, a_n\}$, and let $\Gamma \subset GL(\mathbf{C}^n)$ denote the subgroup of all diagonal matrices of the form $\operatorname{diag}(t^{a_1}, t^{a_2}, \ldots, t^{a_n})$, $t \in \mathbf{C}^*$, where $\mathbf{C}^*$ denotes the multiplicative group of non-zero complex numbers. Show that the invariant ring $\mathbf{C}[x_1, \ldots, x_n]^\Gamma$ is finitely generated as a $\mathbf{C}$-algebra.

(4) Let C[X] denote the ring of polynomial functions on an $n \times n$-matrix $\mathbf{X} = (x_{ij})$ of indeterminates. The general linear group $GL(\mathbf{C}^n)$ acts on C[X] by conjugation, i.e., $\mathbf{X} \mapsto A\mathbf{X}A^{-1}$ for $A \in GL(\mathbf{C}^n)$. The invariant ring $\mathbf{C}[\mathbf{X}]^{GL(\mathbf{C}^n)}$ consists of all polynomial functions which are invariant under the action. Find a fundamental set of invariants.

1.4. Torus invariants and integer programming

Let $n \geq d$ be positive integers, and let $\mathcal{A} = (a_{ij})$ be any integer $n \times d$-matrix of rank d. Integer programming is concerned with the algorithmic study of the *monoid defined by* $\mathcal{A}$:

$$\mathcal{M}_{\mathcal{A}} := \{(v_1, \ldots, v_n) \in \mathbf{Z}^n \setminus \{0\} : \\ v_1, \ldots, v_n \geq 0 \text{ and } (v_1, \ldots, v_n) \cdot \mathcal{A} = 0\}. \quad (1.4.1)$$

We are interested in the following three specific questions:

(a) Feasibility Problem: "Is $\mathcal{M}_{\mathcal{A}}$ nonempty?" If yes, find a vector $v = (v_1, \ldots, v_n)$ in $\mathcal{M}_{\mathcal{A}}$.
(b) Optimization Problem: Given any cost vector $\omega = (\omega_1, \ldots, \omega_n) \in \mathbf{R}_+^n$, find a vector $v = (v_1, \ldots, v_n) \in \mathcal{M}_{\mathcal{A}}$ such that $\langle \omega, v \rangle = \sum_{i=1}^n \omega_i v_i$ is minimized.
(c) Hilbert Basis Problem: Find a finite minimal spanning subset $\mathcal{H}$ in $\mathcal{M}_{\mathcal{A}}$.

By "spanning" in (c) we mean that every $\beta \in \mathcal{M}_{\mathcal{A}}$ has a representation

$$\beta = \sum_{v \in \mathcal{H}} c_v \cdot v, \quad (1.4.2)$$

where the c_v are non-negative integers. It is known (see, e.g., Schrijver 1986, Stanley 1986) that such a set $\mathcal{H}$ exists and is unique. It is called the *Hilbert basis* of $\mathcal{M}_{\mathcal{A}}$. The existence and uniqueness of the Hilbert basis will also follow from our correctness proof for Algorithm 1.4.5 given below.

Example 1.4.1. Let $n = 4, d = 1$. We choose the matrix $\mathcal{A} = (3, 1, -2, -2)^T$ and the cost vector $\omega = (5, 5, 6, 5)$. Our three problems have the following solutions:

(a) $\mathcal{M}_{\mathcal{A}} \neq \emptyset$ because $v = (1, 1, 1, 1) \in \mathcal{M}_{\mathcal{A}}$.
(b) $v = (0, 2, 0, 1) \in \mathcal{M}_{\mathcal{A}}$ has minimum cost $\langle \omega, v \rangle = 15$.
(c) The Hilbert basis of $\mathcal{M}_{\mathcal{A}}$ equals $\mathcal{H} = \{(2, 0, 3, 0), (2, 0, 2, 1), (2, 0, 1, 2),$
    $(2, 0, 0, 3), (1, 1, 2, 0), (1, 1, 1, 1), (1, 1, 0, 2), (0, 2, 0, 1), (0, 2, 1, 0)\}$.

The Hilbert basis problem (c) has a natural translation into the context of invariant theory; see, e.g., Hochster (1972), Wehlau (1991). Using this translation and Gröbner bases theory, we will present algebraic algorithms for solving the problems (a), (b) and (c).

With the given integer $n \times d$-matrix $\mathcal{A}$ we associate a group of diagonal $n \times n$-matrices:

$$\Gamma_\mathcal{A} := \left\{ \operatorname{diag}\left(\prod_{i=1}^d t_i^{a_{1i}}, \prod_{i=1}^d t_i^{a_{2i}}, \ldots, \prod_{i=1}^d t_i^{a_{ni}}\right) : t_1, \ldots, t_d \in \mathbf{C}^* \right\}. \qquad (1.4.3)$$

The matrix group $\Gamma_\mathcal{A}$ is isomorphic to the group $(\mathbf{C}^*)^d$ of invertible diagonal $d \times d$-matrices, which is called the *d-dimensional algebraic torus*. We call $\Gamma_\mathcal{A}$ the *torus defined by* $\mathcal{A}$. In this section we describe an algorithm for computing its invariant ring $\mathbf{C}[x_1, x_2, \ldots, x_n]^{\Gamma_\mathcal{A}}$.

The action of $\Gamma_\mathcal{A}$ maps monomials into monomials. Hence a polynomial $f(x_1, \ldots, x_n)$ is an invariant if and only if each of the monomials appearing in f is an invariant. The invariant monomials are in bijection with the elements of the monoid $\mathcal{M}_\mathcal{A}$.

Lemma 1.4.2.

(a) A monomial $\mathbf{x}^\nu = x_1^{\nu_1} \cdots x_n^{\nu_n}$ is $\Gamma_\mathcal{A}$-invariant if and only if $\nu = (\nu_1, \ldots, \nu_n) \in \mathcal{M}_\mathcal{A}$.
(b) A finite set $\mathcal{H} \subset \mathbf{Z}^n$ equals the Hilbert basis of $\mathcal{M}_\mathcal{A}$ if and only if the invariant ring $\mathbf{C}[x_1, \ldots, x_n]^{\Gamma_\mathcal{A}}$ is minimally generated as a $\mathbf{C}$-algebra by $\{\mathbf{x}^\nu : \nu \in \mathcal{H}\}$.

Proof. The image of $\mathbf{x}^\nu$ under a torus element $\operatorname{diag}(\prod_{i=1}^d t_i^{a_{1i}}, \ldots, \prod_{i=1}^d t_i^{a_{ni}})$ equals

$$(x_1 \prod_{i=1}^d t_i^{a_{1i}})^{\nu_1} \cdots \cdots (x_n \prod_{i=1}^d t_i^{a_{ni}})^{\nu_n} = (x_1^{\nu_1} \cdots x_n^{\nu_n}) \cdot \prod_{i=1}^d t_i^{\sum_{j=1}^n \nu_j a_{ji}}. \qquad (1.4.4)$$

Therefore $\mathbf{x}^\nu$ is invariant under the action of $\Gamma_\mathcal{A}$ if and only if $\sum_{j=1}^n \nu_j a_{ji} = 0$, for $i = 1, \ldots, d$. This is equivalent to $\nu \cdot \mathcal{A} = 0$, which is the defining equation of the monoid $\mathcal{M}_\mathcal{A}$. Part (b) follows from the fact that (1.4.2) translates into $\mathbf{x}^\beta = \prod_{\nu \in \mathcal{H}} (\mathbf{x}^\nu)^{c_\nu}$. ◁

Example 1.4.1 (continued). Let $\Gamma_\mathcal{A}$ be the group of diagonal 4×4-matrices of the form $\operatorname{diag}(t^3, t^1, t^{-2}, t^{-2})$, where $t \in \mathbf{C}^*$. The invariant ring equals $\mathbf{C}[x_1, x_2, x_3, x_4]^{\Gamma_\mathcal{A}} = \mathbf{C}[x_1^2 x_3^3, x_1^2 x_3^2 x_4, x_1^2 x_3 x_4^2, x_1^2 x_4^3, x_1 x_2 x_3^2, x_1 x_2 x_3 x_4, x_1 x_2 x_4^2, x_2^2 x_3, x_2^2 x_4]$. ◁

We first show how to solve the easiest of the three problems. The subsequent Algorithms 1.4.3 and 1.4.4 are due to L. Pottier (1992). For an alternative Gröbner basis approach to integer programming see Conti and Traverso (1991).

1.4. Torus invariants and integer programming

Algorithm 1.4.3 (Integer programming – Feasibility problem (a)).
Input: An integer $n \times d$-matrix $\mathcal{A} = (a_{ij})$.
Output: A vector $(\beta_1, \ldots, \beta_n)$ in the monoid $\mathcal{M}_\mathcal{A}$ if $\mathcal{M}_\mathcal{A} \neq \emptyset$; "INFEASIBLE" otherwise.

1. Compute any reduced Gröbner basis $\mathcal{G}$ for the kernel of the **C**-algebra homomorphism

$$\mathbf{C}[x_1, x_2, \ldots, x_n] \to \mathbf{C}[t_1, \ldots, t_d, t_1^{-1}, \ldots, t_d^{-1}], \quad x_i \mapsto \prod_{j=1}^{d} t_j^{a_{ij}}. \quad (1.4.5)$$

2. Does there exist an element of the form $x_1^{\beta_1} x_2^{\beta_2} \cdots x_n^{\beta_n} - 1$ in $\mathcal{G}$?
   If yes, then output "$(\beta_1, \ldots, \beta_n) \in \mathcal{M}_\mathcal{A}$". If no, then output "INFEASIBLE".

In step 1 of Algorithm 1.4.3 we may encounter negative exponents a_{ij}. In practice these are dealt with as follows. Let t_0 be a new variable, and choose any elimination order $\{t_0, t_1, \ldots, t_d\} \succ \{x_1, \ldots, x_n\}$. Using the additional relation $t_0 t_1 \cdots t_d - 1$, clear the denominators in $x_i - \prod_{j=1}^{d} t_j^{a_{ij}}$, for $i = 1, 2, \ldots, n$. For the resulting $n+1$ polynomials compute a Gröbner basis $\mathcal{G}'$ with respect to $\prec$. Let $\mathcal{G} := \mathcal{G}' \cap \mathbf{C}[x_1, \ldots, x_n]$.

Algorithm 1.4.4 (Integer programming – Optimization problem (b)).

0. Choose a monomial order $\prec$ which *refines* the given cost vector $\omega \in \mathbf{R}_+^n$. By this we mean

$$\forall \alpha, \beta \in \mathbf{N}^n : \langle \alpha, \omega \rangle < \langle \beta, \omega \rangle \implies \mathbf{x}^\alpha \prec \mathbf{x}^\beta.$$

1. Let $\mathcal{G}$ be the reduced Gröbner basis with respect to $\prec$ for the kernel of (1.4.5).
2. Among all polynomials of the form $x_1^{\beta_1} x_2^{\beta_2} \cdots x_n^{\beta_n} - 1$ appearing in $\mathcal{G}$, choose the one which is smallest with respect to $\prec$. Output $(\beta_1, \beta_2, \ldots, \beta_n)$.

Proof of correctness for Algorithms 1.4.3 and 1.4.4. Let I denote the kernel of the map (1.4.5). This is a prime ideal in the polynomial ring $\mathbf{C}[x_1, \ldots, x_n]$, having the generic point $(\prod_{i=1}^{d} t_i^{a_{1i}}, \ldots, \prod_{i=1}^{d} t_i^{a_{ni}})$. By the proof of Lemma 1.4.2, a monomial $\mathbf{x}^\beta$ is invariant under $\Gamma_\mathcal{A}$ if and only if $\mathbf{x}^\beta$ is congruent to 1 modulo I. Therefore, if Algorithm 1.4.3 outputs a vector β, then β must lie in $\mathcal{M}_\mathcal{A}$.

We must show that Algorithm 1.4.3 outputs "INFEASIBLE" only if $\mathcal{M}_\mathcal{A} = \emptyset$. Suppose that $\mathcal{M}_\mathcal{A} \neq \emptyset$ and let $\beta \in \mathcal{M}_\mathcal{A}$. Then $\mathbf{x}^\beta - 1$ lies in the ideal I, and hence the normal form of $\mathbf{x}^\beta$ modulo the Gröbner basis $\mathcal{G}$ equals 1. In each step in the reduction of $\mathbf{x}^\beta$ a monomial reduces to another monomial. In the last step some monomial $\mathbf{x}^\gamma$ reduces to 1. This implies that $\mathbf{x}^\gamma - 1 \in \mathcal{G}$. This is a contradiction to the assumption that the output equals "INFEASIBLE". We conclude that Algorithm 1.4.3 terminates and is correct.

To see the correctness of Algorithm 1.4.4, we suppose that the output vector $\beta = (\beta_1, \ldots, \beta_n)$ is not the optimal solution to problem (b). Then there exists a vector $\beta' = (\beta'_1, \ldots, \beta'_n)$ in $\mathcal{M}_A$ such that $\langle \omega, \beta' \rangle < \langle \omega, \beta \rangle$. Since the monomial order $\prec$ refines ω, the reduction path from $\mathbf{x}^{\beta'}$ to 1 decreases the ω-cost of the occurring monomials. The last step in this reduction uses a relation $\mathbf{x}^\gamma - 1 \in \mathcal{G}$ with $\langle \omega, \gamma \rangle \leq \langle \omega, \beta' \rangle < \langle \omega, \beta \rangle$. This is a contradiction, because $\mathbf{x}^\gamma - 1$ would be chosen instead of $\mathbf{x}^\beta - 1$ in step 2. ◁

Our next algorithm uses $2n + d$ variables $t_1, \ldots, t_d, x_1, \ldots, x_n, y_1, \ldots, y_n$. We fix any elimination monomial order $\{t_1, \ldots, t_d\} \succ \{x_1, \ldots, x_n\} \succ \{y_1, \ldots, y_n\}$. Let J_A denote the kernel of the **C**-algebra homomorphism

$$\mathbf{C}[x_1, x_2, \ldots, x_n, y_1, y_2 \ldots, y_n] \to \mathbf{C}[t_1, \ldots, t_d, t_1^{-1}, \ldots, t_d^{-1}, y_1, \ldots, y_n],$$

$$x_1 \mapsto y_1 \prod_{j=1}^{d} t_j^{a_{1j}}, \ldots, x_n \mapsto y_n \prod_{j=1}^{d} t_j^{a_{nj}}, \ y_1 \mapsto y_1, \ldots, y_n \mapsto y_n.$$

(1.4.6)

Algorithm 1.4.5 (Integer programming – Hilbert basis problem (c)).

1. Compute the reduced Gröbner basis $\mathcal{G}$ with respect to $\prec$ for the ideal J_A.
2. The Hilbert basis $\mathcal{H}$ of $\mathcal{M}_A$ consists of all vectors β such that $\mathbf{x}^\beta - \mathbf{y}^\beta$ appears in $\mathcal{G}$.

Proof of correctness for Algorithm 1.4.5. We first note that J_A is a homogeneous prime ideal and that there is no monomial contained in J_A. By the same reasoning as above, a vector $\beta \in \mathbf{N}^n$ lies in $\mathcal{M}_A$ if and only if the monomial difference $\mathbf{x}^\beta - \mathbf{y}^\beta$ lies in J_A.

We wish to show that the finite subset $\mathcal{H} \subset \mathcal{M}_A$ constructed in step 2 spans the monoid $\mathcal{M}_A$. Suppose this is not the case. Then there exists a minimal (with respect to divisibility) monomial $\mathbf{x}^\beta$ such that $\beta \in \mathcal{M}_A$, but β is not a sum of elements in $\mathcal{H}$. The polynomial $\mathbf{x}^\beta - \mathbf{y}^\beta$ lies in J_A, so it reduces to zero modulo $\mathcal{G}$. By the choice of monomial order, the first reduction step replaces $\mathbf{x}^\beta$ by some monomial $\mathbf{x}^\gamma \mathbf{y}^\delta$, where $\delta = \beta - \gamma$ is non-zero. Therefore

$$\mathbf{x}^\gamma \mathbf{y}^\delta - \mathbf{y}^\beta = \mathbf{y}^\delta (\mathbf{x}^\gamma - \mathbf{y}^\gamma) \in J_A.$$

Since J_A is a prime ideal, not containing any monomials, we conclude that $\mathbf{x}^\gamma - \mathbf{y}^\gamma$ lies in J_A. This implies that γ lies in $\mathcal{M}_A$, and therefore the non-negative vector $\delta = \beta - \gamma$ lies in $\mathcal{M}_A$. By our minimality assumption on β, we have that both δ and γ can be written as sums of elements in $\mathcal{H}$. Therefore $\beta = \gamma + \delta$ can be written as sums of elements in $\mathcal{H}$. This is a contradiction, and the proof is complete. ◁

1.4. Torus invariants and integer programming

Example 1.4.1 (continued). For $\mathcal{A} = (3, 1, -2, -2)^T$, we consider the relations

$$\{x_1 - t^3 y_1, \ x_2 - t y_2, \ t^2 x_3 - y_3, \ t^2 x_4 - y_4\}$$

The reduced Gröbner basis with respect to the lexicographic monomial order $t \succ x_1 \succ x_2 \succ x_3 \succ x_4 \succ y_1 \succ y_2 \succ y_3 \succ y_4$ equals $\mathcal{G} =$

$\{t^3 y_1 - x_1, \ t^2 x_2 y_1 - x_1 y_2, \ t^2 x_3 - y_3, \ t^2 x_4 - y_4, \ t x_1 x_3^2 - y_1 y_3^2,$

$t x_1 x_3 x_4 - y_1 y_3 y_4, \ t x_1 x_4^2 - y_1 y_4^2, \ t x_2^2 y_1 - x_1 y_2^2, \ t x_2 x_3 - y_2 y_3, \ t x_2 x_4 - y_2 y_4,$

$t y_1 y_3 - x_1 x_3, \ t y_1 y_4 - x_1 x_4, \ t y_2 - x_2, \ \underline{x_1^2 x_3^3} - y_1^2 y_3^3, \ \underline{x_1^2 x_3^2 x_4} - y_1^2 y_3^2 y_4,$

$\underline{x_1^2 x_3 x_4^2} - y_1^2 y_3 y_4^2, \ \underline{x_1^2 x_4^3} - y_1^2 y_4^3, \ \underline{x_1 x_2 x_3^2} - y_1 y_2 y_3^2, \ \underline{x_1 x_2 x_3 x_4} - y_1 y_2 y_3 y_4,$

$\underline{x_1 x_2 x_4^2} - y_1 y_2 y_4^2, \ x_1 x_3 y_2 - x_2 y_1 y_3, \ x_1 x_4 y_2 - x_2 y_1 y_4, \ x_1 y_2^3 - x_2^3 y_1,$

$\underline{x_2^2 x_3} - y_2^2 y_3, \ \underline{x_2^2 x_4} - y_2^2 y_4, \ x_3 y_4 - x_4 y_3 \}$

The polynomials not containing the variable t form a Gröbner basis for the ideal $J_\mathcal{A}$. The Hilbert basis of $\mathcal{M}_\mathcal{A}$ consists of the nine underlined monomials.

Exercises

(1) Compute a Hilbert basis for $\mathcal{M}_\mathcal{A}$ where $\mathcal{A} = (4, 1, -2, -3)^T$. Verify your result using the polyhedral methods given in (Stanley 1986: section 4.6).
(2) * Give a bound for the complexity of the Hilbert basis $\mathcal{H}$ in terms of the input data $\mathcal{A}$.
(3) * With an integer $n \times d$-matrix $\mathcal{A}$ we can also associate the monoid

$$\mathcal{M}'_\mathcal{A} := \{\mu \in \mathbf{Z}^d \mid \mathcal{A} \cdot \mu \geq 0\}.$$

Give an algorithm, using Gröbner bases, for computing a Hilbert basis for $\mathcal{M}'_\mathcal{A}$.

(4) * With an integer $n \times d$-matrix $\mathcal{A}$ we can also associate the monoid

$$\mathcal{M}''_\mathcal{A} := \{\mu \in \mathbf{Z}^d \mid \exists v \in \mathbf{Q}^n : v \geq 0 \text{ and } v \cdot \mathcal{A} = \mu\}.$$

Give an algorithm, using Gröbner bases, for computing a Hilbert basis for $\mathcal{M}''_\mathcal{A}$.

2 Invariant theory of finite groups

Let $\mathbf{C}[\mathbf{x}]$ denote the ring of polynomials with complex coefficients in n variables $\mathbf{x} = (x_1, x_2, \ldots, x_n)$. We are interested in studying polynomials which remain invariant under the action of a finite matrix group $\Gamma \subset GL(\mathbf{C}^n)$. The main result of this chapter is a collection of algorithms for finding a finite set $\{I_1, I_2, \ldots, I_m\}$ of fundamental invariants which generate the invariant subring $\mathbf{C}[\mathbf{x}]^\Gamma$. These algorithms make use of the Molien series (Sect. 2.2) and the Cohen–Macaulay property (Sect. 2.3). In Sect. 2.4 we include a discussion of invariants of reflection groups, which is an important classical topic. Sections 2.6 and 2.7 are concerned with applications and special cases.

2.1. Finiteness and degree bounds

We start out by showing that every finite group has "sufficiently many" invariants.

Proposition 2.1.1. *Every finite matrix group $\Gamma \subset GL(\mathbf{C}^n)$ has n algebraically independent invariants, i.e., the ring $\mathbf{C}[\mathbf{x}]^\Gamma$ has transcendence degree n over $\mathbf{C}$.*

Proof. For each $i \in \{1, 2, \ldots, n\}$ we define $P_i := \prod_{\pi \in \Gamma}(x_i \circ \pi - t) \in \mathbf{C}[\mathbf{x}][t]$. Consider $P_i = P_i(t)$ as a monic polynomial in the new variable t whose coefficients are elements of $\mathbf{C}[\mathbf{x}]$. Since P_i is invariant under the action of Γ on the $\mathbf{x}$-variables, its coefficients are also invariant. In other words, P lies in the ring $\mathbf{C}[\mathbf{x}]^\Gamma[t]$.

We note that $t = x_i$ is a root of $P_i(t)$ because one of the $\pi \in \Gamma$ in the definition of P equals the identity. This means that all variables $x_1, x_2, \ldots, x_n$ are algebraically dependent upon certain invariants. Hence the invariant subring $\mathbf{C}[\mathbf{x}]^\Gamma$ and the full polynomial ring $\mathbf{C}[\mathbf{x}]$ have the same transcendence degree n over the ground field $\mathbf{C}$. ◁

The proof of Proposition 2.1.1 suggests that "averaging over the whole group" might be a suitable procedure for generating invariants. This idea can be made precise by introducing the following operator which maps polynomial functions onto their average with respect to the group Γ. The *Reynolds operator* "$*$" is defined as

$$* : \mathbf{C}[\mathbf{x}] \to \mathbf{C}[\mathbf{x}]^\Gamma, \quad f \mapsto f^* := \frac{1}{|\Gamma|} \sum_{\pi \in \Gamma} f \circ \pi.$$

Each of the following properties of the Reynolds operator is easily verified.

Proposition 2.1.2. The Reynolds operator "$*$" has the following properties.

(a) "$*$" is a **C**-linear map, i.e., $(\lambda f + \nu g)^* = \lambda f^* + \nu g^*$ for all $f, g \in \mathbf{C}[\mathbf{x}]$ and $\lambda, \nu \in \mathbf{C}$.
(b) "$*$" restricts to the identity map on $\mathbf{C}[\mathbf{x}]^\Gamma$, i.e., $I = I^*$ for all invariants $I \in \mathbf{C}[\mathbf{x}]^\Gamma$.
(c) "$*$" is a $\mathbf{C}[\mathbf{x}]^\Gamma$-module homomorphism, i.e., $(fI)^* = f^* \cdot I$ for all $f \in \mathbf{C}[\mathbf{x}]$ and $I \in \mathbf{C}[\mathbf{x}]^\Gamma$.

We are now prepared to prove a theorem about the invariant rings of finite groups.

Theorem 2.1.3 (Hilbert's finiteness theorem). *The invariant ring $\mathbf{C}[\mathbf{x}]^\Gamma$ of a finite matrix group $\Gamma \subset GL(\mathbf{C}^n)$ is finitely generated.*

Proof. Let $\mathcal{I}_\Gamma := \langle \mathbf{C}[\mathbf{x}]^\Gamma_+ \rangle$ be the *ideal* in $\mathbf{C}[\mathbf{x}]$ which is generated by all homogeneous invariants of positive degree. By Proposition 2.1.2 (a), every invariant I is a **C**-linear combination of symmetrized monomials $(x_1^{e_1} x_2^{e_2} \ldots x_n^{e_n})^*$. These homogeneous invariants are the images of monomials under the Reynolds operator. This implies that the ideal $\mathcal{I}_\Gamma$ is generated by the polynomials $(x_1^{e_1} x_2^{e_2} \ldots x_n^{e_n})^*$, where $\mathbf{e} = (e_1, e_2, \ldots, e_n)$ ranges over all non-zero, nonnegative integer vectors.

By Hilbert's basis theorem (Corollary 1.2.5), every ideal in the polynomial ring $\mathbf{C}[\mathbf{x}]$ is finitely generated. Hence there exist finitely many *homogeneous* invariants $I_1, I_2, \ldots, I_m$ such that $\mathcal{I}_\Gamma = \langle I_1, I_2, \ldots, I_m \rangle$. We shall now prove that all homogeneous invariants $I \in \mathbf{C}[\mathbf{x}]^\Gamma$ can actually be written as polynomial functions in $I_1, I_2, \ldots, I_m$.

Suppose the contrary, and let I be a homogeneous element of minimum degree in $\mathbf{C}[\mathbf{x}]^\Gamma \setminus \mathbf{C}[I_1, I_2, \ldots, I_m]$. Since $I \in \mathcal{I}_\Gamma$, we have $I = \sum_{j=1}^s f_j I_j$ for some homogeneous polynomials $f_j \in \mathbf{C}[\mathbf{x}]$ of degree less than $\deg(I)$. Applying the Reynolds operator on both sides of this equation we get

$$I = I^* = \Big(\sum_{j=1}^s f_j I_j\Big)^* = \sum_{j=1}^s f_j^* I_j$$

from Proposition 2.1.2. The new coefficients f_j^* are homogeneous invariants whose degree is less than $\deg(I)$. From the minimality assumption on I we get $f_j^* \in \mathbf{C}[I_1, \ldots, I_m]$ and therefore $I \in \mathbf{C}[I_1, \ldots, I_m]$, which is a contradiction to our assumption. This completes the proof of Theorem 2.1.3. ◁

This proof of Theorem 2.1.3 implies the remarkable statement that every *ideal basis* $\{I_1, \ldots, I_m\}$ of $\mathcal{I}_\Gamma$ is automatically an *algebra basis* for $\mathbf{C}[\mathbf{x}]^\Gamma$, i.e., a fundamental system of invariants. Observe also that in this proof the finiteness of the group Γ has not been used until the last paragraph. The only hypothesis on the group Γ which we really needed was the existence of an averaging operator "$*$" which satisfies (a), (b) and (c) in Proposition 2.1.2.

2.1. Finiteness and degree bounds

The finiteness theorem and its proof remain valid for infinite groups Γ which do admit a Reynolds operator with these properties. These groups are called *reductive*. In particular, it is known that every matrix representation Γ of a compact Lie group is reductive. The Reynolds operator of such a compact group Γ is defined by the formula $f^* = \int_\Gamma (f \circ \pi) d\pi$, where $d\pi$ is the Haar probability measure on Γ. For details on reductive groups and proofs of the general finiteness theorem we refer to Dieudonné and Carrell (1971) or Springer (1977).

Let us now return to the case of a finite group Γ. Here the general inconstructive finiteness result of Hilbert can be improved substantially. The following effective version of the finiteness theorem is due to E. Noether (1916).

Theorem 2.1.4 (Noether's degree bound). *The invariant ring $\mathbf{C}[\mathbf{x}]^\Gamma$ of a finite matrix group Γ has an algebra basis consisting of at most $\binom{n+|\Gamma|}{n}$ invariants whose degree is bounded above by the group order $|\Gamma|$.*

Proof. With every vector $\mathbf{e} = (e_1, e_2, \ldots, e_n)$ of nonnegative integers we associate the homogeneous invariant $J_\mathbf{e}(\mathbf{x}) := (x_1^{e_1} x_2^{e_2} \ldots x_n^{e_n})^*$ which is obtained by applying the Reynolds operator to the monomial with exponent vector $\mathbf{e}$. We abbreviate $e := |\mathbf{e}| = e_1 + e_2 + \ldots + e_n$.

Let $u_1, \ldots, u_n$ be a new set of variables, and consider the polynomial

$$S_e(\mathbf{u}, \mathbf{x}) := \{(u_1 x_1 + \ldots + u_n x_n)^e\}^*$$
$$= \frac{1}{|\Gamma|} \sum_{\pi \in \Gamma} [u_1(x_1 \circ \pi) + \ldots + u_n(x_n \circ \pi)]^e$$

in the new variables whose coefficients are polynomials in the old variables $x_1, \ldots, x_n$. The Reynolds operator "$*$" acts on such polynomials by regarding the u_i as constants. By complete expansion of the above expression, we find that the coefficient of $u_1^{e_1} \ldots u_n^{e_n}$ in S_e is equal to the invariant $J_\mathbf{e}$ times a positive integer.

The polynomials S_e are the power sums of the $|\Gamma|$ magnitudes $u_1(x_1 \circ \pi) + \ldots + u_n(x_n \circ \pi)$ where π ranges over Γ. By Proposition 1.1.2, we can express each power sum S_e as a polynomial function in the first $|\Gamma|$ power sums $S_1, S_2, \ldots, S_{|\Gamma|}$. Such a representation of S_e shows that all $\mathbf{u}$-coefficients are actually polynomial functions in the $\mathbf{u}$-coefficients of $S_1, S_2, \ldots, S_{|\Gamma|}$.

This argument proves that the invariants $J_\mathbf{e}$ with $|\mathbf{e}| > |\Gamma|$ are contained in the subring $\mathbf{C}[\{J_\mathbf{e} : |\mathbf{e}| \le |\Gamma|\}]$. We have noticed above that every invariant is a $\mathbf{C}$-linear combination of the special invariants $J_\mathbf{e}$. This implies that

$$\mathbf{C}[\mathbf{x}]^\Gamma = \mathbf{C}[\{J_\mathbf{e} : |\mathbf{e}| \le |\Gamma|\}].$$

The set of integer vectors $\mathbf{e} \in \mathbf{N}^n$ with $|\mathbf{e}| \le |\Gamma|$ has cardinality $\binom{n+|\Gamma|}{n}$. ◁

The following proposition shows that, from the point of view of worst case complexity, the Noether degree bound is optimal.

Proposition 2.1.5. For any two integers $n, p \geq 2$ there exists a p-element matrix group $\Gamma \subset GL(\mathbf{C}^n)$ such that every algebra basis for $\mathbf{C}[\mathbf{x}]^\Gamma$ contains at least $\binom{n+p-1}{n-1}$ invariants of degree p.

Proof. Consider the action of the p-element cyclic group on $\mathbf{C}^n$ given by

$$\Gamma := \left\{ \mathrm{diag}(e^{\frac{2ik\pi}{p}}, e^{\frac{2ik\pi}{p}}, \ldots, e^{\frac{2ik\pi}{p}}) : k = 0, 1, \ldots, p-1 \right\}.$$

We can easily determine the action the Reynolds operator on all monomials:

$$(x_1^{e_1} x_2^{e_2} \cdots x_n^{e_n})^* = \begin{cases} x_1^{e_1} x_2^{e_2} \cdots x_n^{e_n} & \text{if } p \text{ divides } e = e_1 + \ldots + e_n \\ 0 & \text{otherwise.} \end{cases}$$

This shows that the invariant ring $\mathbf{C}[\mathbf{x}]^\Gamma$ is the *Veronese subalgebra* of $\mathbf{C}[\mathbf{x}]$ which is generated by all monomials of total degree p. Clearly, any graded algebra basis for this ring must contain a vector space basis for the $\binom{n+p-1}{n-1}$-dimensional $\mathbf{C}$-vector space of n-variate polynomials of total degree p. ◁

The lower bounds in Proposition 2.1.5 have been shown to hold for essentially all primitive groups Γ by Huffman and Sloane (1979). In spite of these discouraging results, there are many special groups for which the system of fundamental invariants is much smaller. For such groups and for studying properties of invariant rings in general, the technique of "linear algebra plus degree bounds" will not be sufficient, but we will need the refined techniques and algorithms to be developed in the subsequent sections.

Exercises

(1) Determine the invariant rings of all finite subgroups of $GL(\mathbf{C}^1)$, that is, the finite multiplicative subgroups of the complex numbers.
(2) Let $* : \mathbf{C}[x_1, x_2] \to \mathbf{C}[x_1, x_2]^{Z_4}$ be the Reynolds operator of the cyclic group in Example 1.3.2., and consider its restriction to the 5-dimensional vector space of homogeneous polynomials of degree 4. Represent this $\mathbf{C}$-linear map "$*$" by a 5×5-matrix A, and compute the rank, image and kernel of A.
(3) Consider the action of the symmetric group S_4 on $\mathbf{C}[x_{12}, x_{13}, x_{14}, x_{23}, x_{24}, x_{34}]$ by permuting indices of the six variables (subject to the relations $x_{ji} = x_{ij}$). Determine a minimal algebra basis for the ring of invariants. Compare your answer with the bounds in Theorem 2.1.4.
(4) Let $\Gamma \subset GL(\mathbf{C}^n)$ be a finite matrix group and $\mathcal{I} \subset \mathbf{C}[\mathbf{x}]$ an ideal which is fixed by Γ. Show that Γ acts on the quotient ring $\mathbf{C}[\mathbf{x}]/\mathcal{I}$, and give

an algorithm for computing a finite algebra basis for the invariant ring $(\mathbf{C}[\mathbf{x}]/\mathcal{I})^\Gamma$.

2.2. Counting the number of invariants

We continue our discussion with the problem "how many invariants does a given matrix group Γ have?" Such an enumerative question can be made precise as follows. Let $\mathbf{C}[\mathbf{x}]_d^\Gamma$ denote the set of all homogeneous invariants of degree d. The invariant ring $\mathbf{C}[\mathbf{x}]^\Gamma$ is the direct sum of the finite-dimensional $\mathbf{C}$-vector spaces $\mathbf{C}[\mathbf{x}]_d^\Gamma$. By definition, the *Hilbert series* of the graded algebra $\mathbf{C}[\mathbf{x}]^\Gamma$ is the generating function $\Phi_\Gamma(z) = \sum_{d=0}^\infty \dim(\mathbf{C}[\mathbf{x}]_d^\Gamma) z^d$.

The following classical theorem gives an explicit formula for the Hilbert series of $\mathbf{C}[\mathbf{x}]^\Gamma$ in terms of the matrices in Γ. We write *id* for the $n \times n$-identity matrix.

Theorem 2.2.1 (Molien 1897). *The Hilbert series of the invariant ring $\mathbf{C}[\mathbf{x}]^\Gamma$ equals*

$$\Phi_\Gamma(z) = \frac{1}{|\Gamma|} \sum_{\pi \in \Gamma} \frac{1}{\det(id - z\pi)}.$$

Theorem 2.2.1 states in other words that the Hilbert series of the invariant ring is the average of the inverted characteristic polynomials of all group elements. In order to prove this result we need the following lemma from linear algebra.

Lemma 2.2.2. *Let $\Gamma \subset GL(\mathbf{C}^n)$ be a finite matrix group. Then the dimension of the invariant subspace*

$$V^\Gamma = \{\mathbf{v} \in \mathbf{C}^n : \pi \mathbf{v} = \mathbf{v} \text{ for all } \pi \in \Gamma\}$$

is equal to $\frac{1}{|\Gamma|} \sum_{\pi \in \Gamma} \text{trace}(\pi)$.

Proof. Consider the average matrix $P_\Gamma := \frac{1}{|\Gamma|} \sum_{\pi \in \Gamma} \pi$. This linear map is a projection onto the invariant subspace V^Γ. Since the matrix P_Γ defines a projection, we have $P_\Gamma = P_\Gamma^2$, which means that P_Γ has only the eigenvalues 0 and 1. Therefore the rank of the matrix P_Γ equals the multiplicity of its eigenvalue 1, and we find $\dim(V^\Gamma) = \text{rank}(P_\Gamma) = \text{trace}(P_\Gamma) = \frac{1}{|\Gamma|} \sum_{\pi \in \Gamma} \text{trace}(\pi)$. ◁

Proof of Theorem 2.2.1. We write $\mathbf{C}[\mathbf{x}]_d$ for the $\binom{n+d-1}{d}$-dimensional vector space of d-forms in $\mathbf{C}[\mathbf{x}]$. For every linear transformation $\pi \in \Gamma$ there is an induced linear transformation $\pi^{(d)}$ on the vector space $\mathbf{C}[\mathbf{x}]_d$. In this linear algebra notation $\mathbf{C}[\mathbf{x}]_d^\Gamma$ becomes precisely the invariant subspace of $\mathbf{C}[\mathbf{x}]_d$ with respect to the induced group $\{\pi^{(d)} : \pi \in \Gamma\}$ of $\binom{n+d-1}{d} \times \binom{n+d-1}{d}$-matrices.

In order to compute the trace of an induced transformation $\pi^{(d)}$, we identify the vector space $\mathbf{C}^n$ with its linear forms $\mathbf{C}[\mathbf{x}]_1$. Let $\ell_{\pi,1}, \ldots, \ell_{\pi,n} \in \mathbf{C}[\mathbf{x}]_1$ be the eigenvectors of $\pi = \pi^{(1)}$, and let $\rho_{\pi,1}, \ldots, \rho_{\pi,n} \in \mathbf{C}$ denote the corresponding eigenvalues. Note that each matrix $\pi \in \Gamma$ is diagonalizable over $\mathbf{C}$ because it has finite order.

The eigenvectors of $\pi^{(d)}$ are precisely the $\binom{n+d-1}{d}$ d-forms $\ell_{\pi,1}^{d_1} \cdots \ell_{\pi,n}^{d_n}$ where $d_1 + \ldots + d_n = d$. The eigenvalues of $\pi^{(d)}$ are therefore the complex numbers $\rho_{\pi,1}^{d_1} \cdots \rho_{\pi,n}^{d_n}$ where $d_1 + \ldots + d_n = d$. Since the trace of a linear transformation equals the sum of its eigenvalues, we have the equation

$$\operatorname{trace}(\pi^{(d)}) = \sum_{d_1+\ldots+d_n=d} \rho_{\pi,1}^{d_1} \cdots \rho_{\pi,n}^{d_n}.$$

By Lemma 2.2.2, the dimension of the invariant subspace $\mathbf{C}[\mathbf{x}]_d^\Gamma$ equals the average of the traces of all group elements. Rewriting this dimension count in terms of the Hilbert series of the invariant ring, we get:

$$\Phi_\Gamma(z) = \sum_{d=0}^\infty \frac{1}{|\Gamma|} \sum_{\pi \in \Gamma} \Big(\sum_{d_1+\ldots+d_n=d} \rho_{\pi,1}^{d_1} \cdots \rho_{\pi,n}^{d_n} \Big) z^d$$

$$= \frac{1}{|\Gamma|} \sum_{\pi \in \Gamma} \sum_{(d_1,\ldots,d_n)\in\mathbf{N}^n} \rho_{\pi,1}^{d_1} \cdots \rho_{\pi,n}^{d_n} z^{d_1+\ldots+d_n}$$

$$= \frac{1}{|\Gamma|} \sum_{\pi \in \Gamma} \frac{1}{(1-z\rho_{\pi,1}) \cdots (1-z\rho_{\pi,n})}$$

$$= \frac{1}{|\Gamma|} \sum_{\pi \in \Gamma} \frac{1}{\det(id - z\pi)}. \quad \triangleleft$$

In the remainder of this section we illustrate the use of Molien's theorem for computing invariants. For that purpose we need the following general lemma which describes the Hilbert series of a graded polynomial subring of $\mathbf{C}[\mathbf{x}]$.

Lemma 2.2.3. Let $p_1, p_2, \ldots, p_m$ be algebraically independent elements of $\mathbf{C}[\mathbf{x}]$ which are homogeneous of degrees $d_1, d_2, \ldots, d_m$ respectively. Then the Hilbert series of the graded subring $R := \mathbf{C}[p_1, p_2, \ldots, p_m]$ equals

$$H(R, z) := \sum_{n=0}^\infty (\dim_\mathbf{C} R_d) z^d = \frac{1}{(1-z^{d_1})(1-z^{d_2})\ldots(1-z^{d_m})}.$$

Proof. Since the p_i are algebraically independent, the set

$$\{p_1^{i_1} p_2^{i_2} \cdots p_m^{i_m} : i_1, i_2, \ldots, i_m \in \mathbf{N} \text{ and } i_1 d_1 + i_2 d_2 + \ldots + i_m d_m = d\}$$

is a basis for the $\mathbf{C}$-vector space R_d of degree d elements in R. Hence the

2.2. Counting the number of invariants

dimension of R_d equals the cardinality of the set

$$A_d := \{(i_1, i_2, \ldots, i_m) \in \mathbf{N}^m : i_1 d_1 + i_2 d_2 + \ldots + i_m d_m = d\}.$$

The expansion

$$\frac{1}{(1-z^{d_1})(1-z^{d_2})\ldots(1-z^{d_m})} = \frac{1}{(1-z^{d_1})} \cdot \frac{1}{(1-z^{d_2})} \cdots \frac{1}{(1-z^{d_m})}$$

$$= \Big(\sum_{i_1=0}^{\infty} z^{i_1 d_1}\Big)\Big(\sum_{i_2=0}^{\infty} z^{i_2 d_2}\Big)\cdots\Big(\sum_{i_m=0}^{\infty} z^{i_m d_m}\Big)$$

$$= \sum_{d=0}^{\infty} \sum_{(d_1, d_2 \ldots, d_m) \in A_d} z^d = \sum_{d=0}^{\infty} |A_d| z^d$$

proves the claim of Lemma 2.2.3. ◁

The following matrix group had already been considered in Example 1.3.2.

Example 2.2.4.
The invariant ring $\mathbf{C}[x_1, x_2]^{Z_4}$ of the group $\{\pm \begin{pmatrix} 1 & 0 \\ 0 & 1 \end{pmatrix}, \pm \begin{pmatrix} 0 & -1 \\ 1 & 0 \end{pmatrix}\}$ is generated by the invariants $I_1 := x_1^2 + x_2^2$, $I_2 := x_1^2 x_2^2$ and $I_3 := x_1 x_2^3 - x_1^3 x_2$.

Proof. The graded algebra $\mathbf{C}[I_1, I_2, I_3]$ is clearly contained in the graded algebra $\mathbf{C}[x_1, x_2]^{Z_4}$. In order to establish that these two algebras are equal, it suffices that, for each $d \in \mathbf{N}$, their graded components $\mathbf{C}[I_1, I_2, I_3]_d$ and $\mathbf{C}[x_1, x_2]^{Z_4}_d$ have the same finite dimension as $\mathbf{C}$-vector spaces. In other words, we need to show that the Hilbert series of $\mathbf{C}[I_1, I_2, I_3]$ equals the Molien series of the invariant ring.

The Hilbert series $\Phi_{Z_4}(z)$ of $\mathbf{C}[x_1, x_2]^{Z_4}$ can be computed using Molien's Theorem.

$$\Phi_{Z_4}(z)$$
$$= \frac{1}{4}\left[\frac{1}{\begin{vmatrix} 1-z & 0 \\ 0 & 1-z \end{vmatrix}} + \frac{1}{\begin{vmatrix} -1-z & 0 \\ 0 & -1-z \end{vmatrix}} + \frac{1}{\begin{vmatrix} -z & 1 \\ -1 & -z \end{vmatrix}} + \frac{1}{\begin{vmatrix} -z & -1 \\ 1 & -z \end{vmatrix}}\right]$$

$$= \frac{1}{4}\left[\frac{1}{(1-z)^2} + \frac{1}{(1+z)^2} + \frac{2}{1+z^2}\right]$$

$$= \frac{1+z^4}{(1-z^2)(1-z^4)}$$

$$= 1 + z^2 + 3z^4 + 3z^6 + 5z^8 + 5z^{10} + 7z^{12} + 7z^{14} + 9z^{16} + 9z^{18} + \ldots$$

The Hilbert series of $\mathbf{C}[I_1, I_2, I_3]$ can be computed as follows. Using the Gröbner basis method discussed in Sect. 1.2 (see also Subroutine 2.5.3), we find that the algebraic relation $I_3^2 - I_2 I_1^2 + 4I_2^3$ generates the ideal of syzygies among the I_j. This implies that every polynomial $p \in \mathbf{C}[I_1, I_2, I_3]$ can be written uniquely in the form $p(I_1, I_2, I_3) = q(I_1, I_2) + I_3 \cdot r(I_1, I_2)$ where q and r are bivariate polynomials. In other words, the graded algebra in question is decomposed as the direct sum of graded $\mathbf{C}$-vector spaces

$$\mathbf{C}[I_1, I_2, I_3] = \mathbf{C}[I_1, I_2] \oplus I_3 \, \mathbf{C}[I_1, I_2].$$

The first component in this decomposition is a subring generated by algebraically independent homogeneous polynomials. Using Lemma 2.2.3, we find that its Hilbert series equals $\frac{1}{(1-z^2)(1-z^4)}$. Since the degree d elements in $\mathbf{C}[I_1, I_2]$ are in one-to-one correspondence with the degree $d+4$ elements in $I_3 \, \mathbf{C}[I_1, I_2]$, the Hilbert series of the second component equals $\frac{z^4}{(1-z^2)(1-z^4)}$. The sum of these two series equals $\Phi_{Z_4}(z)$, and it is the Hilbert series of $\mathbf{C}[I_1, I_2, I_3]$ because the vector space decomposition is direct. ◁

The method we used in Example 2.2.4 for proving the completeness of a given system of invariants works in general.

Algorithm 2.2.5 (Completeness of fundamental invariants). Suppose we are given a set of invariants $\{I_1, \ldots, I_m\} \subset \mathbf{C}[\mathbf{x}]^\Gamma$. We wish to decide whether this set is complete, i.e., whether the invariant ring $\mathbf{C}[\mathbf{x}]^\Gamma$ equals its subalgebra $R = \mathbf{C}[I_1, \ldots, I_m]$. This is the case if and only if the Hilbert series $H(R, z)$ is equal to the Molien series $\Phi_\Gamma(z)$. Otherwise, we can subtract $H(R, z)$ from the Molien series, and we get $\Phi_\Gamma(z) - H(R, z) = c_d z^d + \textit{higher terms}$, where c_d is some positive integer. From this we conclude that there are c_d linearly independent invariants of degree d which cannot be expressed as polynomials in $I_1, \ldots, I_m$. We may now compute these extra invariants (using the Reynolds operator) and proceed by adding them to the initial set $\{I_1, \ldots, I_m\}$.

Hence our problem is reduced to computing the Hilbert function of a graded subalgebra $\mathbf{C}[I_1, \ldots, I_m] \subset \mathbf{C}[\mathbf{x}]$ which is presented in terms of homogeneous generators. Let $d_j := \deg(I_j)$. Using the Subroutine 2.5.3, we compute any Gröbner basis $\mathcal{G} = \{g_1, \ldots, g_r\}$ for the kernel $\mathcal{I}$ of the map of polynomial rings $\mathbf{C}[y_1, \ldots, y_m] \to \mathbf{C}[x_1, \ldots, x_n]$, $y_i \mapsto I_i(\mathbf{x})$. Then R is isomorphic as a graded $\mathbf{C}$-algebra to $\mathbf{C}[y_1, \ldots, y_m]/\mathcal{I}$ where the degree of each variable y_j is defined to be d_j.

By Theorem 1.2.6, R is isomorphic as a graded $\mathbf{C}$-vector space to $\mathbf{C}[y_1, \ldots, y_m]/\langle \operatorname{init}(g_1), \ldots, \operatorname{init}(g_r) \rangle$. Hence the d-th coefficient $\dim_\mathbf{C}(R_d)$ of the desired Hilbert series $H(R, z)$ equals the number of monomials $y_1^{i_1} y_2^{i_2} \cdots y_m^{i_m}$ with $i_1 d_1 + \ldots + i_m d_m = d$ which are not multiples of any of the monomials $\operatorname{init}(g_1), \ldots, \operatorname{init}(g_r)$. Fast combinatorial algorithms for determining this number are given in Bayer and Stillman (1992) and Bigatti et al. (1992). These algorithms are implemented in the computer algebra systems MACAULAY and COCOA respectively.

2.2. Counting the number of invariants

Example 2.2.6 (A 3-dimensional representation of the dihedral group D_6). Consider the action of the dihedral group $D_6 = \{id, \delta, \delta^2, \delta^3, \delta^4, \delta^5, \sigma, \sigma\delta, \sigma\delta^2, \sigma\delta^3, \sigma\delta^4, \sigma\delta^5\}$ on $\mathbf{C}[x, y, z]$ which is defined by the matrices

$$\sigma = \begin{pmatrix} 1 & 0 & 0 \\ 0 & -1 & 0 \\ 0 & 0 & -1 \end{pmatrix} \quad \text{and} \quad \delta := \begin{pmatrix} 1/2 & -\sqrt{3}/2 & 0 \\ \sqrt{3}/2 & 1/2 & 0 \\ 0 & 0 & 1 \end{pmatrix}$$

By computing the characteristic polynomials of all twelve matrices we obtain

$$\Phi_{D_6}(t) = \frac{1}{12} \sum_{\pi \in D_6} \frac{1}{\det(id - t\pi)}$$

$$= \frac{1}{12}\left(\frac{1}{(1-t)^3} + \frac{2}{(1-t)(t^2-t+1)} + \frac{2}{(1-t)(t^2+t+1)} + \frac{7}{(1-t)(t+1)^2}\right)$$

$$= 1 + 2t^2 + 3t^4 + 5t^6 + t^7 + 7t^8 + 2t^9 + 9t^{10} + 3t^{11} + 12t^{12} + 5t^{13} + 15t^{14} + O(t^{15}).$$

According to Proposition 2.1.1 there exist three algebraically independent invariants. The Molien series $\Phi_{D_6}(t)$ suggests to search for such invariants in degree 2 and 6. Using the Reynolds operator we find

$$P_2 := x^2 + y^2, \quad Q_2 := z^2, \quad P_6 := x^6 - 6x^4y^2 + 9x^2y^4.$$

We can see (e.g., using Gröbner bases) that P_2, Q_2 and P_6 are algebraically independent over $\mathbf{C}$. By Lemma 2.2.3, their subring $R = \mathbf{C}[P_2, Q_2, P_6]$ has the Hilbert series

$$H(R, t) = \frac{1}{(1-t^2)^2(1-t^6)} = 1 + 2t^2 + 3t^4 + 5t^6 + 7t^8 + 9t^{10} + 12t^{12} + \ldots$$

Since $\Phi_{D_6}(t) - H(R, t) = t^7 + 2t^9 + \ldots$ is non-zero, R is a proper subring of $\mathbf{C}[x, y, z]^{D_6}$. We need to find an additional invariant in degree 7. For instance, let
$$P_7 := 3x^5yz - 10x^3y^3z + 3xy^5z.$$

Following Algorithm 2.2.5 we now compute a Gröbner basis $\mathcal{G}$ for the set $\{P_2(x, y, z) - p_2, Q_2(x, y, z) - Q_2, P_6(x, y, z) - p_6, P_7(x, y, z) - p_7\}$, where p_2, q_2, p_6, p_7 are new variables lexicographically smaller than x, y, z. We find

$$\mathcal{G} \cap \mathbf{C}[p_2, q_2, p_6, p_7] = \{p_7^2 - p_2^3 q_2 q_6 + q_2 q_6^2\}.$$

which means that the four invariants satisfy a unique syzygy of degree 14. We conclude that the current subring $R' = \mathbf{C}[P_2, Q_2, P_6, P_7]$ has the Hilbert series

$$H(R', t) = \frac{(1+t^7)}{(1-t^2)^2(1-t^6)}$$
$$= 1 + 2t^2 + 3t^4 + 5t^6 + t^7 + 7t^8 + 2t^9 + 9t^{10} + 3t^{11} + \cdots$$

This series is equal to the Molien series, and hence $\{P_2, Q_2, P_6, P_7\}$ is a complete set of invariants. Every other invariant $I(x, y, z)$ can be expressed as a polynomial function in P_2, Q_2, P_6, P_7 by computing the normal form of $I(x, y, z)$ with respect to $\mathcal{G}$.

In the remainder of this section we present an application of the invariant theory of finite groups to the study of error-correcting codes. Our discussion is based on an expository paper of N. J. A. Sloane (1977), and we refer to that article for details and a guide to the coding theory literature. According to Sloane's "general plan of attack", there are two stages in using invariant theory to solve a problem.

I. Convert the assumptions about the problem (e.g., from coding theory) into algebraic constraints on polynomials (e.g., weight enumerators).
II. Use invariant theory to find *all* possible polynomials satisfying these constraints.

Imagine a noisy telegraph line from Ithaca to Linz, which transmits 0s and 1s. Usually when a 0 is sent from Ithaca it is received as a 0 in Linz, but occasionally a 0 is received as a 1. Similarly a 1 is occasionally received as a 0. The problem is to send a lot of important messages down this line, as quickly and as reliably as possible. The coding theorist's solution is to send certain strings of 0s and 1s, called *code words*.

Consider a simple example: One of two messages will be sent, either YES or NO. The message YES will be encoded into the code word 00000, and NO into 11111. Suppose 10100 is received in Linz. The receiver argues that it is more likely that 00000 was sent (and two errors occurred) than that 11111 was sent (and three errors occurred), and therefore *decodes* 10100 as 00000 = YES. For in some sense 10100 is closer to 00000 than to 11111. To make this precise, define the *Hamming distance* dist($\mathbf{u}, \mathbf{v}$) between two vectors $\mathbf{u} = (u_1, \ldots, u_n)$ and $\mathbf{v} = (v_1, \ldots, v_n)$ to be the number of places where $u_i \neq v_i$. It is easily checked that "dist" is a metric. Then the receiver should decode the received vector as the closest code word, measured in the Hamming distance.

Notice that in the above example two errors were corrected. This is possible because the code words 00000 and 11111 are at distance 5 apart. In general, if d is the minimum Hamming distance between any two code words, then the code can correct $e = [(d-1)/2]$ errors, where $[x]$ denotes the greatest integer not exceeding $[x]$. This motivates the following definition. Let V be the vector space of dimension n over $GF(2)$ consisting of all n-tuples of 0s and 1s. An

2.2. Counting the number of invariants

$[n, k, d]$ *binary code* is a k-dimensional linear subspace $C \subset V$ such that any two code words in C differ in at least d places. Then n is called the *length*, k the *dimension*, and d the *minimum distance* of the code. In a good code n is small (for rapid transmission), k is large (for an efficient code), and d is large (to correct many errors). These are incompatible goals, and coding theory deals with the problem to find best possible compromises.

The *weight* of vector $\mathbf{u} = (u_1, \ldots, u_n)$ is the number of nonzero u_i. Since a code C is a linear space, for any code words $\mathbf{u}, \mathbf{v}$, dist$(\mathbf{u}, \mathbf{v})$ = weight$(\mathbf{u} - \mathbf{v})$ = weight$(\mathbf{w})$ for some $\mathbf{w} \in C$. Therefore the minimum distance d between code words equals the smallest weight of any nonzero code word. The weight enumerator of an $[n, k, d]$ code C is a bivariate polynomial which tells the number of code words of each weight. If C contains a_i code words of weight i, then the *weight enumerator* of C is defined to be

$$W_C(x_1, x_2) := \sum_{i=0}^{n} a_i \, x_1^{n-i} x_2^i.$$

Notice that W_C is a homogeneous polynomial of degree n. The weight enumerator immediately gives the minimum distance d of C. For C always contains the zero code word, giving the leading monomial x_1^n of W_C, and the next non-zero monomial is $a_d \, x_1^{n-d} x_2^d$. As an example consider the $[3, 2, 2]$ code $C_1 = \{000, 011, 101, 110\}$. Its weight enumerator equals $W_{C_1} = x_1^3 + 3x_1 x_2^2$.

Let C be any $[n, k, d]$ code. The *dual code* $C^\perp$ consists of all vectors having zero dot product in $GF(2)$ with every code word of C. It is an $[n, n-k, d']$ code for some d'. E.g., the dual code of C_1 is the $[3, 1, 3]$ code $C_1^\perp = \{000, 111\}$. A *self-dual* code is one for which $C^\perp = C$. In a self-dual code k must be equal to $n/2$, and so n must be even.

Example 2.2.7. The following 16 code words

```
00000000  11101000  01110100  00111010  10011100  01001110
10100110  11010010  11111111  00010111  10001011  11000101
01100011  10110001  01011001  00101101
```

define a self-dual $[8, 4, 4]$ code C_2. Its weight enumerator equals $W_{C_2} = x_1^8 + 14x_1^4 x_2^4 + x_2^8$.

The following theorem relates the weight enumerators of dual pairs of codes. A proof can be found in Sloane (1977: theorem 6).

Theorem 2.2.8. If C is an $[n, k, d]$ binary code with dual code $C^\perp$, then

$$W_{C^\perp}(x_1, x_2) = \frac{1}{2^k} \cdot W_C(x_1 + x_2, x_1 - x_2).$$

The class of self-dual codes is of particular interest in coding theory because here the *decoding* of messages is relatively easy (Sloane, 1977: sect. II.B). It is the study of this class of codes to which invariant theory of finite groups has been applied. The basic observation is the following.

Corollary 2.2.9. Let $W_\mathcal{C}$ be the weight enumerator of a self-dual binary code $\mathcal{C}$. Then
$$W_\mathcal{C}\big((x_1 + x_2)/\sqrt{2},\ (x_1 - x_2)/\sqrt{2}\big) = W_\mathcal{C}(x_1, x_2)$$
$$W_\mathcal{C}(x_1, -x_2) = W_\mathcal{C}(x_1, x_2).$$

Proof. The first of these identities follows from Theorem 2.2.8 and the fact that $W_\mathcal{C}$ is homogeneous of degree $n = 2k$. The second one is equivalent to the fact that every $\mathbf{w} \in \mathcal{C}$ has an even number of 1s, since $\mathbf{w} \cdot \mathbf{w} = 0$. ◁

We rephrase Corollary 2.2.9 in the language of invariant theory. Consider the group D_8 which is generated by the matrices $\frac{1}{\sqrt{2}}\begin{pmatrix} 1 & 1 \\ 1 & -1 \end{pmatrix}$ and $\begin{pmatrix} 1 & 0 \\ 0 & -1 \end{pmatrix}$. It consists of 16 elements, and geometrically speaking, D_8 is the symmetry group of a regular hexagon in the plane.

Corollary 2.2.9'. Let $W_\mathcal{C}$ be the weight enumerator of a self-dual binary code $\mathcal{C}$. Then $W_\mathcal{C}$ is a polynomial invariant of the group D_8.

Proposition 2.2.10. The invariant ring $\mathbf{C}[x_1, x_2]^{D_8}$ is generated by the fundamental invariants $\theta_1 := x_1^2 + x_2^2$ and $\theta_2 := x_1^2 x_2^2 (x_1^2 - x_2^2)^2$.

Corollary 2.2.11. The weight enumerator of every self-dual binary code is a polynomial function in θ_1 and θ_2.

As an example consider the weight enumerator $W_{\mathcal{C}_2} = x_1^8 + 14x_1^4 x_2^4 + x_2^8$ of the self-dual code in Example 2.2.7. We have the representation $W_{\mathcal{C}_2} = \theta_1^4 - 4 \cdot \theta_2$ in terms of fundamental invariants.

One of the main applications of Sloane's approach consisted in proving the nonexistence of certain very good codes. The desired properties (e.g., minimum distance) of the code are expressed in a tentative weight enumerator W, and invariant theory can then be used to show that no such invariant W exists.

Exercises

(1) Compute the Hilbert series of the ring $\mathbf{C}[\sigma_1, \sigma_2, \ldots, \sigma_n]$ of symmetric polynomials.
(2) Consider the subring of $\mathbf{C}[x_1, x_2, x_3, x_4]$ consisting of all polynomials p which satisfy the shift invariance $p(x_1, x_2, x_3, x_4) = p(x_2, x_3, x_4, x_1)$. Find a generating set for this invariant ring and use Molien's theorem to prove the correctness of your result.
(3) The dihedral group D_n acts on $\mathbf{C}[x, y]$ via the symmetries of a regular

n-gon. Show that there exists an invariant θ of degree n such that $\mathbf{C}[x^2+y^2, \theta] = \mathbf{C}[x,y]^{D_n}$. In particular, prove Proposition 2.2.10.
(4) * Let $\Gamma \subset GL(\mathbf{C}^n)$ be a finite matrix group and let $\chi : \Gamma \to \mathbf{C}^*$ be any character.
   (a) Find a generalization of Molien's theorem 2.2.1 for the graded vector space of *relative invariants* $\mathbf{C}[\mathbf{x}]_\chi^\Gamma = \{f \in \mathbf{C}[\mathbf{x}] : f \circ \pi = \chi(\pi) \cdot f\}$.
   (b) Show that $\mathbf{C}[\mathbf{x}]_\chi^\Gamma$ is a finitely generated module over the invariant ring $\mathbf{C}[\mathbf{x}]^\Gamma$, and give an algorithm for computing a set of module generators.
   (c) Find an example where $\mathbf{C}[\mathbf{x}]_\chi^\Gamma$ is not free as a $\mathbf{C}[\mathbf{x}]^\Gamma$-module.
(5) Consider the action of a finite matrix group $\Gamma \subset GL(\mathbf{C}^n)$ on the *exterior algebra* $\wedge^*\mathbf{C}^n = \bigoplus_{d=0}^n \wedge^d \mathbf{C}^n$, and let $(\wedge^*\mathbf{C}^n)^\Gamma = \bigoplus_{d=0}^n (\wedge^d \mathbf{C}^n)^\Gamma$ denote the subalgebra of Γ-invariants. Prove the following anticommutative version of Molien's theorem:

$$\sum_{d=0}^n \dim((\wedge^*\mathbf{C}^n)^\Gamma) z^d = \frac{1}{|\Gamma|} \sum_{\pi \in \Gamma} \det(z\, id + \pi).$$

(6) * Prove the following expression of the Molien series in terms of the character "trace" of the given representation of Γ. This formula is due to Jarić and Birman (1977).

$$\Phi_\Gamma(z) = \frac{1}{|\Gamma|} \sum_{\pi \in \Gamma} \exp\left(\sum_{l=1}^\infty \frac{\text{trace}(\pi^l) z^l}{l} \right).$$

2.3. The Cohen–Macaulay property

In this section we show that invariant rings are Cohen–Macaulay, which implies that they admit a very nice decomposition. Cohen–Macaulayness is a fundamental concept in commutative algebra, and most of its aspects are beyond the scope of this text. What follows is a brief introduction to some basic concepts and properties. For further reading on Cohen–Macaulayness and commutative algebra in general we refer to Atiyah and MacDonald (1969), Kunz (1985), Matsumura (1986), and Hochster and Eagon (1971). I wish to thank Richard Stanley for supplying the elementary proof of Theorem 2.3.1 given below.

Let $R = R_0 \oplus R_1 \oplus R_2 \oplus \ldots$ be a graded $\mathbf{C}$-algebra of dimension n. This means that $R_0 = \mathbf{C}$, $R_i \cdot R_j \subset R_{i+j}$, and that n is the maximal number of elements of R which are algebraically independent over $\mathbf{C}$. This number is the *Krull dimension* of R, abbreviated $\dim(R) := n$. We write $H(R_+)$ for the set of homogeneous elements of positive degree in R. A set $\{\theta_1, \ldots, \theta_n\} \subset H(R_+)$ is said to be a *homogeneous system of parameters* (h. s. o. p.) provided R is finitely generated as a module over its subring $\mathbf{C}[\theta_1, \ldots, \theta_n]$. This implies in particular that $\theta_1, \ldots, \theta_n$ are algebraically independent. A basic result of commutative algebra, known as the *Noether normalization lemma*, implies that an h. s. o. p. for R always exists. See Logar (1990) and Eisenbud and Sturmfels (1992) for discussions of the Noether normalization lemma from the computer algebra point of view. We will need the following result from commutative algebra.

Theorem 2.3.1. Let R be a graded $\mathbf{C}$-algebra, and let $\theta_1, \ldots, \theta_n$ be an h. s. o. p. for R. Then the following two conditions are equivalent.

(a) R is a finitely generated *free* module over $\mathbf{C}[\theta_1, \ldots, \theta_n]$. In other words, there exist $\eta_1, \ldots, \eta_t \in R$ (which may be chosen to be homogeneous) such that

$$R = \bigoplus_{i=1}^{t} \eta_i \mathbf{C}[\theta_1, \ldots, \theta_n]. \tag{2.3.1}$$

(b) For every h. s. o. p. $\phi_1, \ldots, \phi_n$ of R, the ring R is a finitely-generated free $\mathbf{C}[\phi_1, \ldots, \phi_n]$-module.

If condition (a) and therefore (b) holds, then the elements $\eta_1, \ldots, \eta_t$ satisfy (2.3.1) if and only if their images form a $\mathbf{C}$-vector space basis of the quotient algebra $R/\langle \theta_1, \ldots, \theta_n \rangle$.

The proof of Theorem 2.3.1 is based on two lemmas. We recall that a sequence $\theta_1, \ldots, \theta_n$ of elements in R is said to be *regular* if θ_i is not a zero-divisor in $R/\langle \theta_1, \ldots, \theta_{i-1} \rangle$ for $i = 1, 2, \ldots, n$. This is equivalent to R being a free module over its subring $\mathbf{C}[\theta_1, \ldots, \theta_n]$, provided $\theta_1, \ldots, \theta_n$ are algebraically independent.

Lemma 2.3.2. Let R be a graded $\mathbf{C}$-algebra and $a_1, \ldots, a_n$ positive integers.

(a) A set $\{\theta_1, \ldots, \theta_n\} \subset H(R_+)$ is an h. s. o. p. if and only if $\{\theta_1^{a_1}, \ldots, \theta_n^{a_n}\}$ is an h. s. o. p.
(b) A sequence $\theta_1, \ldots, \theta_n \in H(R_+)$ is regular if and only if the sequence $\theta_1^{a_1}, \ldots, \theta_n^{a_n}$ is regular.

Proof. Suppose $\theta_1, \ldots, \theta_n$ are algebraically independent over $\mathbf{C}$. Then the polynomial ring $\mathbf{C}[\theta_1, \ldots, \theta_n]$ is a free module of rank $a_1 a_2 \cdots a_n$ over its subring $\mathbf{C}[\theta_1^{a_1}, \ldots, \theta_n^{a_n}]$. In fact, the set $\{\theta_1^{b_1} \cdots \theta_n^{b_n} \mid 0 \leq b_i < a_i\}$ is a free basis. This implies both (a) and (b). ◁

We also need the following "weak exchange property". For combinatorialists we note that h. s. o. p.'s do not form the bases of a matroid.

Lemma 2.3.3. Let $\phi_1, \ldots, \phi_n$ and $\theta_1, \ldots, \theta_n$ be h. s. o. p.'s of R, with all θ_i of the same degree. Then there exists a $\mathbf{C}$-linear combination $\theta = \lambda_1 \theta_1 + \ldots + \lambda_n \theta_n$ such that $\phi_1, \ldots, \phi_{n-1}, \theta$ is an h. s. o. p.

Proof. The ring $S = R/\langle \phi_1, \ldots, \phi_{n-1} \rangle$ has Krull dimension $\dim(S) = 1$. Let T denote the image of $\mathbf{C}[\theta_1, \ldots, \theta_n]$ in S. Since S is finitely generated as a module over its subring T, we have $\dim(T) = \dim(S) = 1$. By the Noether

2.3. The Cohen–Macaulay property

normalization lemma, there exists a linear combination $\theta = \lambda_1\theta_1 + \ldots + \lambda_n\theta_n$, $\lambda_i \in \mathbf{C}$, which is a parameter for T. Now T is a finitely generated $\mathbf{C}[\theta]$-module, and hence S is a finitely generated $\mathbf{C}[\theta]$-module. Thus θ is a parameter for S, and hence $\phi_1, \ldots, \phi_{n-1}, \theta$ is an h. s. o. p. for R. ◁

Proof of Theorem 2.3.1. Clearly, (b) implies (a). To prove the converse, suppose that $\theta_1, \ldots, \theta_n$ is a regular sequence in R and that $\phi_1, \ldots, \phi_n$ is any h. s. o. p. We need to show that $\phi_1, \ldots, \phi_n$ is a regular sequence. We proceed by induction on $n = \dim(R)$.

$n = 1$: Let $\theta \in H(R_+)$ be regular and $\phi \in H(R_+)$ a parameter. In other words, θ is not a zerodivisor, and R is a finitely generated $\mathbf{C}[\phi]$-module. Suppose ϕ is not regular, and pick an element $u \in H(R_+)$ such that $\phi u = 0$ in R. Thus ϕ lies in the annihilator $\text{Ann}(u) = \{v \in R \mid vu = 0\}$. Since $\phi \in \text{Ann}(u)$ is a parameter for the 1-dimensional ring R, the quotient ring $R/\text{Ann}(u)$ is zero-dimensional. Hence $\theta^m \in \text{Ann}(u)$ for some $m \in \mathbf{N}$. This means that θ^m is a zero-divisor and hence not regular. This is a contradiction to Lemma 2.3.2, because θ was assumed to be regular.

$n-1 \to n$: By Lemma 2.3.2, we may assume that $\theta_1, \ldots, \theta_n$ are of the same degree. Choose θ as in Lemma 2.3.3, and suppose (after relabeling if necessary) that $\theta_1, \ldots, \theta_{n-1}, \theta$ are linearly independent over $\mathbf{C}$. Then $\theta_1, \ldots, \theta_{n-1}, \theta$ is a regular sequence in R, and consequently $\theta_1, \ldots, \theta_{n-1}$ is a regular sequence in the $(n-1)$-dimensional quotient algebra $S := R/\langle\theta\rangle$.

By the choice of θ, the set $\{\phi_1, \ldots, \phi_{n-1}\}$ is an h. s. o. p. for S. Applying the induction hypothesis to S, we conclude that $\phi_1, \ldots, \phi_{n-1}$ is regular for S and therefore $\phi_1, \ldots, \phi_{n-1}, \theta$ is regular for R. In particular, θ is a non-zerodivisor in the 1-dimensional ring $R/\langle\phi_1, \ldots, \phi_{n-1}\rangle$. Applying the induction hypothesis again, we find that the parameter ϕ_n is also a non-zerodivisor in $R/\langle\phi_1, \ldots, \phi_{n-1}\rangle$. Hence $\phi_1, \ldots, \phi_n$ is a regular sequence in R. This completes the proof of the implication from (a) to (b).

For the second part of the statement we rewrite the $\mathbf{C}$-linear decomposition (2.3.1) as

$$R = \left(\bigoplus_{i=1}^{t} \eta_i\, \mathbf{C}\right) \oplus \left(\bigoplus_{(i_1,\ldots,i_n)\in\mathbf{N}^n\setminus\{0\}} \bigoplus_{i=1}^{t} \eta_i\, \theta_1^{i_1}\cdots\theta_n^{i_n}\, \mathbf{C}\right).$$

The claim follows from the fact that the second summand is the ideal $\langle\theta_1, \ldots, \theta_n\rangle$. ◁

A graded $\mathbf{C}$-algebra R satisfying the conditions (a) and (b) in Theorem 2.3.1 is said to be *Cohen–Macaulay*. The decomposition (2.3.1) is called a *Hironaka decomposition* of the Cohen–Macaulay algebra R. Once we know an explicit Hironaka decomposition for R, then it is easy to read off the Hilbert series of R. The following formula is a direct consequence of Lemma 2.2.3.

Corollary 2.3.4. Let R be an n-dimensional graded Cohen–Macaulay algebra with Hironaka decomposition (2.3.1). Then the Hilbert series of R equals

$$H(R, z) = \left(\sum_{i=1}^{t} z^{\deg \eta_i}\right) \Big/ \prod_{j=1}^{n}(1 - z^{\deg \theta_j}).$$

We now come to the main result of this section. Theorem 2.3.5 first appeared in Hochster and Eagon (1971) although it was apparently part of the "folklore" of commutative algebra before that paper appeared.

Theorem 2.3.5. *The invariant ring $\mathbf{C}[\mathbf{x}]^\Gamma$ of a finite matrix group $\Gamma \subset GL(\mathbf{C}^n)$ is Cohen–Macaulay.*

Proof. Consider the polynomial ring $\mathbf{C}[\mathbf{x}]$ as a module over the invariant subring $\mathbf{C}[\mathbf{x}]^\Gamma$. We have seen in the proof of Proposition 2.1.1 that every coordinate function x_i satisfies a *monic* equation with coefficients in $\mathbf{C}[\mathbf{x}]^\Gamma$. This implies that $\mathbf{C}[\mathbf{x}]$ is finitely generated as a $\mathbf{C}[\mathbf{x}]^\Gamma$-module. Note that the set $U := \{f \in \mathbf{C}^\Gamma : f^* = 0\}$ of polynomials which are mapped to zero by the Reynolds operator is also a $\mathbf{C}[\mathbf{x}]^\Gamma$-module. We can write the full polynomial ring as the direct sum $\mathbf{C}[\mathbf{x}] = \mathbf{C}[\mathbf{x}]^\Gamma \oplus U$ of $\mathbf{C}[\mathbf{x}]^\Gamma$-modules.

By the Noether normalization lemma, there exists an h. s. o. p. $\theta_1, \ldots, \theta_n$ for $\mathbf{C}[\mathbf{x}]^\Gamma$. Since $\mathbf{C}[\mathbf{x}]$ is *finite* over $\mathbf{C}[\mathbf{x}]^\Gamma$ as observed above, and since $\mathbf{C}[\mathbf{x}]^\Gamma$ is finite over the subring $\mathbf{C}[\theta_1, \ldots, \theta_n]$, it follows that $\mathbf{C}[\mathbf{x}]$ is also finite over $\mathbf{C}[\theta_1, \ldots, \theta_n]$. Hence $\theta_1, \ldots, \theta_n$ is an h. s. o. p. also for $\mathbf{C}[\mathbf{x}]$.

Taking the coordinate functions $x_1, \ldots, x_n$ as an h. s. o. p. for the polynomial ring $\mathbf{C}[\mathbf{x}]$, we see that $\mathbf{C}[\mathbf{x}]$ is Cohen–Macaulay. From the implication "(a) $\Rightarrow$ (b)" of Theorem 2.3.1 we get that $\mathbf{C}[\mathbf{x}]$ is a finitely generated free $\mathbf{C}[\theta_1, \ldots, \theta_n]$-module.

From the module decomposition $\mathbf{C}[\mathbf{x}] = \mathbf{C}[\mathbf{x}]^\Gamma \oplus U$ we obtain a decomposition

$$\mathbf{C}[\mathbf{x}]/\langle\theta_1, \ldots, \theta_n\rangle = \mathbf{C}[\mathbf{x}]^\Gamma/\langle\theta_1, \ldots, \theta_n\rangle \oplus U/(\theta_1 U + \ldots + \theta_n U)$$
$$f + \sum h_i \theta_i \mapsto f^* + \sum h_i^* \theta_i + (f - f^*) + \sum(h_i^* - h_i)\theta_i$$

of finite-dimensional $\mathbf{C}$-vector spaces. We can choose a homogeneous $\mathbf{C}$-basis $\bar{\eta}_1, \ldots, \bar{\eta}_t, \bar{\eta}_{t+1}, \ldots, \bar{\eta}_s$ for $\mathbf{C}[\mathbf{x}]/\langle\theta_1, \ldots, \theta_n\rangle$ such that $\bar{\eta}_1, \ldots, \bar{\eta}_t$ is a $\mathbf{C}$-basis for $\mathbf{C}[\mathbf{x}]^\Gamma/\langle\theta_1, \ldots, \theta_n\rangle$ and $\bar{\eta}_{t+1}, \ldots, \bar{\eta}_s$ is a $\mathbf{C}$-basis for $U/(\theta_1 U + \ldots + \theta_n U)$. Lift $\bar{\eta}_1, \ldots, \bar{\eta}_t$ to homogeneous elements $\eta_1, \ldots, \eta_t$ of $\mathbf{C}[\mathbf{x}]^\Gamma$, and lift $\bar{\eta}_{t+1}, \ldots, \bar{\eta}_s$ to homogeneous elements $\eta_{t+1}, \ldots, \eta_s$ of U. By the last part of Theorem 2.3.1, $\mathbf{C}[\mathbf{x}] = \bigoplus_{i=1}^{s} \eta_i \mathbf{C}[\theta_1, \ldots, \theta_n]$. This implies the desired Hironaka decomposition

$$\mathbf{C}[\mathbf{x}]^\Gamma = \bigoplus_{i=1}^{t} \eta_i \, \mathbf{C}[\theta_1, \ldots, \theta_n] \qquad (2.3.2)$$

which shows that the invariant ring $\mathbf{C}[\mathbf{x}]^\Gamma$ is Cohen–Macaulay. ◁

2.3. The Cohen–Macaulay property

In the following we shall see that the Hironaka decomposition (2.3.2) promised by Theorem 2.3.5 is a useful way of representing the invariant ring of a finite matrix group Γ. In this representation every invariant $I(\mathbf{x})$ can be written uniquely as

$$I(\mathbf{x}) = \sum_{i=1}^{t} \eta_i(\mathbf{x}) \cdot p_i\big(\theta_1(\mathbf{x}), \ldots, \theta_n(\mathbf{x})\big) \qquad (2.3.3)$$

where $p_1, p_2, \ldots, p_t$ are suitable n-variate polynomials. In particular, we have that $\{\theta_1, \ldots, \theta_n, \eta_1, \ldots, \eta_t\}$ is a set of fundamental invariants for Γ. The polynomials θ_i in the h. s. o. p. are called *primary invariants* while the η_j are called *secondary invariants*. We abbreviate the respective degrees with $d_i := \deg(\theta_i)$ and $e_j := \deg(\eta_j)$.

Note that for a given group Γ there are many different Hironaka decompositions. Also the degrees of the primary and secondary invariants are not unique. For instance, take $\Gamma = \{1\} \subset GL(\mathbf{C}^1)$, then we have

$$\mathbf{C}[x]^\Gamma = \mathbf{C}[x] = \mathbf{C}[x^2] \oplus x\, \mathbf{C}[x^2] = \mathbf{C}[x^3] \oplus x\, \mathbf{C}[x^3] \oplus x^2\, \mathbf{C}[x^3] = \ldots.$$

But there is also a certain uniqueness property. Suppose that we already know the primary invariants or at least their degrees d_i, $i = 1, \ldots, n$. Then the number t of secondary invariants can be computed from the following explicit formula. In the algebraic language of the proof of Theorem 2.3.5 the integer t is the rank of the invariant ring $\mathbf{C}[\mathbf{x}]^\Gamma$ as a free $\mathbf{C}[\theta_1, \ldots, \theta_n]$-module. Moreover, also the degrees $e_1, \ldots, e_t$ of the secondary invariants are uniquely determined by the numbers $d_1, \ldots, d_n$.

Proposition 2.3.6. *Let $d_1, d_2, \ldots, d_n$ be the degrees of a collection of primary invariants of a matrix group Γ. Then*

(a) *the number of secondary invariants equals*

$$t = \frac{d_1 d_2 \ldots d_n}{|\Gamma|},$$

(b) *the degrees (together with their multiplicities) of the secondary invariants are the exponents of the generating function*

$$\phi_\Gamma(z) \cdot \prod_{i=1}^{n} (1 - z^{d_j}) = z^{e_1} + z^{e_2} + \ldots + z^{e_t}.$$

Proof. We equate the formula for the Hilbert series of a Cohen–Macaulay algebra given in Corollary 2.3.4 with Molien's formula (Theorem 2.2.1) for the

Hilbert series $\Phi_\Gamma(z)$ of the invariant ring $\mathbf{C}[\mathbf{x}]^\Gamma$:

$$\frac{1}{|\Gamma|} \sum_{\pi \in \Gamma} \frac{1}{\det(id - z\pi)} = \left(\sum_{i=1}^{t} z^{e_i}\right) / \prod_{j=1}^{n}(1 - z^{d_j}). \qquad (2.3.4)$$

Multiplying both sides of (2.3.4) with $(1-z)^n$, we get

$$\frac{1}{|\Gamma|} \sum_{\pi \in \Gamma} \frac{(1-z)^n}{\det(id - z\pi)} = \left(\sum_{i=1}^{t} z^{e_i}\right) / \prod_{j=1}^{n}(1 + z + z^2 + \ldots + z^{d_j - 1}). \qquad (2.3.5)$$

We now take the limit $z \to 1$ in (2.3.5). The expressions $\frac{(1-z)^n}{\det(id - z\pi)}$ all converge to zero except for one summand where π equals the identity matrix. For that summand we get 1, and hence the left hand side of (2.3.5) converges to $1/|\Gamma|$. On the right hand side we get $t/d_1 d_2 \ldots d_n$. The resulting identity $t/d_1 d_2 \ldots d_n = 1/|\Gamma|$ proves statement (a). The statement (b) follows directly from Eq. (2.3.4). ◁

Now it is really about time for a concrete example which casts some light on the abstract discussion on the last few pages.

Example 2.3.7. Consider the matrix group

$$\Gamma = \left\{ \begin{pmatrix} 1 & 0 & 0 \\ 0 & 1 & 0 \\ 0 & 0 & 1 \end{pmatrix}, \begin{pmatrix} 0 & 1 & 0 \\ -1 & 0 & 0 \\ 0 & 0 & -1 \end{pmatrix}, \begin{pmatrix} -1 & 0 & 0 \\ 0 & -1 & 0 \\ 0 & 0 & 1 \end{pmatrix}, \begin{pmatrix} 0 & -1 & 0 \\ 1 & 0 & 0 \\ 0 & 0 & -1 \end{pmatrix} \right\}.$$

This is a three dimensional representation of the cyclic group of order 4. Its invariant ring equals

$$\mathbf{C}[x_1, x_2, x_3]^\Gamma = \{f \in \mathbf{C}[x_1, x_2, x_3] : f(x_1, x_2, x_3) = f(-x_2, x_1, -x_3)\}.$$

By Molien's theorem the invariant ring has the Hilbert series

$$\Phi_\Gamma(z) = \tfrac{1}{4} \cdot \left[\frac{1}{(1-z)^3} + \frac{2}{(1+z)(1+z^2)} + \frac{1}{(1+z)^2(1-z)} \right]$$

$$= \frac{z^3 + z^2 - z + 1}{(1+z)^2(1+z^2)(1-z)^3}$$

$$= 1 + 2z^2 + 2z^3 + 5z^4 + 4z^5 + 8z^6 + 8z^7 + 13z^8 + 12z^9 + 18z^{10} + \ldots$$

The following three invariants

$$\theta_1 := x_1^2 + x_2^2, \quad \theta_2 := x_3^2, \quad \theta_3 := x_1^4 + x_2^4,$$

are algebraically independent and they have no common roots except the zero

2.3. The Cohen–Macaulay property

vector. This means that $\mathbf{C}[x_1, x_2, x_3]$ is a finitely generated free $\mathbf{C}[\theta_1, \theta_2, \theta_3]$-module. By the arguments in our proof of Theorem 2.3.5, also the invariant ring $\mathbf{C}[\theta_1, \theta_2, \theta_3]^\Gamma$ is then a finitely generated free $\mathbf{C}[\theta_1, \theta_2, \theta_3]$-module, which means that $\theta_1, \theta_2, \theta_3$ can serve as primary invariants.

Proposition 2.3.6 tells us that we need to find four secondary invariants $\eta_1, \eta_2, \eta_3, \eta_4$ whose degrees e_1, e_2, e_3, e_4 are computed by the formula

$$z^{e_1} + z^{e_2} + z^{e_3} + z^{e_4} = \Phi_\Gamma(z) \cdot (1 - z^{d_1})(1 - z^{d_2})(1 - z^{d_3})$$
$$= \frac{(z^3 + z^2 - z + 1)(1 - z^2)^2(1 - z^4)}{(1 + z)^2(1 + z^2)(1 - z)^3}$$
$$= 1 + 2 \cdot z^3 + z^4.$$

We can simply read off $e_1 = 1$, $e_2 = e_3 = 3$, $e_4 = 4$. Now we can apply the Reynolds operator

$$* : f \mapsto \tfrac{1}{4}[f(x_1, x_2, x_3) + f(-x_2, x_1, -x_3) + f(-x_1, -x_2, x_3)$$
$$+ f(x_2, -x_1, -x_3)]$$

to all monomials of degree 3 and 4, and we obtain the desired secondary invariants

$$\eta_1 := 1, \quad \eta_2 := x_1 x_2 x_3, \quad \eta_3 := x_1^2 x_3 - x_2^2 x_3, \quad \eta_4 := x_1^3 x_2 - x_1 x_2^3.$$

Using the Gröbner basis methods of Sects. 1.2 and 2.5 (or by hand calculations) we finally verify that there does not exist a non-trivial polynomial relation of the form $\sum_{i=1}^{4} \eta_i \, p_i(\theta_1, \theta_2, \theta_3) = 0$. Therefore the invariant ring has the Hironaka decomposition

$$\mathbf{C}[x_1, x_2, x_3]^\Gamma = \mathbf{C}[\theta_1, \theta_2, \theta_3] \oplus \eta_2 \mathbf{C}[\theta_1, \theta_2, \theta_3] \oplus \eta_3 \mathbf{C}[\theta_1, \theta_2, \theta_3]$$
$$\oplus \eta_3 \mathbf{C}[\theta_1, \theta_2, \theta_3].$$

Exercises

(1) Prove that the algebra $A := \mathbf{C}[x_1, x_2]/\langle x_1 x_2 \rangle$ is Cohen–Macaulay, and find a Hironaka decomposition for A. (Hint: Try $\theta = x_1 + x_2$.)
Prove that $B := \mathbf{C}[x_1, x_2]/\langle x_1^2 x_2, x_1 x_2^2 \rangle$ is *not* Cohen–Macaulay. (Hint: Every homogeneous element of positive degree in B is a zero-divisor.) Compare the Hilbert functions of both algebras.

(2) Consider the six invariants $\eta_2^2, \eta_2 \eta_3, \eta_2 \eta_4, \eta_3^2, \eta_3 \eta_4, \eta_4^2$ in Example 2.3.7, and compute their Hironaka decompositions

$$\eta_i \eta_j \to p_{ij1}(\theta_1, \theta_2, \theta_3) + \eta_2 \cdot p_{ij2}(\theta_1, \theta_2, \theta_3) + \eta_3 \cdot p_{ij3}(\theta_1, \theta_2, \theta_3)$$
$$+ \eta_4 \cdot p_{ij4}(\theta_1, \theta_2, \theta_3).$$

Using your results, find a rewriting rule for computing the Hironaka decomposition of an arbitrary invariant $T = I(\theta_1, \theta_2, \theta_3, \eta_2, \eta_3, \eta_4)$.
(3) * Let $\Gamma \subset GL(\mathbf{C}^n)$ be a matrix group and $H \subset \Gamma$ any subgroup. Show that $\mathbf{C}[\mathbf{x}]^H$ is a free module of rank $[\Gamma : H]$ over its subring $\mathbf{C}[\mathbf{x}]^\Gamma$. How can you compute a free module basis?
(4) Let R be the subring of $\mathbf{C}[x_1, x_2, x_3, x_4]$ spanned by all monomials $x_1^{a_1} x_2^{a_2} x_3^{a_3} x_4^{a_4}$ with

$$a_1 + 3a_2 + a_3 \equiv 0 \pmod{4} \quad \text{and} \quad 4a_1 - a_3 + 2a_4 \equiv 0 \pmod{5}.$$

(a) Show that R is the invariant ring of a finite abelian group Γ.
(b) Compute the Hilbert series of the graded ring R.
(c) Compute a Hironaka decomposition for R.

2.4. Reflection groups

In view of Theorem 2.3.5 it is natural to ask under what circumstances does the Hironaka representation $\mathbf{C}[\mathbf{x}]^\Gamma = \bigoplus_{i=1}^t \eta_i \, \mathbf{C}[\theta_1, \ldots, \theta_n]$ have a particularly simple or interesting form. In this section we discuss the simplest possibility of all, namely the case $\mathbf{C}[\mathbf{x}]^\Gamma = \mathbf{C}[\theta_1, \ldots, \theta_n]$ when the invariant ring is generated by n algebraically independent invariants. We have seen in Sect. 1.1 that this happens for the symmetric group S_n of $n \times n$ permutation matrices. In Exercise 2.2.3 we have seen that also the invariant ring of the symmetry group of a regular n-gon is isomorphic to a polynomial ring in two variables.

The main theorem in this section characterizes those matrix groups whose invariants form a polynomial ring. In order to state this theorem we need two definitions. A matrix or linear transformation $\pi \in GL(\mathbf{C}^n)$ is called a *reflection* if precisely one eigenvalue of π is not equal to one. Actually, these reflections are what some authors call "generalized reflections" or "pseudo-reflections". The "usual" hyperplane reflections in $\mathbf{R}^n$ are those reflections whose n-th eigenvalue is equal to -1. A finite subgroup $\Gamma \subset GL(\mathbf{C}^n)$ is said to be a *reflection group* if Γ is generated by reflections.

Theorem 2.4.1 (Shephard–Todd–Chevalley theorem). *The invariant ring $\mathbf{C}[\mathbf{x}]^\Gamma$ of a finite matrix group $\Gamma \subset GL(\mathbf{C}^n)$ is generated by n algebraically independent homogeneous invariants if and only if Γ is a reflection group.*

It is important to note that being a reflection group is *not* a property of the abstract group underlying Γ but it depends on the specific n-dimensional representation. For instance, the 2-dimensional representation of the dihedral group D_6 as the symmetry group of a hexagon is a reflection group (and its invariant ring is a polynomial ring by Exercise 2.2.3), while the 3-dimensional representation of D_6 considered in Example 2.2.6 is not a reflection group (and its invariant ring is not a polynomial ring). See also Example 2.4.6.

Theorem 2.4.1 was first proved for real reflection groups by Shephard and Todd (1954), and subsequently generalized to the complex case by Chevalley

2.4. Reflection groups

(1955) and Serre (Stanley 1979b). Shephard and Todd explicitly determined all finite subgroups generated by reflections, and they verified the if-part of Theorem 2.4.1 by explicitly computing their invariant rings $\mathbf{C}[\mathbf{x}]^\Gamma$.

The proof of the if-part to be presented here follows the exposition of Chevalley's proof given in Grove and Benson (1985). Let $\sigma \in GL(\mathbf{C}^n)$ be any reflection. Then the kernel of the linear transformation $\sigma - id$ is a hyperplane H_σ in $\mathbf{C}^n$. Let L_σ denote the linear polynomial whose zero set is the hyperplane H_σ.

Lemma 2.4.2. For all polynomials $f \in \mathbf{C}[\mathbf{x}]$, the linear polynomial L_σ is a divisor of $\sigma f - f$.

Proof. Given $\mathbf{v} \in \mathbf{C}^n$ with $L_\sigma(\mathbf{v}) = 0$, then we have

$$\mathbf{v} \in H_\sigma \Rightarrow \sigma \mathbf{v} = \mathbf{v} \Rightarrow f(\sigma \mathbf{v}) - f(\mathbf{v}) = 0 \Rightarrow (\sigma f - f)(\mathbf{v}) = 0.$$

Since the linear polynomial L_σ is irreducible, Hilbert's Nullstellensatz implies that $\sigma f - f$ is a multiple of L_σ. ◁

In the following let $\Gamma \subset GL(\mathbf{C}^n)$ be a finite reflection group. Let $\mathcal{I}_\Gamma$ denote the ideal in $\mathbf{C}[\mathbf{x}]$ which is generated by all homogeneous invariants of positive degree.

Proposition 2.4.3. Let $h_1, h_2, \ldots, h_m \in \mathbf{C}[\mathbf{x}]$ be homogeneous polynomials, let $g_1, g_2, \ldots, g_m \in \mathbf{C}[\mathbf{x}]^\Gamma$ be invariants, and suppose that $g_1 h_1 + g_2 h_2 + \ldots + g_m h_m = 0$. Then either $h_1 \in \mathcal{I}_\Gamma$, or g_1 is contained in the ideal $\langle g_2, \ldots, g_m \rangle$ in $\mathbf{C}[\mathbf{x}]$.

Proof. We proceed by induction on the degree of h_1. If $h_1 = 0$, then $h_1 \in \mathcal{I}_\Gamma$. If $\deg(h_1) = 0$, then h_1 is a constant and hence $g_1 \in \langle g_2, \ldots, g_m \rangle$. We may therefore assume $\deg(h_1) > 0$ and that the assertion is true for smaller degrees. Suppose that $g_1 \notin \langle g_2, \ldots, g_m \rangle$.

Let $\sigma \in \Gamma$ be any reflection. Then

$$\sum_{i=1}^m g_i \cdot (\sigma h_i) = \sigma \left(\sum_{i=1}^m g_i \cdot h_i \right) = \sigma(0) = 0.$$

By Lemma 2.4.2, we can write $\sigma h_i = h_i + L_\sigma \cdot \tilde{h}_i$, where $\tilde{h}_i$ is a homogeneous polynomial of degree $\deg(h_i) - 1$. We get

$$0 = \sum_{i=1}^m g_i \cdot (h_i + L_\sigma \cdot \tilde{h}_i) = L_\sigma \cdot \sum_{i=1}^m g_i \tilde{h}_i,$$

and consequently $g_1 \tilde{h}_1 + g_2 \tilde{h}_2 + \ldots + g_m \tilde{h}_m = 0$. By the induction hypothesis, we have $\tilde{h}_1 \in \mathcal{I}_\Gamma$, and therefore $\sigma h_1 - h_1 = \tilde{h}_1 \cdot L_\sigma \in \mathcal{I}_\Gamma$.

Now let $\pi = \sigma_1 \sigma_2 \ldots \sigma_l$ be an arbitrary element of Γ, written as a product of reflections. Since the ideal $\mathcal{I}_\Gamma$ is invariant under the action of Γ,

$$\pi h_1 - h_1 = \sum_{i=1}^{l}(\sigma_1\ldots\sigma_i\sigma_{i+1}h_1 - \sigma_1\ldots\sigma_i h_1)$$

$$= \sum_{i=1}^{l}(\sigma_1\ldots\sigma_i)(\sigma_{i+1}h_1 - h_1) \in \mathcal{I}_\Gamma.$$

This implies $h_1^* - h_1 \in \mathcal{I}_\Gamma$ and consequently $h_1 \in \mathcal{I}_\Gamma$. ◁

Proof of Theorem 2.4.1 (if-part). By Hilbert's basis theorem (Corollary 1.2.5), there exists a finite set $\{f_1, f_2, \ldots, f_m\} \subset \mathbf{C}[\mathbf{x}]$ of homogeneous invariants which generates the ideal $\mathcal{I}_\Gamma$. With the same argument as in the proof of Theorem 2.1.3 this automatically implies

$$\mathbf{C}[\mathbf{x}]^\Gamma = \mathbf{C}[f_1, f_2, \ldots, f_m].$$

Suppose now that m is minimal with this property, i.e., no smaller set of homogeneous invariants generates $\mathcal{I}_\Gamma$. We need to prove that $m = n$, or, equivalently, that the invariants $f_1, f_2, \ldots, f_m$ are algebraically independent over $\mathbf{C}$.

Our proof is by contradiction. Suppose there exists a non-zero polynomial $g \in \mathbf{C}[y_1, y_2, \ldots, y_m]$ such that $g(f_1, f_2, \ldots, f_m) = 0$ in $\mathbf{C}[\mathbf{x}]$. We may assume that g is of minimal degree and that all monomials $x_1^{i_1} x_2^{i_2} \ldots x_n^{i_n}$ occurring (before cancellation) in the expansion of $g(f_1, f_2, \ldots, f_m)$ have the same degree $d := i_1 + i_2 + \ldots + i_n$.

For $i = 1, 2, \ldots, m$ consider the invariant

$$g_i := \left(\frac{\partial g}{\partial y_i}\right)(f_1, f_2, \ldots, f_m) \in \mathbf{C}[\mathbf{x}]^\Gamma.$$

Each g_i is either 0 or of degree $d - \deg f_i$. Since $g(y_1, \ldots, y_m)$ is not a constant, there exists an i with $\left(\frac{\partial g}{\partial y_i}\right)(y_1, \ldots, y_m) \neq 0$, and hence $g_i \neq 0$, by the choice of g.

Let $\mathcal{J}$ denote the ideal in $\mathbf{C}[\mathbf{x}]$ generated by $\{g_1, g_2, \ldots, g_m\}$, and relabel if necessary so that $\mathcal{J}$ is generated by $\{g_1, \ldots, g_k\}$ but no proper subset. For $i = k+1, \ldots, m$ write $g_i = \sum_{j=1}^{k} h_{ij} g_j$ where h_{ij} is either 0 or homogeneous of degree $\deg(g_i) - \deg(g_j) = \deg(f_j) - \deg(f_i)$. We have

$$0 = \frac{\partial}{\partial x_s}[g(f_1, f_2, \ldots, f_m)]$$

$$= \sum_{i=1}^{m} g_i \frac{\partial f_i}{\partial x_s}$$

2.4. Reflection groups

$$= \sum_{i=1}^{k} g_i \frac{\partial f_i}{\partial x_s} + \sum_{i=k+1}^{m} \Big(\sum_{j=1}^{k} h_{ij} g_j\Big) \frac{\partial f_i}{\partial x_s}$$

$$= \sum_{i=1}^{k} g_i \Big(\frac{\partial f_i}{\partial x_s} + \sum_{j=k+1}^{m} h_{ji} \frac{\partial f_j}{\partial x_s}\Big).$$

Since $g_1 \notin \langle g_2, \ldots, g_k \rangle$, Proposition 2.4.3 implies

$$\frac{\partial f_1}{\partial x_s} + \sum_{j=k+1}^{m} h_{j1} \frac{\partial f_j}{\partial x_s} \in \mathcal{I}_\Gamma \quad \text{for } s = 1, 2, \ldots, n.$$

Multiplying with x_s and summing over s, we can apply Euler's formula to find

$$\sum_{s=1}^{n} x_s \frac{\partial f_1}{\partial x_s} + \sum_{j=k+1}^{m} h_{j1} \sum_{s=1}^{n} x_s \frac{\partial f_j}{\partial x_s}$$

$$= (\deg f_1) f_1 + \sum_{j=k+1}^{m} h_{j1} (\deg f_j) f_j$$

$$\in \langle x_1, \ldots, x_n \rangle \mathcal{I}_\Gamma$$

$$\subset \langle x_1 f_1, \ldots, x_n f_1 \rangle + \langle f_2, \ldots, f_m \rangle.$$

All monomials in this polynomial are of degree $\deg(f_1)$, and therefore

$$(\deg f_1) f_1 + \sum_{j=k+1}^{m} h_{j1} (\deg f_j) f_j \in \langle f_2, \ldots, f_m \rangle.$$

The last expression implies $f_1 \in \langle f_2, \ldots, f_m \rangle$, which is a contradiction to the minimality of m. This completes the proof of the "if"-part of Theorem 2.4.1. ◁

Our proof of "only-if"-direction follows Stanley (1979b). It is based on some interesting generating function techniques. In what follows we do *not* assume any longer that Γ is a reflection group.

Lemma 2.4.4. Let r be the number of reflections in a finite matrix group $\Gamma \subset GL(\mathbf{C}^n)$. Then the Laurent expansion of the Molien series about $z = 1$ begins

$$\Phi_\Gamma(z) = \frac{1}{|\Gamma|}(1-z)^{-n} + \frac{r}{2|\Gamma|}(1-z)^{-n+1} + O\big((1-z)^{-n+2}\big).$$

Proof. Recall from Theorem 2.2.1 the representation

$$\Phi(z) = \frac{1}{|\Gamma|} \sum_{\pi \in \Gamma} \det(id - z\pi)^{-1}.$$

The only term $\det(id - z\pi)^{-1}$ in this sum to have a pole of order n at $z = 1$ is the term $(1-z)^{-n}$ corresponding to the identity matrix in Γ. If $\det(id - z\pi)^{-1}$ has a pole of order $n-1$ at $z=1$, then π is a reflection and

$$\det(id - z\pi)^{-1} = (1-z)^{-n+1}(1 - \det\pi \cdot z)^{-1}.$$

Hence the coefficient of $(1-z)^{-n+1}$ in the Laurent expansion of $\Phi_\Gamma(z)$ equals

$$\frac{1}{|\Gamma|}\sum_\sigma (1 - \det\sigma)^{-1}$$

where the sum ranges over all reflections σ in Γ. Since σ is a reflection if and only if σ^{-1} is a reflection, we conclude

$$2 \cdot \sum_\sigma \frac{1}{1-\det\sigma} = \sum_\sigma \left(\frac{1}{1-\det\sigma} + \frac{1}{1-(\det\sigma)^{-1}}\right) = \sum_\sigma 1 = r,$$

completing the proof. ◁

Corollary 2.4.5. Let $\Gamma \subset GL(\mathbf{C}^n)$ be a finite matrix group whose invariant ring $\mathbf{C}[\mathbf{x}]^\Gamma$ is generated by n algebraically independent homogeneous invariants $\theta_1, \ldots, \theta_n$ where $d_i := \deg\theta_i$. Let r be the number of reflections in Γ. Then

$$|\Gamma| = d_1 d_2 \ldots d_n \quad \text{and} \quad r = d_1 + d_2 + \ldots + d_n - n.$$

Proof. By Lemma 2.2.3, we have

$$\Phi_\Gamma(z) = \frac{1}{(1-z^{d_1})}\frac{1}{(1-z^{d_2})} \cdots \frac{1}{(1-z^{d_n})}.$$

The Laurent expansion of this power series about $z=1$ begins

$$\Phi_\Gamma(z) = \frac{1}{d_1 d_2 \ldots d_n}(1-z)^{-n} + \frac{d_1 + d_2 + \ldots + d_n - n}{2 d_1 d_2 \ldots d_n}(1-z)^{-n+1} + O((1-z)^{-n+2}).$$

Comparing with Lemma 2.4.4 completes the proof. ◁

Proof of Theorem 2.4.1 (only-if part). Suppose that $\mathbf{C}[\mathbf{x}]^\Gamma = \mathbf{C}[\theta_1, \ldots, \theta_n]$ with $\deg(\theta_i) = d_i$. Let H be the subgroup of Γ generated by all reflections in Γ. Then by the if-part of Theorem 2.4.1, we have

$$\mathbf{C}[\mathbf{x}]^H = \mathbf{C}[\psi_1, \ldots, \psi_n],$$

2.4. Reflection groups

where $\deg(\psi_j) = e_j$. Clearly $\mathbf{C}[\mathbf{x}]^\Gamma \subseteq \mathbf{C}[\mathbf{x}]^H$, so each θ_i is a polynomial function in the ψ's.

Since the θ's and the ψ's are both algebraically independent, the Jacobian determinant $\det(\partial\theta_i/\partial\psi_j)$ is non-zero. Hence there exists a permutation π with

$$\frac{\partial\theta_{\pi(1)}}{\partial\psi_1} \frac{\partial\theta_{\pi(2)}}{\partial\psi_2} \cdots \frac{\partial\theta_{\pi(n)}}{\partial\psi_n} \neq 0.$$

This means that ψ_i actually appears in $\theta_{\pi(i)} = \theta_{\pi(i)}(\psi_1, \ldots, \psi_n)$, and consequently $e_i = \deg \psi_i \leq d_{\pi(i)} = \deg \theta_{\pi(i)}$.

Let r be the number of reflections in Γ and therefore in H. By Corollary 2.4.6, we have

$$r = \sum_{i=1}^n (d_i - 1) = \sum_{i=1}^n (d_{\pi(i)} - 1) = \sum_{i=1}^n (e_i - 1).$$

Since $e_i \leq d_{\pi(i)}$, we have $e_i = d_{\pi(i)}$, so again by Corollary 2.4.5 we have $|\Gamma| = d_1 d_2 \ldots d_n = e_1 e_2 \ldots e_n = |H|$, and hence $H = \Gamma$. ◁

The "only-if" part is useful in that it proves that most invariant rings are not polynomial rings.

Example 2.4.6 (Twisted symmetric polynomials). Let S_n denote the set of permutation matrices in $GL(\mathbf{C}^n)$, and consider its representation $\Gamma := \{\mathrm{sign}(\sigma) \cdot \sigma : \sigma \in S_n\}$. We call the elements of the invariant ring $\mathbf{C}[\mathbf{x}]^\Gamma$ *twisted symmetric polynomials*. Note that a homogeneous polynomial f is twisted symmetric if and only if $f \circ \sigma = \mathrm{sign}(\sigma)^{\deg f} \cdot f$ for all permutations σ. Theorem 2.4.1 implies that the ring $\mathbf{C}[\mathbf{x}]^\Gamma$ is not a polynomial ring. For instance, for $n = 3$ we have the Hironaka decomposition

$$\mathbf{C}[x_1, x_2, x_3] = \mathbf{C}[\theta_1, \theta_2, \theta_3] \oplus \eta \, \mathbf{C}[\theta_1, \theta_2, \theta_3],$$

where $\theta_1 := x_1^2 + x_2^2 + x_3^2$, $\theta_2 := x_1 x_2 + x_1 x_3 + x_2 x_3$, $\theta_3 := x_1^2 x_2 + x_2^2 x_3 + x_3^2 x_1 - x_2^2 x_1 - x_1^2 x_3 - x_3^2 x_1$ and $\eta := x_1^4 + x_2^4 + x_3^4$.

Exercises

(1) Consider the full symmetry group $\Gamma \subset GL(\mathbf{R}^3)$ of any of the five Platonic solids. (The five Platonic solids are the tetrahedron, the octahedron, the cube, the icosahedron, and the dodecahedron.)
   (a) Show that Γ is a reflection group.
   (b) Find three algebraically independent invariants $\theta_1, \theta_2, \theta_3$ which generate the invariant ring $\mathbf{C}[x, y, z]^\Gamma$.
   (c) How are the degrees of $\theta_1, \theta_2, \theta_3$ related to the order of the group Γ? How are they related to the face numbers of the polytope in question?
   (d) Find an explicit formula which expresses each symmetrized monomial

$(x^i y^j z^k)^*$ as a polynomial function in the fundamental invariants $\theta_1, \theta_2, \theta_3$.

(2) Determine the Molien series and a Hironaka decomposition for the ring of twisted symmetric polynomials in n variables for $n \geq 4$ (cf. Example 2.4.6). Can you generalize your results to the case where "sign" is replaced by an arbitrary character of the symmetric group?

2.5. Algorithms for computing fundamental invariants

In this section we present algorithms for computing a fundamental set of invariants for any finite matrix group Γ. Our input will be a black box which evaluates the Reynolds operator "*" of Γ, and our output will be a set of primary and secondary invariants as in Sect. 2.3. The reason for this assumption is that the knowledge of the full group Γ might not be necessary to compute the invariants: it suffices to know the Reynolds operator.

We begin with a description of six commutative algebra subroutines based on Buchberger's method. The first four of these algorithms are well-known, and they are discussed in practically every introduction to Gröbner bases theory. It is those four subroutines which we will apply later in this section. The other two subroutines 2.5.5 and 2.5.6 are perhaps a little less known. These are included here because they are quite useful for working with Cohen–Macaulay rings such as invariant rings of finite groups.

Whenever the monomial order is left unspecified, any monomial order will work for the Gröbner bases computation in question. The two most frequently used monomial orders are the *purely lexicographical order* "$>_{\text{pl}}$" and the *reverse lexicographical order* "$>_{\text{rl}}$". These are defined as follows. We assume that an order is given on the variables, say $x_1 > x_2 > \ldots > x_n$. We then put $\mathbf{x}^\alpha >_{\text{pl}} \mathbf{x}^\beta$ if there exists i, $1 \leq i \leq n$, such that $\alpha_j = \beta_j$ for all $j < i$, and $\alpha_i > \beta_i$. In contrast to "$>_{\text{pl}}$", the reverse lexicographic order "$>_{\text{rl}}$" is a linear extension of the natural grading on $\mathbf{C}[\mathbf{x}]$. We define $\mathbf{x}^\alpha >_{\text{rl}} \mathbf{x}^\beta$ if $\sum \alpha_i > \sum \beta_i$, or if $\sum \alpha_i = \sum \beta_i$ and there exists i, $1 \leq i \leq n$, such that $\alpha_j = \beta_j$ for all $j > i$, and $\alpha_i < \beta_i$.

Subroutine 2.5.1 (Radical containment).
Input: $f_1, f_2, \ldots, f_m, g \in \mathbf{C}[\mathbf{x}]$.
Question: Let $I := \langle f_1, \ldots, f_m \rangle$. Does g lie in $\text{Rad}(I)$, the radical of I?
Solution: Let G be a Gröbner basis of $\langle f_1, f_2, \ldots, f_m, gz - 1 \rangle$, where z is a new variable. Then $g \in \text{Rad}(I)$ if and only if $1 \in G$.

Subroutine 2.5.2 (Solvability of homogeneous equations).
Input: *Homogenous* polynomials $f_1, f_2, \ldots, f_m \in \mathbf{C}[\mathbf{x}]$.
Question: Is there a *non-zero* vector $\mathbf{a} \in \mathbf{C}^n$ such that $f_1(\mathbf{a}) = f_2(\mathbf{a}) = \ldots = f_m(\mathbf{a}) = 0$.
Solution: Compute a Gröbner basis G of the ideal $I := \langle f_1, f_2, \ldots, f_m \rangle$. We have $\text{Rad}(I) = \langle x_1, x_2, \ldots, x_n \rangle$ (i.e., there is no non-zero solution) if and only

2.5. Algorithms for computing fundamental invariants

if a monomial of the form $x_i^{j_i}$ occurs among the initial monomials in G for every i, for $1 \leq i \leq n$.

Subroutine 2.5.3 (Algebraic dependence).
Input: A set $F := \{f_1, f_2, \ldots, f_m\} \subset \mathbf{C}[\mathbf{x}]$, considered as subset of the field of rational functions $\mathbf{C}(\mathbf{x})$.
Questions: Is F algebraically dependent over $\mathbf{C}$? If so, find an m-variate polynomial P such that $P(f_1, f_2, \ldots, f_m) = 0$ in $\mathbf{C}(\mathbf{x})$.
Solution: Introduce m new "slack" variables $\mathbf{y} := (y_1, \ldots, y_m)$, and compute a Gröbner basis G of $\{f_1 - y_1, f_2 - y_2, \ldots, f_m - y_m\}$ with respect to the purely lexicographical order induced from $x_1 > \ldots > x_n > y_1 > \ldots > y_m$. Let $G' := G \cap \mathbf{C}[\mathbf{y}]$. Then F is algebraically independent if and only if $G' = \emptyset$. On the other hand, if $P(\mathbf{y}) \in G'$, then $P(f_1, \ldots, f_m) = 0$ in $\mathbf{C}[\mathbf{x}]$.

Subroutine 2.5.4 (Subring containment).
Input: $f_1, f_2, \ldots, f_m, g \in \mathbf{C}[\mathbf{x}]$.
Question: Is g contained in the subring $\mathbf{C}[f_1, \ldots, f_m]$ of $\mathbf{C}[\mathbf{x}]$? If so, find an m-variate polynomial Q such that $g = Q(f_1, f_2, \ldots, f_m)$ in $\mathbf{C}[\mathbf{x}]$.
Solution: Compute the Gröbner basis G as in Subroutine 2.5.3, and let $Q \in \mathbf{C}[\mathbf{x}, \mathbf{y}]$ be the unique normal form of g with respect to G. Then $g \in \mathbf{C}[f_1, \ldots, f_m]$ if and only if Q is contained in $\mathbf{C}[\mathbf{y}]$. In that case we have the identity $g = Q(f_1, f_2, \ldots, f_m)$ in $\mathbf{C}[\mathbf{x}]$.

Subroutine 2.5.5 (Hironaka decomposition of a Cohen–Macaulay subring).
Input: Homogeneous polynomials $f_1, f_2, \ldots, f_m \in \mathbf{C}[\mathbf{x}]$, generating the ideal I.
Question: Decide whether $R = \mathbf{C}[\mathbf{x}]/I$ is a d-dimensional Cohen–Macaulay ring, and, if so, construct a *Hironaka decomposition* as in (2.3.1).
Solution:

1. Pick a random $n \times d$ matrix $(a_{ij})_{1 \leq i \leq n, 1 \leq j \leq d}$ over $\mathbf{C}$, and abbreviate

$$\theta_1 := \sum_{i=1}^{n} a_{i1} x_i, \quad \theta_2 := \sum_{i=1}^{n} a_{i2} x_i, \quad \ldots, \quad \theta_d := \sum_{i=1}^{n} a_{id} x_i.$$

2. Introduce d new variables $\mathbf{z} := (z_1, \ldots, z_d)$. Compute a reduced Gröbner basis $\mathcal{G}$ with respect to reverse lexicographic order induced from $z_1 < z_2 < \ldots < z_d < x_1 < x_2 < \ldots < x_n$ for the ideal

$$J := I + \langle \theta_1 - z_1, \theta_2 - z_2, \ldots, \theta_d - z_d \rangle \quad \text{in } \mathbf{C}[\mathbf{x}, \mathbf{z}].$$

3. Does the initial monomial of some element in $\mathcal{G}$ contain a new variable z_i? If so; STOP: R is not a free $\mathbf{C}[\theta_1, \ldots, \theta_d]$-module. Otherwise, proceed with Step 4.

4. Let $\mathbf{F}$ be the set of $\alpha \in \mathbf{N}^n$ such that $\mathbf{x}^\alpha$ is standard (i.e., not a multiple of the initial monomial of some element in $\mathcal{G}$). If $\mathbf{F}$ is infinite (i.e., $\exists i \; \forall s \; \forall g \in \mathcal{G} : x_i^s \neq \text{init}(g)$), then STOP: R is an infinite-dimensional free $\mathbf{C}[\theta_1, \ldots, \theta_d]$-

module. If **F** is finite, then R is a d-dimensional Cohen–Macaulay ring having the Hironaka decomposition

$$R = \bigoplus_{\alpha \in \mathbf{F}} \mathbf{x}^\alpha \, \mathbf{C}[\theta_1, \theta_2, \ldots, \theta_d].$$

Subroutine 2.5.6 (Normal form with respect to a Hironaka decomposition).
Input: Generators $\theta_1, \ldots, \theta_n, \eta_1, \ldots, \eta_t \in \mathbf{C}[\mathbf{x}]$ of a Cohen–Macaulay subring R with Hironaka decomposition as in Theorem 2.3.1 (1).
Question: Decide whether a given polynomial $f \in \mathbf{C}[\mathbf{x}]$ lies in the subring R, and if so, find the unique representation

$$f(\mathbf{x}) = \sum_{i=1}^{t} \eta_i(\mathbf{x}) \cdot p_i\big(\theta_1(\mathbf{x}), \ldots, \theta_n(\mathbf{x})\big). \qquad (2.5.1)$$

Solution: Introduce $n+t$ new "slack" variables $(\mathbf{y}, \mathbf{z}) := (y_1, \ldots, y_n, z_1, \ldots, z_t)$, and compute a Gröbner basis $\mathcal{G}$ of $\{\theta_1 - y_1, \ldots, \theta_n - y_n, \eta_1 - z_1, \ldots, \eta_t - z_t\}$ with respect to the following monomial order "$\succ$" on $\mathbf{C}[\mathbf{x}, \mathbf{y}, \mathbf{z}]$. We define $\mathbf{x}^\alpha \mathbf{y}^\beta \mathbf{z}^\gamma \succ \mathbf{x}^{\alpha'} \mathbf{y}^{\beta'} \mathbf{z}^{\gamma'}$ if $\mathbf{x}^\alpha > \mathbf{x}^{\alpha'}$ in the *purely* lexicographic order, or else if $\mathbf{y}^\beta > \mathbf{y}^{\beta'}$ in the *degree* lexicographic order, or else if $\mathbf{z}^\gamma > \mathbf{z}^{\gamma'}$ in the *purely* lexicographic order.

Then $f \to_{\mathcal{G}} \sum_{i=1}^{t} z_i \cdot p_i(y_1, \ldots, y_n)$ if and only if the identity (2.5.1) holds. Note that $\mathcal{G}$ contains in particular those rewriting relations $\eta_i \eta_j - \sum_{k=1}^{t} z_i \cdot q_{ijk}(y_1, \ldots, y_n)$ which express the Hironaka decompositions of all quadratic monomials in the η's.

We now come to the problem of computing a fundamental set of invariants for a given finite matrix group $\Gamma \subset GL(\mathbf{C}^n)$. Our algorithm will be set up so that it generates an explicit Hironaka decomposition for the invariant ring $\mathbf{C}[\mathbf{x}]^\Gamma$.

In the following we will assume that the group Γ is presented by its Reynolds operator

$$* : \mathbf{C}[\mathbf{x}] \to \mathbf{C}[\mathbf{x}]^\Gamma$$

$$f \mapsto f^* := \frac{1}{|\Gamma|} \sum_{\sigma \in \Gamma} \sigma(f).$$

We recall from Proposition 2.1.2 that the Reynolds operator "$*$" is a $\mathbf{C}[\mathbf{x}]^\Gamma$-module homomorphism and that the restriction of "$*$" to $\mathbf{C}[\mathbf{x}]^\Gamma$ is the identity.

In the course of our computation we will repeatedly call the function "$*$", irrespective of how this function is implemented. One obvious possibility is to store a complete list of all group elements in Γ, but this may be infeasible in some instances. The number of calls of the Reynolds operator is a suitable measure for the running time of our algorithm. As far as asymptotic worst case complexity is concerned, Proposition 2.1.5 implies that also in this measure we will not be able to beat Noether's bound (Theorem 2.1.4).

Let us mention parenthetically that the approach presented here generalizes

2.5. Algorithms for computing fundamental invariants

directly to infinite reductive algebraic groups, provided the Reynolds operator "*" and the ideal of the nullcone are given effectively. The *nullcone* of any matrix group is defined as the set of common zeros of all invariants. Finding a defining set for the nullcone is generally easier than computing a fundamental set of invariants. For the case of the general linear group $\Gamma = GL(\mathbf{C}^n)$ we will discuss this in detail in Chap. 4.

Here we are concerned with a finite group Γ, and the following lemma states that in this case the nullcone consists only of the origin. Equivalently, the ideal of the nullcone equals the *irrelevant ideal* $M := \langle x_1, x_2, \ldots, x_n \rangle$.

Lemma 2.5.7. *Let $\Gamma \subset GL(\mathbf{C}^n)$ be any finite matrix group, and let I^Γ denote the ideal in $\mathbf{C}[\mathbf{x}]$ generated by all homogeneous invariants of positive degree. Then $\mathrm{Rad}(I^\Gamma) = M$.*

Proof. Each homogeneous polynomial of positive degree lies in the irrelevant ideal M. Therefore we need only show the reverse inclusion $M \subseteq \mathrm{Rad}(I^\Gamma)$. In view of Hilbert's Nullstellensatz, it is sufficient to show that the variety of I^Γ in $\mathbf{C}^n$ equals the variety of M, which is the origin. We will do so by constructing, for an arbitrary nonzero vector $\mathbf{a} \in \mathbf{C}^n$, a suitable invariant in I^Γ which does not vanish at $\mathbf{a}$.

Let $\mathbf{a} \in \mathbf{C}^n \setminus \{0\}$. Since every matrix σ in the group Γ is invertible, the orbit $\Gamma\mathbf{a} = \{\sigma\mathbf{a} \in \mathbf{C}^n \mid \sigma \in \Gamma\}$ does not contain the origin. The orbit $\Gamma\mathbf{a}$ is a finite subset of $\mathbf{C}^n$, and therefore it is Zariski closed. This means there exists a polynomial function $f \in \mathbf{C}[\mathbf{x}]$ such that $f(0) = 0$ and $f(\sigma\mathbf{a}) = 1$ for all $\sigma \in \Gamma$.

We apply the Reynolds operator to the polynomial f, and we obtain an invariant f^* which lies in I^Γ because $f^*(0) = 0$. On the other hand we have $f^*(\mathbf{a}) = \frac{1}{|\Gamma|} \sum_{\sigma \in \Gamma} f(\sigma\mathbf{a}) = 1$. Hence the point $\mathbf{a}$ does not lie in the variety of I^Γ. This completes the proof of Lemma 2.5.7. ◁

We will now present the basic algorithm for computing a Hironaka decomposition of $\mathbf{C}[\mathbf{x}]^\Gamma$. In Algorithm 2.5.8 we do not make use of the Molien series techniques in Sects. 2.2 and 2.3. A more practical variant based upon the Molien series will be presented in Algorithm 2.5.14.

We fix any monomial order $m_1 < m_2 < m_3 < m_4 < \ldots$ which refines the partial order given by the total degree on the set of monomials in $\mathbf{C}[\mathbf{x}]$.

Algorithm 2.5.8.
Input: The Reynolds operator $* : \mathbf{C}[\mathbf{x}] \to \mathbf{C}[\mathbf{x}]^\Gamma$ of a finite subgroup Γ of $GL(\mathbf{C}^n)$.
Output: A Hironaka decomposition for the invariant ring $\mathbf{C}[\mathbf{x}]^\Gamma$.

0. Let $t := 1$ and $\mathcal{Q} := \emptyset$.
1. Repeat $t := t + 1$ until $m_t^* \notin \mathrm{Rad}(\langle \mathcal{Q} \rangle)$ (using Subroutine 2.5.1).
2. Let $\mathcal{Q} := \mathcal{Q} \cup \{m_t^*\}$. If $\mathrm{Rad}(\langle \mathcal{Q} \rangle) \neq M$ then go to step 1 (using Subroutine 2.5.2).

3. If $\mathcal{Q}$ has cardinality n
   3.1. then $\mathcal{P} := \mathcal{Q}$;
   3.2. else modify $\mathcal{Q}$ to an algebraically independent set $\mathcal{P}$ of invariants with Rad($\langle\mathcal{P}\rangle$) = M (using Subroutine 2.5.3; see Subroutine 2.5.10).
4. Write $\mathcal{P} = \{\theta_1, \theta_2, \ldots, \theta_n\}$, let $\mathcal{S} := \{1\}$, $t := 0$, and set bound := $\sum_{i=1}^{n} \text{degree}(\theta_i) - n$.
5. Let $t := t + 1$. If degree(m_t) > bound then STOP. At this point $\mathcal{P}$ and $\mathcal{S}$ are primary and secondary invariants respectively, and their union generates $\mathbf{C}[\mathbf{x}]^\Gamma$ as a ring.
6. Test whether m_t^* lies in the $\mathbf{C}[\mathcal{P}]$-module generated by $\mathcal{S}$ (see Exercise (3) below). If no, then $\mathcal{S} := \mathcal{S} \cup \{m_t^*\}$. Go to 5.

We will explain the details of Algorithm 2.5.8 and prove its correctness along the way.

Theorem 2.5.9. Algorithm 2.5.8 terminates with finite sets $\mathcal{P} = \{\theta_1, \theta_2, \ldots, \theta_n\}$ (the *primary* invariants) and $\mathcal{S} = \{\eta_1, \eta_2 \ldots, \eta_t\}$ (the *secondary* invariants, where $\eta_1 = 1$) such that the invariant ring $\mathbf{C}[\mathbf{x}]^\Gamma$ is a free $\mathbf{C}[\mathcal{P}]$-module with basis $\mathcal{S}$. In other words, for any $f \in \mathbf{C}[\mathbf{x}]^\Gamma$, there exist *unique* polynomials $f_1, \ldots, f_t \in \mathbf{C}[\mathbf{x}]$ such that

$$f = \sum_{i=1}^{t} f_i(\theta_1, \ldots, \theta_n) \cdot \eta_i.$$

Thus we have the Hironaka decomposition $\mathbf{C}[\mathbf{x}]^\Gamma = \bigoplus_{i=1}^{t} \eta_i \, \mathbf{C}[\mathcal{P}]$.

The steps 0 to 3 in Algorithm 2.5.8 generate the primary invariants. These form an h. s. o. p. for $\mathbf{C}[\mathbf{x}]^\Gamma$. By Theorem 2.3.5, the invariant ring is a finitely generated free module over the subring generated by this h. s. o. p. A free module basis over this subring is then constructed in steps 4 to 6.

In steps 1 and 2 we generate a sequence of homogeneous invariants whose variety gets smaller and smaller. In step 2 we will in practice first check whether the symmetrized monomial m_t^* is zero. Only if m_t^* is nonzero we will employ Subroutine 2.5.1 to test radical containment. This entire process will terminate once this variety consists of the origin only, and termination is guaranteed by Lemma 2.5.7.

After the completion of step 2 we have a set $\mathcal{Q}$ of invariants whose variety equals the nullcone, namely, the origin. Moreover, the degree of the polynomial in $\mathcal{Q}$ will be lexicographically optimal with respect to this property, since we proceed one degree level at a time. The homogeneous ideal generated by $\mathcal{Q}$ contains $\{x_1^{d_1}, \ldots, x_n^{d_n}\}$ for some $d_1, \ldots, d_n$. This implies that the polynomial ring $\mathbf{C}[\mathbf{x}]$ is finitely generated as a $\mathbf{C}[\mathcal{Q}]$-module, and therefore the invariant ring $\mathbf{C}[\mathbf{x}]^\Gamma$ is finitely generated a $\mathbf{C}[\mathcal{Q}]$-module. This implies that $\mathcal{Q}$ contains at least n elements. If $\mathcal{Q}$ contains precisely n elements, then $\mathcal{Q}$ is an h. s. o. p for $\mathbf{C}[\mathbf{x}]$ and hence also for $\mathbf{C}[\mathbf{x}]^\Gamma$. In this case (step 3.1) we choose $\mathcal{P} = \mathcal{Q}$ as the set of primary invariants. If $\mathcal{Q}$ contains more than n elements, then we proceed as follows.

2.5. Algorithms for computing fundamental invariants

Subroutine 2.5.10 (Creating a homogeneous system of parameters). Suppose $\mathcal{Q} = \{q_1, q_2, \ldots, q_m\}$ with $m > n$ and let $d_j := \mathrm{degree}(q_j)$. Let d denote the greatest common divisor of $d_1, d_2, \ldots, d_m$. We now choose an $n \times m$-matrix (a_{ij}) of complex numbers such that the ideal generated by the polynomials $p_i := \sum_{j=1}^{m} a_{ij} q_j^{d/d_j}$, $i = 1, 2, \ldots, n$, has the irrelevant ideal M as its radical. It follows from Noether's normalization lemma that every sufficiently generic matrix (a_{ij}) will have this property. In practice it is of course desirable to choose (a_{ij}) as sparse as possible. See Eisenbud and Sturmfels (1992) for a systematic approach to maintaining sparseness during this process.

We now discuss the remaining steps in Algorithm 2.5.8. Upon entering step 4, we are given an explicit h. s. o. p. $\mathcal{P} = \{\theta_1, \ldots, \theta_n\}$ for the invariant ring. In steps 5 and 6 we determine a set of symmetrized monomials which forms a free basis for $\mathbf{C}[\mathbf{x}]^\Gamma$ as a $\mathbf{C}[\mathcal{P}]$-module. In step 4 we assign a degree upper bound for the possible symmetrized monomials to be considered. The correctness of steps 4, 5 and 6 follows from Theorem 2.3.1 and the validity of this degree bound.

Lemma 2.5.11. Let $\mathcal{P} = \{\theta_1, \theta_2 \ldots, \theta_n\}$ be a set of algebraically independent invariant generators of I^Γ. Then there exists a finite set of invariants $\mathcal{S}$ of degree bounded by $\sum_{i=1}^{n} \mathrm{degree}(\theta_i) - n$ such that $\mathbf{C}[\mathbf{x}]^\Gamma$ is a free $\mathbf{C}[\mathcal{P}]$-module with basis $\mathcal{S}$.

Proof. Let $d_1, \ldots, d_n$ be the degrees of $\theta_1, \ldots, \theta_n$, and let d be the maximum degree occurring in any system of secondary invariants. By Proposition 2.3.6 (b), the Molien series satisfies an identity

$$\Phi_\Gamma(z) \prod_{i=1}^{n} (1 - z^{d_j}) = p_d(z),$$

where p_d is a polynomial of degree d. We multiply both sides of this identity with the polynomial

$$q(z) := \prod_{\pi \in \Gamma} \det(id - z\pi).$$

The right hand side $p_d(z)q(z)$ is a polynomial of degree $d+n|\Gamma|$. The expression $q(z)\Phi_\Gamma(z)$ is a polynomial of degree at most $n|\Gamma| - n$. Therefore the left hand side is a polynomial of degree at most $d_1 + \ldots + d_n + n|\Gamma| - n$. This implies the desired inequality $d \leq d_1 + \ldots + d_n - n$, and it completes the proof of Lemma 2.5.11 and Theorem 2.5.9. ◁

It can happen that the degree d of a system of primary invariants generated as in Subroutine 2.5.10 exceeds the cardinality of the group Γ. The following approach due to Dade provides an alternative method for computing a system of primary invariants all of whose degrees are divisors of the group order.

Subroutine 2.5.12 (Dade's algorithm for generating primary invariants).
Input: The Reynolds operator $* : \mathbf{C}[\mathbf{x}] \to \mathbf{C}[\mathbf{x}]^\Gamma$ of a finite subgroup Γ of $GL(\mathbf{C}^n)$.
Output: An h. s. o. p. $\{\theta_1, \ldots, \theta_n\}$ for $\mathbf{C}[\mathbf{x}]^\Gamma$ having the property that degree(θ_i) divides $|\Gamma|$ for $i = 1, \ldots, n$.
  For i from 1 to n do
- Choose a linear form $\ell_i(\mathbf{x})$ on $\mathbf{C}^n$ which does not vanish on the subspace defined by the linear forms $\ell_1 \circ \sigma_1, \ldots, \ell_{i-1} \circ \sigma_{i-1}$, for any choice of matrices $\sigma_1, \ldots, \sigma_{i-1}$ in Γ.
- Let θ_i denote the product over the set $\{\ell_i \circ \sigma : \sigma \in \Gamma\}$.

Note that the required choice of the linear forms ℓ_i is always possible since $\mathbf{C}$ is an infinite field. All necessary computations are using only linear algebra. To check correctness, we first observe that the cardinality of the set $\{\ell_i \circ \sigma : \sigma \in \Gamma\}$ divides the order of the group Γ, and hence so does the degree of the homogeneous invariants θ_i. By construction, the zero set of the linear forms $\ell_1 \circ \sigma_1, \ldots, \ell_n \circ \sigma_n$ consists only of the origin, for any choice of matrices $\sigma_1, \ldots, \sigma_n$ in Γ. This shows that the set of common zeros of the homogeneous invariants $\theta_1, \ldots, \theta_n$ consists only of the origin. This implies, as above, that $\theta_1, \ldots, \theta_n$ is an h. s. o. p. for $\mathbf{C}[\mathbf{x}]^\Gamma$.

The method described in Subroutine 2.5.12 can also be rephrased as follows. Let $\mathbf{u} = (u_1, \ldots, u_n)$ be a new set of variables. We define the *Chow form* of the matrix group Γ to be the polynomial

$$R(\mathbf{u}, \mathbf{x}) := \prod_{\sigma \in \Gamma} \langle \mathbf{u}, \sigma \mathbf{x} \rangle,$$

where $\langle, \rangle$ denotes the usual scalar product. This polynomial can be expanded as $R(\mathbf{u}, \mathbf{x}) = \sum_\alpha r_\alpha(\mathbf{x}) \mathbf{u}^\alpha$, where $\alpha \in$ ranges over all non-negative integer vectors whose coordinates sum to $|\Gamma|$.

Proposition 2.5.13. *A system of primary invariants can be obtained by taking n sufficiently generic $\mathbf{C}$-linear combinations of the coefficients $r_\alpha(\mathbf{x})$ of the Chow form $R(\mathbf{u}, \mathbf{x})$.*

Proof. By construction, the polynomials $r_\alpha(\mathbf{x})$ are homogeneous invariants having the same degree $|\Gamma|$. It therefore suffices to show that their common zero set consists only of the origin. Suppose $\mathbf{a} \in \mathbf{C}^n$ is a common zero of the r_α. Then $R(\mathbf{u}, \mathbf{a}) = \prod_{\sigma \in \Gamma} \langle \mathbf{u}, \sigma \mathbf{a} \rangle$ vanishes identically as a polynomial in $\mathbf{C}[\mathbf{u}]$. Hence there is an invertible matrix $\sigma \in \Gamma$ such that $\langle \mathbf{u}, \sigma \mathbf{a} \rangle = 0$ in $\mathbf{C}[\mathbf{u}]$. But this implies $\sigma \mathbf{a} = 0$ and consequently $\mathbf{a} = 0$. ◁

In practice we will usually be able to precompute the Molien series of the group Γ. Naturally, we will then use this information to speed up all computations. We close this section with the following general algorithm which summarizes most of the techniques we have discussed so far in this chapter.

2.5. Algorithms for computing fundamental invariants

Algorithm 2.5.14 (Computing the invariants of a finite matrix group).
Input: The Reynolds operator $^* : \mathbf{C}[\mathbf{x}] \to \mathbf{C}[\mathbf{x}]^\Gamma$ of a finite subgroup Γ of $GL(\mathbf{C}^n)$.
Output: A Hironaka decomposition for the invariant ring $\mathbf{C}[\mathbf{x}]^\Gamma$.

0. Compute the Molien series $\Phi(z)$ of Γ as a rational function in z.
1. Choose a system $\mathcal{P} = \{\theta_1, \ldots, \theta_n\}$ of primary invariants for Γ (using any of the procedures suggested in 2.5.8, 2.5.10, 2.5.12, or 2.5.13). Abbreviate $d_i := \mathrm{degree}(\theta_i)$.
2. Compute the polynomial

$$\Phi_\Gamma(z) \cdot \prod_{i=1}^n (1 - z^{d_j}) = c_1 z^{e_1} + c_2 z^{e_2} + \ldots + c_r z^{e_r},$$

where $c_1, \ldots, c_r$ are positive integers.
3. Using the Reynolds operator "$*$", find c_i linearly independent invariants $\eta_1, \ldots, \eta_{c_i}$ of degree e_i which do not lie in the ideal generated by $\theta_1, \ldots, \theta_n$, for $i = 1, \ldots, r$.

In step 1 we will use the information provided by a partial expansion of the Molien series $\Phi_\Gamma(z)$ to skip those degree levels which have no invariants whatsoever. In step 3 we can consider the ideal $\langle \theta_1, \ldots, \theta_n \rangle$ either in $\mathbf{C}[\mathbf{x}]^\Gamma$ or in $\mathbf{C}[\mathbf{x}]$. It seems reasonable to precompute a Gröbner basis $\mathcal{G}$ for $\langle \theta_1, \ldots, \theta_n \rangle$ after step 1. Then the ideal membership tests in step 3 amount to a normal form reduction with respect to $\mathcal{G}$.

Exercises

(1) Verify the correctness of Subroutines 2.5.5 and 2.5.6 for the rings discussed in Exercises 2.3. (1).
(2) Let $\Gamma \subset GL(\mathbf{C}^n)$ be a finite matrix group and fix $\mathbf{a} \in \mathbf{C}^n \setminus \{0\}$. Give an algorithm for computing a homogeneous invariant $I \in \mathbf{C}[\mathbf{x}]^\Gamma$ such that $I(\mathbf{a}) \neq 0$.
(3) Explain how the test in step 6 of Algorithm 2.5.14 can be implemented using Gröbner bases. (Hint: Use a monomial order like in Subroutine 2.5.6).
(4) * Implement some version of Algorithm 2.5.14 in your favorite computer algebra system (e.g., MAPLE, MATHEMATICA, ...)
(5) Let Γ be any subgroup of the group S_n of $n \times n$-permutation matrices. Show that there exists a system of secondary invariants whose degree does not exceed $\binom{n}{2}$.
(6) Let Γ be the cyclic subgroup of order 210 of the group S_{17} of 17×17-permutation matrices which is generated by the permutation (in cycle notation)

$$\sigma = (1, 2)(3, 4, 5)(6, 7, 8, 9, 10)(11, 12, 13, 14, 15, 16, 17).$$

(a) Give an example of an invariant of Γ which is not invariant under any permutation group $\Gamma' \subseteq S_{17}$ which properly contains Γ.
   (b) Compute the Molien series $\Phi_\Gamma(z)$ of this matrix group.
   (c) Identify a system of primary invariants of degree at most 7.
   (d) Show that the degree of the secondary invariants is at most 45. Is this estimate sharp? What is the number of secondary invariants?
(7) * Compile a list of all (isomorphism classes of) five-dimensional representations of the symmetric group S_3. Compute an explicit Hironaka decomposition for the invariant ring of each representation.

2.6. Gröbner bases under finite group action

In the preceding sections we have seen how algorithms from computer algebra can be used to solve problems in the invariant theory of finite groups. In the following the reverse point of view is taken: we wish to illustrate the application of invariant-theoretic methods for solving typical problems in computer algebra. Our main attention will be on computing with ideals and varieties which are fixed by some finite matrix group.

Consider the problem of finding the zeros of an ideal $I \subset \mathbf{C}[\mathbf{x}]$ which is presented in terms of generators. A standard solution method consists in computing a lexicographic Gröbner basis for I, from which the zeros can be "read off". It is known that this computation is generally very time-consuming. Moreover, it has been observed in practice that the running time is particularly bad if the given set of generators for I happens to be invariant under some finite group action. This is unsatisfactory because many polynomial systems arising from applications *do* have symmetries. It is our first goal to show how invariant theory can be used to compute a Gröbner basis which respects all symmetries.

The problem of solving systems of polynomial equations with symmetry has recently been addressed by Gatermann (1990). In this work the author shows how to substantially simplify and solve symmetric polynomial systems using representation theory of finite groups. The invariant-theoretic ideas to be presented in this section may lead to useful additions to the representation-theoretic algorithms introduced by Gatermann.

We fix a finite group $\Gamma \subset GL(\mathbf{C}^n)$ of $n \times n$-matrices. The set of Γ-orbits in $\mathbf{C}^n$ is denoted $\mathbf{C}^n/\Gamma$ and called the *orbit space* of Γ. We have an induced action of Γ on the coordinate ring of $\mathbf{C}^n$, which is the polynomial ring $\mathbf{C}[\mathbf{x}]$ in n complex variables $\mathbf{x} = (x_1, x_2, \ldots, x_n)$. The invariant subring $\mathbf{C}[\mathbf{x}]^\Gamma$ consists of all polynomials which are fixed under the action of Γ.

By Hilbert's finiteness theorem, there exists a finite set $\{I_1(\mathbf{x}), I_2(\mathbf{x}), \ldots, I_r(\mathbf{x})\}$ of fundamental invariants which generates the invariant ring $\mathbf{C}[\mathbf{x}]^\Gamma$. In geometric terms, the choice of these invariants amounts to choosing an embedding of the orbit space $\mathbf{C}^n/\Gamma$ as an algebraic subvariety into affine r-space $\mathbf{C}^r$. The equations defining the *orbit variety* $\mathbf{C}^n/\Gamma$ in $\mathbf{C}^r$ are the *syzygies* or algebraic relations among the $I_j(\mathbf{x})$.

We have seen in Sect. 2.5 how to use Gröbner bases for computing fundamental invariants and their syzygies. This preprocessing will be done once for

2.6. Gröbner bases under finite group action

the given group Γ. In the special case where Γ equals the symmetric group S_n of $n \times n$-permutation matrices this preprocessing is taken care of by Theorem 1.2.7, and Subroutine 2.6.1 is unnecessary.

Subroutine 2.6.1 (Preprocessing a fixed group Γ). Let Γ be any finite matrix group. We first compute a fundamental set of invariants $\{I_1(\mathbf{x}), I_2(\mathbf{x}), \ldots, I_r(\mathbf{x})\}$ as in the preceding sections. We then compute a Gröbner basis $\mathcal{G}_0$ for the ideal generated by

$$\{I_1(\mathbf{x}) - y_1, I_2(\mathbf{x}) - y_2, \ldots, I_r(\mathbf{x}) - y_r\} \quad \text{in } \mathbf{C}[x_1, x_2, \ldots, x_n, y_1, y_2, \ldots, y_r]$$

with respect to the lexicographic monomial order induced from $x_1 \succ \ldots \succ x_n \succ y_1 \succ \ldots \succ y_r$ (cf. Subroutine 2.5.3). Then the set $\mathcal{G}_0 \cap \mathbf{C}[y_1, y_2, \ldots, y_r]$ is a Gröbner basis for the ideal J defining the orbit variety $\mathcal{V}(J) = \mathbf{C}^n/\Gamma \hookrightarrow \mathbf{C}^r$.

Let $\mathcal{F} = \{f_1(\mathbf{x}), f_2(\mathbf{x}), \ldots, f_m(\mathbf{x})\}$ be a set of polynomials which is invariant under the action of Γ, i.e., $\forall \pi \in \Gamma \ \forall i \ \exists j : f_i \circ \pi = f_j$. Then its ideal $I = \langle \mathcal{F} \rangle$ is invariant under the action of Γ on $\mathbf{C}[\mathbf{x}]$, and its variety $\mathcal{V}(\mathcal{F}) = \mathcal{V}(I)$ is invariant under the action of Γ on $\mathbf{C}^n$. When applying Gröbner bases to study $\mathcal{V}(I)$, usually the following happens.

(a) One starts with *symmetric* input data $\mathcal{F}$.
(b) The Buchberger algorithm applied to $\mathcal{F} \subset K[\mathbf{x}]$ breaks all symmetries, and one gets a Gröbner basis $\mathcal{G}$ which is *not symmetric* at all.
(c) The *symmetric* variety $\mathcal{V}(I)$ is computed from the asymmetric polynomial set $\mathcal{G}$.

Invariant theory enables us, at least in principle, to replace step (b) by a Gröbner basis computation which preserves all symmetries. Since the variety $\mathcal{V}(I) \subset \mathbf{C}^r$ is invariant under the action of Γ, we can define the *relative orbit variety* $\mathcal{V}(I)/\Gamma$ whose points are the Γ-orbits of zeros of I.

We find that $\mathcal{V}(I)/\Gamma$ is an algebraic subvariety of $\mathbf{C}^n/\Gamma$, and therefore it is an algebraic subvariety of $\mathbf{C}^r$. In order to preserve the symmetry in step (b), we propose to compute a Gröbner basis for the relative orbit variety $\mathcal{V}(I)/\Gamma$ rather than for $\mathcal{V}(I)$ itself. Once we have gotten such a "symmetric Gröbner basis", it is not difficult to reconstruct properties of $\mathcal{V}(I)$ from the knowledge of $\mathcal{V}(I)/\Gamma$.

Algorithm 2.6.2 (Computing the relative orbit variety). Let $\mathcal{G}_0$ and "$\succ$" be as in Subroutine 2.6.1. The computation of a Gröbner basis for the ideal of the relative orbit variety $\mathcal{V}(I)/\Gamma$ in $\mathbf{C}[y_1, y_2, \ldots, y_r]$ works as follows.

- Compute a Gröbner basis $\mathcal{G}_1$ for $\mathcal{F} \cup \mathcal{G}_0$ with respect to the elimination order "$\succ$". Then $\mathcal{G}_2 := \mathcal{G}_1 \cap \mathbf{C}[y_1, y_2, \ldots, y_r]$ is a Gröbner basis for the ideal of $\mathcal{V}(I)/\Gamma$.
- Each point $\tilde{\mathbf{y}}$ in $\mathcal{V}(I)/\Gamma$ gives rise to a unique Γ-orbit in $\mathcal{V}(I)$. Such an orbit is a subset of $\mathbf{C}^n$ of cardinality $\leq |\Gamma|$. The points in the orbit corresponding to $\tilde{\mathbf{y}}$ can be computed by substituting the coordinates of $\tilde{\mathbf{y}} = (\tilde{y}_1, \tilde{y}_2, \ldots, \tilde{y}_r)$

$\in \mathbf{C}^r$ for the variables $y_1, y_2, \ldots, y_r$ in the precomputed Gröbner basis $\mathcal{G}_0$. The desired orbit equals the subvariety of $\mathbf{C}^r$ which is defined by the specialized Gröbner basis $\mathcal{G}_0(\tilde{\mathbf{y}}) \subset \mathbf{C}[\mathbf{x}]$.

Example 2.6.3. Let $n = 3$ and consider the set of polynomials $\mathcal{F} = \{f_1, f_2, f_3\} \subset \mathbf{C}[x_1, x_2, x_3]$ where

$$f_1(\mathbf{x}) = x_1^2 + x_2^2 + x_3^2 - 1$$
$$f_2(\mathbf{x}) = x_1^2 x_2 + x_2^2 x_3 + x_3^2 x_1 - 2x_1 - 2x_2 - 2x_3$$
$$f_3(\mathbf{x}) = x_1 x_2^2 + x_2 x_3^2 + x_3 x_1^2 - 2x_1 - 2x_2 - 2x_3.$$

The Gröbner basis $\mathcal{G}$ of the ideal $I = \langle \mathcal{F} \rangle$ with respect to the purely lexicographic order induced from $x_1 > x_2 > x_3$ equals

$$\{ \underline{50750x_1} + 50750x_2 + 54x_3^{11} + 585x_3^9 + 1785x_3^7 + 17580x_3^5 + 28695x_3^3$$
$$+ 32797x_3, \quad \underline{5800x_2^2} + 1740x_3^5 x_2 + 1740x_3^3 x_2 + 4060x_2 x_3 + 27x_3^{10}$$
$$+ 9345x_3^2 - 3825x_3^4 + 1110x_3^6 + 75x_3^8 - 3684, \quad \underline{420x_3^6 x_2} - 420x_3^4 x_2$$
$$+ 2940x_3^2 x_2 - 560x_2 + 9x_3^{11} + 45x_3^9 + 210x_3^7 + 165x_3^5 + 1335x_3^3 - 268x_3,$$
$$\underline{9x_3^{12}} - 18x_3^{10} + 315x_3^8 - 465x_3^6 + 1860x_3^4 - 1353x_3^2 + 196 \}.$$

From the underlined initial monomials we see that $\mathbf{C}[x_1, x_2, x_3]/I$ is a $\mathbf{C}$-vector space of dimension 18 (cf. Theorem 1.2.6). This implies that the variety $\mathcal{V}(I)$ consists of 18 points in affine 3-space $\mathbf{C}^3$, possibly counting multiplicities.

The input set $\mathcal{F}$ is invariant with respect to the symmetric group S_3 of 3×3-permutation matrices. The invariant ring $\mathbf{C}[x_1, x_2, x_3]^{S_3}$ is the ring of symmetric polynomials, and it is generated, for instance, by the elementary symmetric functions

$$I_1(\mathbf{x}) = x_1 + x_2 + x_3, \quad I_2(\mathbf{x}) = x_1 x_2 + x_1 x_3 + x_2 x_3, \quad I_3(\mathbf{x}) = x_1 x_2 x_3.$$

By Theorem 1.2.7, the preprocessed Gröbner basis for $\{I_1(\mathbf{x}) - y_1, I_2(\mathbf{x}) - y_2, I_3(\mathbf{x}) - y_3\}$ in the lexicographic order "$\succ$" induced from $x_1 \succ x_2 \succ x_3 \succ y_1 \succ y_2 \succ y_3$ equals

$$\mathcal{G}_0 = \{\underline{x_1} + x_2 + x_3 - y_1, \; \underline{x_2^2} + x_2 x_3 + x_3^2 - x_2 y_1 - x_3 y_1 + y_2, \; \underline{x_3^3} - x_3^2 y_1 + x_3 y_2 - y_3\}.$$

We now compute the Gröbner basis for the orbit variety $\mathcal{V}(I)/S_3$ as in Algorithm 2.6.2, and we find

$$\mathcal{G}_2 = \{\underline{8260 y_1} + 9 y_3^5 - 87 y_3^3 + 5515 y_3, \; \underline{1475 y_2} + 9 y_3^4 - 264 y_3^2 + 736,$$
$$\underline{27 y_3^6} - 513 y_3^4 + 33849 y_3^2 - 784\}.$$

2.6. Gröbner bases under finite group action

The orbit variety $\mathcal{V}(I)/S_3$ consists of six points, where each point $\tilde{\mathbf{y}} = (\tilde{y}_1, \tilde{y}_2, \tilde{y}_3) \in \mathcal{V}(I)/S_3$ corresponds to a three element S_3-orbit in $\mathcal{V}(I)$. If we wish to explicitly compute each individual orbit $\{(\tilde{x}_{\pi(1)}, \tilde{x}_{\pi(2)}, \tilde{x}_{\pi(3)}) : \pi \in S_3\}$, we may do so by factoring the cubic polynomial

$$t^3 - \tilde{y}_1 t^2 + \tilde{y}_2 t - \tilde{y}_3 = (t - \tilde{x}_1)(t - \tilde{x}_2)(t - \tilde{x}_3)$$

in terms of radicals over the rationals. Note that the Gröbner basis $\mathcal{G}_2$ is not only symmetric (after $y_i \mapsto p_i(\mathbf{x})$) but it is also simpler than the "ordinary" Gröbner basis $\mathcal{G}$. In this example the computation time for $\mathcal{G}_2$ is roughly equal to the computation time for $\mathcal{G}$.

It is a natural question whether each Γ-invariant ideal $I \subset \mathbf{C}[\mathbf{x}]$ can be generated by a suitable set of invariants. As stated, the answer to this question is "no". For instance, consider the action of the symmetric group S_2 by permuting the variables in $\mathbf{C}[x, y]$. The irrelevant ideal $I = \langle x, y \rangle$ is invariant under S_2 but this ideal cannot be generated by symmetric polynomials. For, the ideal I' in $\mathbf{C}[x, y]$ generated by *all* symmetric polynomials in I is the proper subideal $I' = \langle x+y, xy \rangle = \langle x^2, y^2, x+y \rangle$. Note, however, that the radical of I' equals I. It is true in general that each Γ-invariant ideal has a subideal with the same radical which is generated by a collection of invariants.

Proposition 2.6.4. Let $I \subset \mathbf{C}[\mathbf{x}]$ be a Γ-invariant ideal, and let I' be the subideal which is generated by all invariants in I. Then $\text{Rad}(I') = \text{Rad}(I)$.

Proof. Since $I' \subseteq I$, we clearly have $\text{Rad}(I') \subseteq \text{Rad}(I)$. By Hilbert's Nullstellensatz, it suffices to show that the variety $\mathcal{V}(I')$ is contained in the variety $\mathcal{V}(I)$. Let $\mathbf{a} \in \mathcal{V}(I')$ and $f \in I$. We need to show that $f(\mathbf{a}) = 0$.

We first note that $f \circ \sigma$ lies in the ideal I for all $\sigma \in \Gamma$. Now consider the polynomial

$$\prod_{\sigma \in \Gamma}(z - f(\sigma \mathbf{x})) = z^{|\Gamma|} + \sum_{j=0}^{|\Gamma|-1} p_j(\mathbf{x}) z^j,$$

where z is a new variable. Each coefficient p_j is a linear combination of $f \circ \sigma$, $\sigma \in \Gamma$, and hence p_j lies in I. Moreover, as in the proof of Proposition 1.1.1, we see that p_j lies in the invariant ring $\mathbf{C}[\mathbf{x}]^\Gamma$. Hence each p_j lies in the subideal I', and therefore $p_j(\mathbf{a}) = 0$. This implies $\prod_{\sigma \in \Gamma}(z - f(\sigma \mathbf{a})) = z^{|\Gamma|}$, and hence $f(\mathbf{a}) = 0$. ◁

Suppose we are given a set of generators $\mathcal{F}$ for an invariant ideal I as above. Then a set of invariant generators for its subideal I' can be computed using Algorithm 2.6.2. Thus we have a method for computing a set of invariant equations for any invariant variety.

In practice we will often be interested in the case where I is a zero-dimensional radical ideal, which means that $\mathcal{V}(I)$ is the union of a finite number of orbits $\Gamma\mathbf{a}$ in $\mathbf{C}^n$. We will often find that the ideal I' of the relative orbit variety is a radical ideal, in which case we have the equality $I' = I$, and the given ideal I is indeed generated by a set of invariants.

Example 2.6.5 (Computing invariant equations for an invariant variety). Consider the two polynomials

$$f(x, y) := 256y^{16} - 1536y^{14} + 3648y^{12} - 4416y^{10} + 3136y^8 - 1504y^6$$
$$+ 508y^4 - 92y^2 + 15$$

$$g(x, y) := 26233x^2 + 95744y^{14} - 572224y^{12} + 1340864y^{10} - 1538288y^8$$
$$+ 913824y^6 - 287484y^4 + 84717y^2 - 33671.$$

These polynomials form a Gröbner basis for their ideal $I := \langle f, g \rangle \subset \mathbf{C}[x, y]$ with respect to the lexicographic monomial order induced from $x \succ y$. Using the corresponding normal form reduction, it can be verified that the transformed polynomials $f(x+y, x-y)$ and $g(x+y, x-y)$ also lie in the ideal I. Moreover, we see that $f(x, y) = f(x, -y)$ and $g(x, y) = g(x, -y)$. These considerations show that the ideal I is invariant under the dihedral group D_8 which is generated by the matrices $\frac{1}{\sqrt{2}}\begin{pmatrix} 1 & 1 \\ 1 & -1 \end{pmatrix}$ and $\begin{pmatrix} 1 & 0 \\ 0 & -1 \end{pmatrix}$.

By Proposition 2.2.10, the invariant ring $\mathbf{C}[x, y]^{D_8}$ is generated by the invariants

$$a(x, y) = x^2 + y^2 \quad \text{and} \quad b(x, y) = x^8 + 14x^4y^4 + y^8.$$

Let us apply the preprocessing of Subroutine 2.6.1 with respect to the lexicographic monomial order induced from $x \succ y \succ A \succ B$. As the result we obtain the Gröbner basis

$$\mathcal{G}_0 = \{\underline{x^2} + y^2 - A, \ \underline{16y^8} - 32Ay^6 + 20A^2y^4 - 4A^3y^2 + A^4 - B\}$$

for a generic D_8-orbit.

We now apply Algorithm 2.6.2 to compute the relative orbit variety $\mathcal{V}(I)/D_8$ (in the embedding into $\mathbf{C}^2$ defined by the fundamental invariants $a(x, y)$ and $b(x, y)$). This results in the Gröbner basis

$$\mathcal{G}_2 = \{\underline{A} - B - 1, \ \underline{B^2} - B\}.$$

We see that $\mathcal{V}(I)/D_8$ consists of the two points $(\tilde{A} = 1, \tilde{B} = 0)$ and $(\tilde{A} = 2, \tilde{B} = 1)$. Each point corresponds to a regular orbit (i.e., having cardinality 16) of the dihedral group D_8. For each individual orbit we get a Gröbner basis by substituting $A = \tilde{A}$ and $B = \tilde{B}$ in $\mathcal{G}_0$. Using the notation of Proposition 2.6.4,

2.6. Gröbner bases under finite group action

we conclude that the ideal I is equal to I' and that it has the invariant decomposition

$$I = \langle x^2 + y^2 - 1, \, x^8 + 14x^4y^4 + y^8 \rangle \cap \langle x^2 + y^2 - 2, \, x^8 + 14x^4y^4 + y^8 - 1 \rangle.$$

We now return to our general discussion with the following question. What happens if our Algorithm 2.6.2 is applied to a set of polynomials $\mathcal{F}$ which is *not* invariant under Γ? We will see that this operation corresponds to taking the image of the variety $\mathcal{V}(\mathcal{F})$ under the group Γ.

Proposition 2.6.6. Let $I \subset \mathbf{C}[\mathbf{x}]$ be any ideal and let I' be the subideal which is generated by all Γ-invariants in I. Then $\mathcal{V}(I') = \Gamma \cdot \mathcal{V}(I)$, the image of $\mathcal{V}(I)$ under Γ.

Proof. We need to show that a point $\mathbf{a} \in \mathbf{C}^n$ lies in $\mathcal{V}(I')$ if and only if its orbit $\Gamma \mathbf{a}$ intersects $\mathcal{V}(I)$. For the "if"-direction suppose that $\sigma \mathbf{a} \in \mathcal{V}(I)$ for some group element $\sigma \in \Gamma$, and consider any $f \in I'$. Since $I' \subseteq I$, we have $f(\sigma \mathbf{a}) = 0$, and since f is an invariant we conclude $f(\mathbf{a}) = f(\sigma \mathbf{a}) = 0$.

For the "only if"-direction suppose that $\Gamma \mathbf{a} \cap \mathcal{V}(I) = \emptyset$. By Hilbert's Nullstellensatz, there exists a polynomial $g \in I$ which is identically 1 on the finite set $\Gamma \mathbf{a}$. Now consider the invariant polynomial $f(\mathbf{x}) := \prod_{\sigma \in \Gamma} g(\sigma \mathbf{x})$. By construction, f lies in the ideal I' but we have $f(\mathbf{a}) = 1$. Therefore $\mathbf{a} \notin \mathcal{V}(I')$, which completes the proof. ◁

Note that in the following algorithm the group Γ is presented only by the output of our preprocessing Subroutine 2.6.1. Neither the explicit group elements nor the Reynolds operator is needed at all. The correctness of Algorithm 2.6.7 is a direct corollary of Proposition 2.6.6.

Algorithm 2.6.7 (Computing the image of a variety under a finite group).
Input: Fundamental invariants $I_1(\mathbf{x}), \ldots, I_r(\mathbf{x})$ and the preprocessed Gröbner basis $\mathcal{G}_0$ of a finite matrix group $\Gamma \subset GL(\mathbf{C}^n)$. Any finite set of polynomials $\mathcal{F} \subset \mathbf{C}[\mathbf{x}]$.
Output: A finite set $\mathcal{H} \subset \mathbf{C}[\mathbf{x}]^\Gamma$ such that $\mathcal{V}(\mathcal{H}) = \Gamma \cdot \mathcal{V}(\mathcal{F})$ in $\mathbf{C}^n$.

- Compute a Gröbner basis $\mathcal{G}_1$ for $\mathcal{F} \cup \mathcal{G}_0$ with respect to the elimination order $\mathbf{x} \succ \mathbf{y}$ on $\mathbf{C}[\mathbf{x}, \mathbf{y}]$.
- Then $\mathcal{G}_2 := \mathcal{G}_1 \cap \mathbf{C}[\mathbf{y}]$ is a Gröbner basis for the subvariety $(\Gamma \cdot \mathcal{V}(\mathcal{F}))/\Gamma$ of $\mathbf{C}^r$.
- Let $\mathcal{H} \subset \mathbf{C}[\mathbf{x}]$ be the set obtained from $\mathcal{G}_2$ by substituting $y_j \mapsto I_j(\mathbf{x})$.

Example 2.6.8 (Computing the image of a variety under the symmetric group S_3). The invariant ring of the group S_3 of 3×3-permutation matrices is generated by the elementary symmetric functions $a := x + y + z$, $b := xy + xz + yz$, $c := xyz$. We will give six examples of sets of polynomials $\mathcal{F} \subset \mathbf{C}[x, y, z]$ and

$\mathcal{H} \subset \mathbf{C}[a, b, c]$ such that the variety of $\mathcal{H}$ is the image of the variety of $\mathcal{F}$ under the action of S_3. In each case $\mathcal{H}$ is computed from $\mathcal{F}$ using Algorithm 2.6.7.

(1) $\mathcal{F} = \{x - 1, y - 2, z - 3\}$; $\mathcal{H} = \{c - 6, b - 11, a - 6\}$.
(2) $\mathcal{F} = \{xy\}$; $\mathcal{H} = \{c\}$.
(3) $\mathcal{F} = \{x - 1\}$; $\mathcal{H} = \{a - b + c - 1\}$.
(4) $\mathcal{F} = \{xy - z\}$; $\mathcal{H} = \{2ac + c - c^2 - 2cb - b^2 + ca^2\}$.
(5) $\mathcal{F} = \{xy - 2, z^2 - 3\}$; $\mathcal{H} = \{c^2 + c - 12, -bc - c + 6a\}$.
(6) $\mathcal{F} = \{xy^2 - 2, z^2 - 3\}$; $\mathcal{H} = \{c^6 + 4c^5 + 4c^4 - 12c^2b^2 + 144cb - 432,\ c^9 + 8c^8 + 24c^7 + 32c^6 + 16c^5 + 216c^4b + 864c^3b + 864c^2b - 864c^2 - 144cb^4 + 864b^3 + 5184a\}$.

Exercises

(1) In Example 2.6.8, give a geometric description of all six varieties and their S_3-images.
(2) Let $I \subset \mathbf{C}[\mathbf{x}]$ be a radical ideal which is invariant under the action of a finite matrix group Γ, and suppose that the stabilizer in Γ of each root of I is trivial. Is it true that then I is generated by a set of invariant polynomials?
(3) * Consider the ideals I and I' in Proposition 2.6.6.
   (a) Show that if I is principal then I' is principal.
   (b) Compare I and I' with respect to some typical numerical invariants such as dimension, multiplicity, minimal number of generators, ...
(4) Let $\Gamma \subset GL(\mathbf{R}^3)$ be the the symmetry group of a regular cube $\mathcal{C} := \{(x, y, z) \in \mathbf{R}^3 : -1 \leq x, y, z \leq 1\}$. Following Algorithm 2.6.2, compute the relative orbit variety X/Γ for the following Γ-invariant subvarieties X of $\mathbf{C}^3$.
   $X =$ the eight vertices of $\mathcal{C}$,
   $X =$ the midpoints of the twelve edges of $\mathcal{C}$,
   $X =$ the centroids of the six faces of $\mathcal{C}$,
   $X =$ the union of the planes spanned by the six faces of $\mathcal{C}$,
   $X =$ the union of the four diagonal lines of $\mathcal{C}$,
   $X =$ the "superellipsoid" $\mathcal{V}(x^4 + y^4 + z^4 - 1)$.

2.7. Abelian groups and permutation groups

By a *permutation group* we mean a subgroup Γ of the group S_n of $n \times n$-permutation matrices. This section deals with specific techniques and algorithms for computing invariants of abelian groups and of permutation groups. We begin our discussion with an example which lies in the intersection of these two classes.

Example 2.7.1 (The cyclic group of order 5). Consider the cyclic subgroup of S_5 generated by the cycle $\sigma = (12345)$, and let $\Gamma \subset GL(\mathbf{C}^5)$ denote the corresponding cyclic permutation group. We abbreviate the scaled image of a

2.7. Abelian groups and permutation groups

monomial $x_1^i x_2^j x_3^k x_4^l x_5^m$ under the Reynolds operator of this group by

$$J_{ijklm} := x_1^i x_2^j x_3^k x_4^l x_5^m + x_2^i x_3^j x_4^k x_5^l x_1^m + x_3^i x_4^j x_5^k x_1^l x_2^m + x_4^i x_5^j x_1^k x_2^l x_3^m + x_5^i x_1^j x_2^k x_3^l x_4^m.$$

By Noether's theorem 2.1.4, the invariant ring

$$\mathbf{C}[\mathbf{x}]^{\Gamma} = \{f \in \mathbf{C}[\mathbf{x}] : f(x_1, x_2, x_3, x_4, x_5) = f(x_2, x_3, x_4, x_5, x_1)\}$$

is generated by the set of invariants $\{J_{ijklm} : 0 \leq i+j+k+l+m \leq 5\}$. The Molien series of Γ equals

$$\Phi_{\Gamma}(z) = \frac{1 + z^2 + 3z^3 + 4z^4 + 6z^5 + 4z^6 + 3z^7 + z^8 + z^{10}}{(1-z)(1-z^2)(1-z^3)(1-z^4)(1-z^5)}$$

$$= 1 + z + 3z^2 + 7z^3 + 14z^4 + 26z^5 + 42z^6 + 66z^7 + 99z^8$$
$$+ 143z^9 + 201z^{10} + 273z^{11} + \ldots$$

By the results of Sect. 2.3 we know that the invariant ring $\mathbf{C}[\mathbf{x}]^{\Gamma}$ is a free module over the subring of symmetric polynomials $\mathbf{C}[\mathbf{x}]^{S_5} = \mathbf{C}[\sigma_1, \sigma_2, \sigma_3, \sigma_4, \sigma_5]$. The rank of this module equals 24, which is the index of Γ in S_5. We can read off the degrees in any free basis from the above presentation of the Molien series: there is one generator in degree 0, one generator in degree 2, three generators in degree 3, etc.

Nevertheless it is quite hard to compute a system of 24 secondary invariants using the general methods of Sect. 2.5. For instance, we may start by choosing the secondary invariants J_{11000} in degree 2, we then continue by choosing J_{21000}, J_{11100} and J_{11010} in degree 3 etc. .... During this process we need to guarantee that none of the chosen invariants lies in the submodule generated by the previously chosen ones.

We will instead pursue an alternative approach to $\mathbf{C}[\mathbf{x}]^{\Gamma}$ which is based on the fact that Γ is an abelian group. It is known (cf. Proposition 2.7.2) that the matrices in any finite abelian subgroup $\Gamma \subset GL(\mathbf{C}^n)$ can be diagonalized simultaneously. For a cyclic group this diagonalization process is essentially equivalent to the *discrete Fourier transform*. Let $\omega \in \mathbf{C}$ be any primitive fifth root of unity. We think of ω as a formal variable subject to the relation $\omega^4 + \omega^3 + \omega^2 + \omega^1 + 1 = 0$. We perform a linear change of variables by setting

$$y_0 = x_1 + x_2 + x_3 + x_4 + x_5$$
$$y_1 = x_1 + \omega^4 x_2 + \omega^3 x_3 + \omega^2 x_4 + \omega x_5$$
$$y_2 = x_1 + \omega^3 x_2 + \omega x_3 + \omega^4 x_4 + \omega^2 x_5$$
$$y_3 = x_1 + \omega^2 x_2 + \omega^4 x_3 + \omega x_4 + \omega^3 x_5$$
$$y_4 = x_1 + \omega x_2 + \omega^2 x_3 + \omega^3 x_4 + \omega^4 x_5.$$

It is easy to see that the cyclic group Γ consists of diagonal matrices with respect to this new basis. More precisely, we have $y_i \circ \sigma = \omega^i \cdot y_i$ for $i = 0, 1, \ldots, 4$. This implies the following presentation of the invariant ring

$$\mathbf{C}[\mathbf{x}]^\Gamma = \mathrm{span}_{\mathbf{C}}\{y_0^{i_0} y_1^{i_1} y_2^{i_2} y_3^{i_3} y_4^{i_4} \mid i_1 + 2i_2 + 3i_3 + 4i_4 \equiv 0 \pmod{5}\}.$$

Each linear homogeneous congruence, such as $i_1 + 2i_2 + 3i_3 + 4i_4 \equiv 0 \pmod{5}$, is equivalent to a linear homogeneous diophantine equation, such as $i_1 + 2i_2 + 3i_3 + 4i_4 - 5i_5 = 0$.

The problem of solving linear equations over the nonnegative integers is an important problem in combinatorics and in mathematical programming. We refer to Schrijver (1986) for a general introduction to integer programming and to Stanley (1986: section 4.6) for a combinatorial introduction to linear diophantine equations. An algorithm using Gröbner basis was presented in Sect. 1.4.

In our example we find a minimal generating set of eleven lattice points for the solution monoid of the given congruence equation. From this we obtain the following presentation of the invariant ring as a monomial subalgebra of $\mathbf{C}[y_0, y_1, y_2, y_3, y_4]$:

$$\mathbf{C}[\mathbf{x}]^\Gamma = \mathbf{C}[y_0,\ y_1 y_4,\ y_2 y_3,\ y_1 y_2^2,\ y_1^2 y_3,\ y_2 y_4^2,\ y_3^3 y_4,\ y_1^5,\ y_2^5,\ y_3^5,\ y_4^5].$$

A canonical choice of five primary invariants is given by taking the pure powers of the coordinate functions. Indeed, the invariant ring $\mathbf{C}[\mathbf{x}]^\Gamma$ is a free module of rank 125 over its subring $\mathbf{C}[y_0, y_1^5, y_2^5, y_3^5, y_4^5]$. The degree generating function for a free basis equals

$$\Phi_\Gamma(z) \cdot (1-z)(1-z^5)^4 = 1 + 2z^2 + 4z^3 + 7z^4 + 8z^5 + 16z^6 + 16z^7 + 17z^8$$
$$+ 16z^9 + 16z^{10} + 8z^{11} + 7z^{12} + 4z^{13} + 2z^{14} + z^{16}.$$

The unique generator in degree 16 equals $y_1^4 y_2^4 y_3^4 y_4^4$, the two generators in degree 14 are $y_1^3 y_2^4 y_3^4 y_4^3$ and $y_1^4 y_2^3 y_3^3 y_4^4$, etc. ...

It is an easy task to express an arbitrary invariant $I(y_0, y_1, y_2, y_3, y_4)$ in terms of the eleven basic invariants. For each monomial $y_0^{i_0} y_1^{i_1} y_2^{i_2} y_3^{i_3} y_4^{i_4}$ which occurs in the expansion of $I(y_0, y_1, y_2, y_3, y_4)$ satisfies the congruence $i_1 + 2i_2 + 3i_3 + 4i_4 \equiv 0 \pmod{5}$ and is therefore a product of some of the eleven basic monomials.

In practice it may be preferable to work in the old variables x_1, x_2, x_3, x_4, x_5 rather than in the transformed variables y_0, y_1, y_2, y_3, y_4. In order to do so, we may express the eleven fundamental invariants in terms of the symmetrized monomials J_{ijklm}:

$$y_0 = J_{10000},$$

$$y_1 y_4 = J_{20000} - J_{11000} + (-J_{11000} + J_{10100})\omega^2 + (-J_{11000} + J_{10100})\omega^3,$$

2.7. Abelian groups and permutation groups

$$y_2 y_3 = J_{20000} - J_{10100} + (J_{11000} - J_{10100})\omega^2 + (J_{11000} - J_{10100})\omega^3,$$

$$\begin{aligned} y_1 y_2^2 = & J_{30000} - J_{21000} - 2J_{10200} + 2J_{11010} + (J_{12000} - J_{21000} + 2J_{20100} \\ & - 2J_{10200})\omega + (2J_{12000} - J_{21000} - 2J_{11100} - J_{10200} + 2J_{11010})\omega^2 \\ & + (J_{21000} + J_{20100} - 2J_{11100} - 2J_{10200} + 2J_{11010})\omega^3, \end{aligned}$$

$$\begin{aligned} y_1^2 y_3 = & J_{30000} - 2J_{21000} - J_{20100} + 2J_{11100} + (2J_{12000} - 2J_{21000} - J_{20100} \\ & + J_{10200})\omega + (2J_{11100} - J_{21000} - J_{20100} + 2J_{10200} - 2J_{11010})\omega^2 \\ & + (-2J_{21000} + J_{12000} + J_{20100} + 2J_{11100} - 2J_{11010})\omega^3, \ldots \text{etc.} \ldots \end{aligned}$$

From this we get an easy algorithm for rewriting any invariant $I(x_1, x_2, x_3, x_4, x_5)$ in terms of symmetrized monomials J_{ijklm} of degree at most 5. First replace the x_i's by y_j's via

$$5x_1 = y_0 + y_1 + y_2 + y_3 + y_4$$

$$5x_2 = (y_3 - y_4)\omega^3 + (y_2 - y_4)\omega^2 + (y_1 - y_4)\omega + y_0 - y_4$$

$$5x_3 = (y_4 - y_2)\omega^3 + (y_1 - y_2)\omega^2 + (y_3 - y_2)\omega + y_0 - y_2$$

$$5x_4 = (y_1 - y_3)\omega^3 + (y_4 - y_3)\omega^2 + (y_2 - y_3)\omega + y_0 - y_3$$

$$5x_5 = (y_2 - y_1)\omega^3 + (y_3 - y_1)\omega^2 + (y_4 - y_1)\omega + y_0 - y_1.$$

Write the result as a polynomial function $I = I'(y_0, y_1 y_4, y_2 y_3, \ldots, y_4^5)$ in the eleven basic invariants. Finally, substitute the above expansions in terms of J_{ijklm} into I'. ◁

We now consider the case of an arbitrary abelian subgroup Γ of $GL(\mathbf{C}^n)$. The following result from linear algebra is well known. Its proof follows from the fact that each matrix of finite order can be diagonalized over $\mathbf{C}$ and that the centralizer of the subgroup of diagonal matrices in $GL(\mathbf{C}^n)$ equals the diagonal matrices themselves.

Proposition 2.7.2. A finite matrix group $\Gamma \subset GL(\mathbf{C}^n)$ is abelian if and only if there exists a linear transformation $T_\Gamma \in GL(\mathbf{C}^n)$ such that the conjugate group $T_\Gamma \cdot \Gamma \cdot T_\Gamma^{-1}$ consists only of diagonal matrices.

It is important to note that this theorem is false over the real or rational numbers. As we have seen in Example 2.7.1, it can happen that each matrix in Γ has rational entries but the diagonalization matrix T_Γ has entries in some algebraic extension of the rationals.

For a given abelian group Γ we can use algorithms from representation theory to compute the matrix T_Γ. In the following we will assume that this preprocessing of Γ has been done and that our input consists of an abelian matrix group Γ together with its diagonalization matrix T_Γ. The following algorithm shows that the invariant theory of finite abelian groups is equivalent to the study of linear homogeneous congruences.

Algorithm 2.7.3 (Computing fundamental invariants for an abelian group).
Input: A set of generating matrices $\Omega_1, \ldots, \Omega_m$ for an abelian group $\Gamma \subset GL(\mathbf{C}^n)$, and a matrix T_Γ which simultaneously diagonalizes $\Omega_1, \ldots, \Omega_m$.
Output: A finite algebra basis for the invariant ring $\mathbf{C}[\mathbf{x}]^\Gamma$.

1. Introduce new variables $\mathbf{y} = (y_1, \ldots, y_n)$ via $\mathbf{y} = T_\Gamma \mathbf{x}$.
2. For $i = 1, \ldots, m$ write $T_\Gamma \Omega_i T_\Gamma^{-1} = \operatorname{diag}(\omega_{i1}, \omega_{i2}, \ldots, \omega_{in})$. Let d_{ij} denote the order of the complex number ω_{ij}, and let g_i denote the order of the matrix Ω_i.
3. Consider the system of m linear homogeneous congruences

$$d_{i1}\mu_1 + d_{i2}\mu_2 + \ldots + d_{in}\mu_n \equiv 0 \pmod{g_i} \quad i = 1, 2, \ldots, m. \tag{2.7.1}$$

Compute a finite generating set $\mathcal{H}$ for the solution monoid of this system. Then

$$\mathbf{C}[\mathbf{x}]^\Gamma = \mathbf{C}[\mathbf{y}]^{T_\Gamma \Gamma T_\Gamma^{-1}} = \mathbf{C}\bigl[y_1^{\mu_1} y_2^{\mu_2} \cdots y_n^{\mu_n} : \mu = (\mu_1, \mu_2, \ldots, \mu_n) \in \mathcal{H}\bigr].$$

We need to make a few comments about Algorithm 2.7.3. In step 2 the *order* of the complex number ω_{ij} is the smallest positive integer d_{ij} such that $\omega_{ij}^{d_{ij}} = 1$. The order g_i of the matrix Ω_i then simply equals the least common multiple of $d_{i1}, d_{i2}, \ldots, d_{in}$.

The generating set $\mathcal{H}$ in step 3 is a *Hilbert basis* of the monoid $\mathcal{F}$. It has the property that every $\mu = (\mu_1, \ldots, \mu_n) \in \mathcal{F}$ is a $\mathbf{Z}_+^n$-linear combination of elements in $\mathcal{H}$. The Hilbert basis can be computed using Algorithm 1.4.5.

Let us briefly discuss the structure of the monoid $\mathcal{F}$ and its Hilbert basis $\mathcal{H}$. For the statement and derivation of Corollary 2.7.4 we shall assume that the reader is familiar with the terminology in Stanley (1986: section 4.6). For any $j \in \{1, \ldots, n\}$ let e_j denote the least common multiple of $d_{1j}, d_{2j}, \ldots, d_{mj}$. Then the scaled j-th coordinate vector $(0, \ldots, 0, e_j, 0, \ldots, 0)$ lies in $\mathcal{F}$, or, equivalently, the monomial $y_j^{e_j}$ is an invariant. This shows that $\mathcal{F}$ is a *simplicial monoid* with set of *quasi-generators* $Q = \{(0, \ldots, 0, e_j, 0, \ldots, 0) : j = 1, \ldots, n\}$.

Consider the "descent set" $D(\mathcal{F}) = \{\mu = (\mu_1, \ldots, \mu_n) \in \mathcal{F} : \mu_1 < e_1, \ldots, \mu_n < e_n\}$. By Stanley (1986: lemma 4.6.7), every $\mu \in \mathcal{F}$ can be written uniquely as the sum of an element in $D(\mathcal{F})$ and a $\mathbf{Z}_+$-linear combination of Q. This implies that the monomials corresponding to Q and $D(\mathcal{F})$ for a system of primary and secondary invariants.

2.7. Abelian groups and permutation groups

Corollary 2.7.4. The invariant ring of the abelian group Γ has the Hironaka decomposition

$$\mathbf{C}[\mathbf{x}]^\Gamma = \mathbf{C}[\mathbf{y}]^{T_\Gamma \Gamma T_\Gamma^{-1}} = \bigoplus_{\mu \in D(\mathcal{F})} \mathbf{C}[y_1^{e_1}, y_2^{e_2}, \ldots, y_n^{e_n}] \cdot y_1^{\mu_1} y_2^{\mu_2} \cdots y_n^{\mu_n}.$$

In Example 2.7.1 this decomposition consists of five primary invariants and $|D(\mathcal{F})| = 125$ secondary invariants. From those 130 invariants we were able to choose a subset of eleven invariants which generate $\mathbf{C}[\mathbf{x}]^\Gamma$ as an algebra. Also in the general setting of Corollary 2.7.4 such a simplification is possible: We can always find a subset $\mathcal{H}$ of $Q \cup D(\mathcal{F})$ which is a *minimal Hilbert basis* for $\mathcal{F}$.

We now come to the study of invariants of permutation groups. As a motivation we discuss an application to classical Galois theory. Suppose we are given a polynomial $f(z) = z^n + a_{n-1} z^{n-1} + \ldots + a_1 z + a_0$ with coefficients in the field $\mathbf{Q}$ of rational numbers. Suppose further that the roots of f are labeled $x_1, \ldots, x_n \in \mathbf{C}$. Such a labeling defines a representation of the Galois group of f as a subgroup Γ of $S_n \subset GL(\mathbf{C}^n)$.

Problem 2.7.5. Suppose the Galois group Γ of $f(z)$ is a solvable group for which an explicit solvable series is given. How can we express the roots $x_1, \ldots, x_n$ in terms of radicals in the coefficients $a_0, \ldots, a_{n-1}$?

Only few text books in algebra provide a satisfactory answer to this question. One notable exception is Gaal (1988) where section 4.5 explains "How to solve a solvable equation". We will here illustrate how Gröbner bases in conjunction with invariant theory of permutation groups can be used to solve Problem 2.7.5. In our discussion the Galois group is part of the input, and it is our objective to compute a formula in terms of radicals which works for *all* polynomials with that Galois group. We are *not* proposing to use invariants for computing the Galois group Γ in the first place. For the problem of computing Galois groups and its complexity we refer to (Landau 1985). We illustrate our approach to Problem 2.7.5 for the case of a general cubic.

Example 2.7.6 (Automatic derivation of Cardano's formula). Consider an arbitrary univariate cubic polynomial

$$p(z) = z^3 + a_2 z^2 + a_1 z + a_0 = (z - x_1)(z - x_2)(z - x_3)$$

with both coefficients and roots in the complex numbers. We wish to synthesize a general formula which expresses the x_i in terms of the a_j. To that end we view $\mathbf{x} = (x_1, x_2, x_3)$ as indeterminates and a_2, a_1, a_0 as generic constants, and we consider the ideal I in $\mathbf{C}[\mathbf{x}]$ which is generated by $x_1 + x_2 + x_3 + a_2$, $x_1 x_2 + x_1 x_3 + x_2 x_3 - a_1$, $x_1 + x_2 + x_3 + a_0$.

Let $\Gamma \subset S_3$ denote the alternating group of even permutations. In order to preprocess its invariant ring, we introduce a new variable ω subject to the relation $\omega^2 + \omega + 1 = 0$ and we set

$$y_0 = x_1 + x_2 + x_3, \quad y_1 = x_1 + \omega^2 x_2 + \omega x_3, \quad y_2 = x_1 + \omega x_2 + \omega^2 x_3.$$

The inverse relations are conveniently encoded in the Gröbner basis

$$\mathcal{G}_0 = \{\underline{3x_1} - y_0 - y_1 - y_2, \underline{3x_2} - y_0 - y_1\omega + y_2\omega + y_2,$$
$$\underline{3x_3} - y_0 + y_1\omega + y_1 - y_2\omega, \underline{\omega^2} + \omega + 1\}.$$

Using Algorithm 2.7.3 we find the four generators $u_0 = y_0, u_{12} = y_1 y_2, u_{13} = y_1^3$ and $u_{23} = y_2^3$ for the invariant ring $\mathbf{C}[\mathbf{x}]^\Gamma$. In this situation our preprocessing Algorithm 2.6.1 would generate the Gröbner basis

$$\mathcal{G}_1 = \{\underline{y_0} - u_0, \underline{y_1^3} - u_{13}, \underline{y_1^2 u_{12}} - y_2 u_{13}, \underline{y_1 y_2} - u_{12}, \underline{y_1 u_{12}^2} - y_2^2 u_{13},$$
$$\underline{y_1 u_{23}} - y_2^2 u_{12}, \underline{y_2^3} - u_{23}, \underline{u_{12}^3} - u_{13} u_{23}\}.$$

This means we embed the orbit variety $\mathbf{C}^3/\Gamma$ into $\mathbf{C}^4$ as the hypersurface $u_{12}^3 - u_{23} u_{13} = 0$.

The ideal I which encodes the roots of f is clearly invariant under the action of Γ. Let us compute the relative orbit variety $\mathcal{V}(I)/\Gamma$. To this end we first transform coordinates in the input equations

$$x_1 + x_2 + x_3 + a_2 = y_0 + a_2$$

$$x_1 x_2 + x_1 x_3 + x_2 x_3 - a_1 = \frac{1}{3} y_0^2 - \frac{1}{3} y_1 y_2 - a_1$$

$$x_1 x_2 x_3 + a_0 = \frac{1}{27} y_0^3 + \frac{1}{27} y_1^3 + \frac{1}{27} y_2^3 - \frac{1}{9} y_0 y_1 y_2 + a_0,$$

and then we apply Algorithm 2.6.2 to find the Gröbner basis

$$\mathcal{G}_2 = \{\underline{u_{12}^3} - u_{23} u_{13}, \underline{u_0} + a_2, \underline{u_{12}} + 3a_1 - a_2^2,$$
$$\underline{u_{13}} + u_{23} + 27a_0 - 9a_1 a_2 + 2a_2^3,$$
$$\underline{u_{23}^2} + 27 u_{23} a_0 - 9 u_{23} a_1 a_2 + 2 u_{23} a_2^3 - 27 a_1^3 + 27 a_1^2 a_2^2 - 9 a_1 a_2^4 + a_2^6\}.$$

Now the combination of the Gröbner bases $\mathcal{G}_2, \mathcal{G}_1$ and $\mathcal{G}_0$ provides an explicit formula for the x_i in terms of the a_j. We first solve for u_{23}, u_{13}, u_{12} and u_0 in $\mathcal{G}_2$ which involves the extraction of a square root. We substitute the result into $\mathcal{G}_1$ and we solve for y_2, y_1, y_0 which involves the extraction of a cube root. We finally get the roots x_1, x_2, x_3 as linear combinations as of y_2, y_1, y_0 by substituting into $\mathcal{G}_0$. The resulting formula is *Cardano's formula* for cubics.

When does an irreducible cubic polynomial f have the alternating group Γ

as its Galois group? This is the case if and only if the relative orbit variety $\mathcal{V}(I)/\Gamma$ is reducible into $[S_3 : \Gamma] = 2$ components over the ground field. This can be checked by factoring the fifth element of $\mathcal{G}_2$ as a polynomial in u_{23}. This polynomial factors if and only if its discriminant is a square, and in this case f has Galois group Γ. ◁

In order to solve Problem 2.7.5 in the general case we may proceed as follows. We will only sketch the basic ideas and leave it as a challenging exercise for the reader to work out the details. For doing so it may be helpful to consult Gaal (1988: section 4.6).

Let $\Gamma = \Gamma_1 > \Gamma_2 > \ldots > \Gamma_{k-1} > \Gamma_k = (id)$ be a composition series of the given permutation group $\Gamma \subset S_n$. This means that each factor group Γ_i/Γ_{i+1} is cyclic of prime order p_i. For each occurring prime number p_i we introduce a primitive p_i-th root of unity ω_i. We may assume that the polynomial $f(z) = \sum a_i z^i$ has distinct roots $x_1, x_2, \ldots, x_n$. Let $\mathcal{G}_0$ be the Gröbner basis constructed in Theorem 1.2.7 for the vanishing ideal $I \subset \mathbf{C}[\mathbf{x}]$ of the S_n-orbit of the point $(x_1, \ldots, x_n)$ in $\mathbf{C}^n$.

- Choose a system of fundamental invariants $I_1(\mathbf{x}), I_2(\mathbf{x}) \ldots, I_m(\mathbf{x})$ for the tentative Galois group Γ. Using the methods of Sect. 2.6 we compute a Gröbner basis for the relative orbit variety $\mathcal{V}(I)/\Gamma$ with respect to the embedding defined by the I_j.
- The Galois group of f is equal to Γ (or a subgroup thereof) if and only if the relative orbit variety $\mathcal{V}(I)/\Gamma$ factors completely over the ground field. In this case we can factor the ideal of $\mathcal{V}(I)/\Gamma$ as an intersection of $[S_n : \Gamma]$ maximal ideals of the form $\langle I_1 - c_1, I_2 - c_2, \ldots, I_m - c_m \rangle$ where the c_i are rational expressions in the a_j. This gives us a decomposition of the ideal I as an intersection of ideals of the form $\langle I_1(\mathbf{x}) - c_1, I_2(\mathbf{x}) - c_2, \ldots, I_m(\mathbf{x}) - c_m \rangle$ in $\mathbf{C}[\mathbf{x}]$.
- Compute the invariant ring $\mathbf{C}[\mathbf{x}]^{\Gamma_i}$ for each intermediate group Γ_i. It follows from Theorem 2.3.5 that $\mathbf{C}[\mathbf{x}]^{\Gamma_{i+1}}$ is a free module of rank p_i over $\mathbf{C}[\mathbf{x}]^{\Gamma_i}$. The cyclic group Γ_i/Γ_{i+1} acts on this module. We can diagonalize the action via a linear change of variables as in Examples 2.7.1 and 2.7.6 (involving the primitive root of unity ω_i). As the result we obtain an element $\xi_i \in \mathbf{C}[\mathbf{x}]^{\Gamma_{i+1}}$ on which a generator of the cyclic group Γ_i/Γ_{i+1} acts via $\xi_i \mapsto \omega_i \cdot \xi_i$. This implies that $\xi_i^{p_i}$ lies in the subring $\mathbf{C}[\mathbf{x}]^{\Gamma_i}$ and that $\{1, \xi_i, \xi_i^2, \ldots, \xi_i^{p_i-1}\}$ is a free basis for $\mathbf{C}[\mathbf{x}]^{\Gamma_{i+1}}$ as a $\mathbf{C}[\mathbf{x}]^{\Gamma_i}$- module.
- Now each element of $\mathbf{C}[\mathbf{x}]^{\Gamma_{i+1}}$ can be expressed by extracting one p_i-th root and by rational operations in terms of elements of $\mathbf{C}[\mathbf{x}]^{\Gamma_i}$. We iterate this process from $i = 1, \ldots, k - 1$ to get an expression for $x_1, \ldots, x_n$ in terms of $c_1, \ldots, c_m$ which involves only rational operations, p_i-th roots of unity ω_i and extracting p_i-th roots.

Next we study the invariant ring of an arbitrary permutation group $\Gamma \subset S_n$. A good choice of primary invariants for Γ consists in the elementary symmetric functions $\sigma_1, \sigma_2, \ldots, \sigma_n$, or any other algebra basis for the ring of symmetric polynomials $\mathbf{C}[\mathbf{x}]^{S_n}$. It is therefore natural to study the structure of $\mathbf{C}[\mathbf{x}]^\Gamma$ as a module over $\mathbf{C}[\mathbf{x}]^{S_n}$. Our first theorem is a direct consequence of the results in Sect. 2.3.

Theorem 2.7.6. *The invariant ring $\mathbf{C}[\mathbf{x}]^\Gamma$ is a free module of rank $t := n!/|\Gamma|$ over the subring of symmetric polynomials. If $\eta_1, \eta_2, \ldots, \eta_t$ is any free module basis then*

$$\sum_{i=1}^{t} z^{\deg(\eta_i)} = \Phi_\Gamma(z) \cdot (1-z)(1-z^2) \cdots (1-z^n).$$

The computation of the Molien series Φ_Γ of a permutation group is facilitated by the following easy observation. The *cycle type* of a permutation σ is the integer vector $\ell(\sigma) = (\ell_1, \ell_2, \ldots, \ell_n)$ where ℓ_i counts the number of cycles of length i in the cycle decomposition of σ.

Remark 2.7.7. The characteristic polynomial of a permutation matrix σ can be read off from its cycle type $\ell(\sigma) = (\ell_1, \ell_2, \ldots, \ell_n)$ via the formula $\det(1-z\sigma) = \prod_{i=1}^{n}(1-z^i)^{\ell_i}$.

A system of secondary invariants $\{\eta_1, \eta_2, \ldots, \eta_t\}$ as in Theorem 2.7.6 is called a *basic set* for Γ. Finding explicit basic sets for permutation groups is an important problem in algebraic combinatorics. A large number of results on this subject are due to Garsia and Stanton (1984), with more recent extensions by Reiner (1992). In what follows we explain the basic ideas underlying the algebraic combinatorics approach.

Let us begin with the seemingly trivial case of the trivial permutation group $\Gamma = \{id\}$. Its invariant ring equals the full polynomial ring $\mathbf{C}[\mathbf{x}]$. By Theorem 2.7.6, $\mathbf{C}[\mathbf{x}]$ is a free module of rank $n!$ over the symmetric polynomials, and finding a basic set for $\{id\}$ means finding $n!$ module generators for $\mathbf{C}[\mathbf{x}]$ over $\mathbf{C}[\mathbf{x}]^{S_n}$. The Hilbert series of the polynomial ring equals $\Phi_{\{id\}}(z) = (1-z)^{-n}$, and so we get the following formula for the degree generating function of any basic set:

$$\frac{(1-z)(1-z^2)(1-z^3) \cdots (1-z^n)}{(1-z)^n}$$
$$= (1+z)(1+z+z^2) \cdots (1+z+z^2+\ldots+z^{n-1}).$$

Let c_i denote the coefficient of z^i in the expansion of this product. This number has the following two combinatorial interpretations; see Stanley (1986: corollary 1.3.10).

Proposition 2.7.8. *The number c_i of elements of degree i in a basic set for $\Gamma = \{id\}$*

(a) *equals the cardinality of the set $\{(\nu_1, \nu_2, \ldots, \nu_n) \in \mathbf{Z}^n : 0 \leq \nu_i < i, \nu_1 + \ldots + \nu_n = i\}$.*
(b) *equals the number of permutations $\pi \in S_n$ having precisely i inversions.*

2.7. Abelian groups and permutation groups

There are natural basic sets associated with both combinatorial interpretations. With each permutation $\pi = (\pi_1, \pi_2, \ldots, \pi_n)$ in S_n we associate its *descent monomial* $m(\pi)$. This is defined as the product of all monomials $x_{\pi_1} x_{\pi_2} \cdots x_{\pi_i}$ where $1 \leq i < n$ and $\pi_i > \pi_{i+1}$. For instance, the descent monomial of the permutation $\pi = (2, 1, 4, 5, 3)$ equals $m(\pi) = x_2(x_2 x_1 x_4 x_5) = x_1 x_2^2 x_4 x_5$. It is a remarkable combinatorial fact that the number of descent monomials of degree i equals the number of permutations $\pi \in S_n$ having precisely i inversions. For a proof see Stanley (1986: corollary 4.5.9).

Theorem 2.7.9.

(a) The set of monomials $x_1^{\nu_1} x_2^{\nu_2} \cdots x_n^{\nu_n}$ with $0 \leq \nu_i < i$ is basic.
(b) The set of descent monomials $m(\pi)$, $\pi \in S_n$, is basic.

Proof. By the dimension count of Proposition 2.7.8, it is sufficient in both cases to show that the given monomials span $\mathbf{C}[\mathbf{x}]$ as a $\mathbf{C}[\mathbf{x}]^{S_n}$-module. Let us see that part (a) is an easy corollary to Theorem 1.2.7. Consider any $p(\mathbf{x}) \in \mathbf{C}[\mathbf{x}]$. Its normal form with respect to the Gröbner basis $\mathcal{G}$ is an expression of the form $\sum_{0 \leq \nu_i < i} q(\mathbf{y}) x_1^{\nu_1} x_2^{\nu_2} \cdots x_n^{\nu_n}$. We may now replace each slack variable y_i by the corresponding elementary symmetric function $\sigma_i(x_1, \ldots, x_n)$ to get a presentation of $p(\mathbf{x})$ as a $\mathbf{C}[\mathbf{x}]^{S_n}$-linear combination of the basic set in question. For the proof of part (b) we refer to Garsia and Stanton (1984). ◁

We can construct a basic set for an arbitrary permutation group Γ as follows. Let $\mathcal{I}$ denote the ideal in $\mathbf{C}[\mathbf{x}]$ spanned by all symmetric polynomials with zero constant term. A Gröbner basis for $\mathcal{I}$ can be read off from Theorem 1.2.7. This means we can easily compute and decide linear independence in the $n!$-dimensional vector space $V := \mathbf{C}[\mathbf{x}]/\mathcal{I}$. Since the ideal $\mathcal{I}$ is Γ-invariant, we get an action of Γ on V.

Corollary 2.7.10. A set of Γ-invariants $\mathcal{C} \subset \mathbf{C}[\mathbf{x}]^\Gamma$ is basic if and only if its image modulo $\mathcal{I}$ is a vector space basis for the invariant subspace V^Γ.

Corollary 2.7.10 is a direct consequence of Theorem 2.3.1. We get the following general method for producing basic sets of permutation groups.

Algorithm 2.7.11 (Constructing secondary invariants for a permutation group). Let $\mathcal{B}$ be either of the two basic sets in Theorem 2.7.9. Its image $\mathcal{B}^*$ under the Reynolds operator of Γ spans the vector space V^Γ. Using Gröbner basis normal form modulo $\mathcal{I}$ we can now find a subset $\mathcal{C}$ of $\mathcal{B}^*$ which is a $\mathbf{C}$-linear basis for V^Γ.

Example 2.7.12 (The dihedral group of order 5). Let $\mathcal{I}$ denote the ideal in $\mathbf{C}[\mathbf{x}] = \mathbf{C}[x_1, x_2, x_3, x_4, x_5]$ generated by the elementary symmetric polynomials. By Theorem 1.2.7, the lexicographic Gröbner basis for $\mathcal{I}$ equals

$$\mathcal{G} = \{\underline{x_1} + x_2 + x_3 + x_4 + x_5,\ \underline{x_4^4} + x_4^3 x_5 + x_4^2 x_5^2 + x_4 x_5^3 + x_5^4,$$
$$\underline{x_3^3} + x_3^2 x_4 + x_3^2 x_5 + x_3 x_4^2 + x_3 x_4 x_5 + x_3 x_5^2 + x_4^3 + x_4^2 x_5 + x_4 x_5^2 + x_5^3,$$
$$\underline{x_2^2} + x_2 x_3 + x_2 x_4 + x_2 x_5 + x_3^2 + x_3 x_4 + x_3 x_5 + x_4^2 + x_4 x_5 + x_5^2,\ \underline{x_5^5}\}.$$

Consider the dihedral group $\Gamma \subset S_5$ which is generated by the permutations (12345) and (12)(35). Let us apply Algorithm 2.7.11 to determine a basic set $\mathcal{C}$ for Γ. We first compute the Molien series $\Phi_\Gamma(z)$ and the degree generating function for $\mathcal{C}$:

$$\Phi_\Gamma(z)(1-z)(1-z^2)(1-z^3)(1-z^4)(1-z^5)$$
$$= 1 + z^2 + z^3 + 2z^4 + 2z^5 + 2z^6 + z^7 + z^8 + z^{10}.$$

Note that $|\mathcal{C}| = 12$ equals the index of Γ in S_5. Consider the basic set $\mathcal{B} = \{x_2^i x_3^j x_4^k x_5^l : 0 \le i < j < k < l \le 5\}$ from Theorem 2.7.9, and let $\mathcal{B}^*$ be its image under the Reynolds operator for Γ. Using normal form reduction with respect to $\mathcal{G}$, we can determine the image of all 120 elements in $\mathcal{B}^*$ modulo $\mathcal{I}$. Let $\mathcal{C}$ be the subset of the following ten symmetrized monomials. These are linearly independent modulo $\mathcal{I}$ because their normal forms have distinct initial terms.

$$(x_4 x_5)^* \to_{\mathcal{G}} \underline{2x_2 x_3} - 2x_2 x_5 + 2x_3^2 + 4x_3 x_4 + 2x_4^2 + 2x_4 x_5$$

$$(x_4 x_5^2)^* \to_{\mathcal{G}} -\underline{x_2 x_3 x_4} + x_2 x_4 x_5 - x_3^2 x_5 - x_3 x_4 x_5 - x_3 x_5^2 + x_4^3 + x_4^2 x_5$$
$$+ x_4 x_5^2$$

$$(x_4 x_5^3)^* \to_{\mathcal{G}} -\underline{x_2 x_3 x_4^2} - x_2 x_3 x_4 x_5 - 2x_2 x_3 x_5^2 - x_2 x_4^3 - x_2 x_4^2 x_5$$
$$- 2x_2 x_4 x_5^2 - \ldots$$

$$(x_3 x_4 x_5^2)^* \to_{\mathcal{G}} -\underline{x_2 x_3^2 x_4} + x_2 x_3^2 x_5 + x_2 x_3 x_4 x_5 + x_2 x_3 x_5^2 + x_2 x_4^3$$
$$+ 2x_2 x_4^2 x_5 + \ldots$$

$$(x_4 x_5^4)^* \to_{\mathcal{G}} \underline{x_2 x_3 x_4^2 x_5} + 2x_2 x_3 x_4 x_5^2 + x_2 x_4^3 x_5 + x_2 x_4^2 x_5^2 + x_2 x_4 x_5^3$$
$$- x_2 x_5^4 + \ldots$$

$$(x_3 x_4 x_5^3)^* \to_{\mathcal{G}} \underline{x_2 x_3^2 x_4^2} - x_2 x_3^2 x_5^2 - 2x_2 x_3 x_4^2 x_5 - 4x_2 x_3 x_4 x_5^2 - 2x_2 x_3 x_5^3 - \ldots$$

$$(x_3 x_4 x_5^4)^* \to_{\mathcal{G}} -\underline{x_2 x_3^2 x_4^2 x_5} + x_2 x_3^2 x_4 x_5^2 + 2x_2 x_3 x_4^2 x_5^2 + x_2 x_3 x_4 x_5^3$$
$$- x_2 x_3 x_5^4 + \ldots$$

$$(x_2 x_3 x_4 x_5^3)^* \to_{\mathcal{G}} -\underline{x_2 x_3 x_4^2 x_5^2} + x_2 x_3 x_4 x_5^3 + x_2 x_3 x_5^4 - x_2 x_4^3 x_5^2 - x_2 x_4^2 x_5^3$$
$$+ x_2 x_4 x_5^4 - \ldots$$

2.7. Abelian groups and permutation groups

$$(x_3x_4^2x_5^4)^* \to_{\mathcal{G}} -x_2x_3^2x_4^2x_5^2 - 2x_2x_3^2x_4x_5^3 - 2x_2x_3^2x_5^4 - 2x_2x_3x_4x_5^4$$
$$+ 2x_2x_4^3x_5^3 + \ldots$$
$$(x_3^2x_4^2x_5^4)^* \to_{\mathcal{G}} -2x_2x_3^2x_4^2x_5^3 - 2x_2x_3^2x_4x_5^4 - 4x_2x_3x_4^2x_5^4 - 2x_2x_4^3x_5^4 + \ldots$$
$$(x_2x_3^2x_4^3x_5^4)^* \to_{\mathcal{G}} \underline{10x_2x_3^2x_4^3x_5^4}$$

This proves that this set $\mathcal{C}$ of symmetrized monomials is basic for Γ.

The work of Garsia and Stanton (1984) provides an explicit combinatorial construction of basic sets for a large class of important permutation groups. This class includes the Young subgroups which are defined as follows. Let $T = \{t_1 < t_2 < \ldots < t_k\}$ be any ordered subset of $[1, n-1]$. We define Γ_T to be subgroup consisting all permutations π which fix each of the intervals $[t_i, t_{i+1} - 1]$, $i = 1, \ldots, k$. (Here $t_{k+1} := n$.) We call Γ_T the *Young permutation group* associated with the *descent set* T. As an abstract group Γ_T is isomorphic to the product of symmetric groups $S_{t_{i+1}-t_i}$, $i = 1, \ldots, k$.

Let C_T denote the set of all permutations π having the property that each of the sequences $(t_i, t_i + 1, t_i + 2, \ldots, t_{i+1} - 1)$ appears in increasing order in $(\pi_1, \pi_2, \pi_3, \ldots, \pi_n)$. The set C_T is clearly a system of representatives for the cosets of Γ_T, and hence its cardinality equals the cardinality of any basic set for Γ_T.

Theorem 2.7.13 (Garsia and Stanton 1984). *The set of symmetrized descent monomials $\{(m(\pi))^* : \pi \in C_T\}$ is a basic set for Γ_T.*

For the proof of this theorem we refer to the article of Garsia and Stanton. It is based on shellability of posets and the theory of Stanley–Reisner rings.

Exercises

(1) Let $\Gamma_T = S_2 \times S_3$ be the Young subgroup of S_5 associated with $T = \{2\}$.
   (a) Apply Theorem 2.7.13 to compute a basic for this permutation group.
   (b) Give an algorithm for rewriting an arbitrary invariant in $\mathbb{C}[x_1, x_2, x_3, x_4, x_5]^{\Gamma_T}$ in terms of these basic invariants.
(2) * Describe an algorithm for solving any fifth degree polynomial with cyclic Galois group in terms of radicals. Use Example 2.7.1 and the methods of Example 2.7.6. Apply your algorithm to the polynomial $f(z) = z^5 - 10z^4 + 40z^3 - 80z^2 + 80z - 29$. Describe an extension for solving quintics with dihedral Galois group (cf. Example 2.7.12).
(3) Determine a minimal algebra basis for the ring of invariants of the cyclic permutation group of order prime p.

3 Bracket algebra and projective geometry

According to the general philosophy outlined in Sect. 1.3, analytic geometry deals with those properties of vectors and matrices which are invariant with respect to some group of linear transformations. Applying this program to projective geometry, one is lead in a natural way to the study of the bracket algebra.

In Sects. 3.1 and 3.2 we present the two "Fundamental Theorems" of classical invariant theory from the point of view of computer algebra. These results will subsequently be used to derive algebraic tools and algorithms for projective geometry. In the last two sections of this chapter we apply bracket algebra to the study of invariants of binary forms, and in particular, we prove Gordan's finiteness theorem for binary forms.

3.1. The straightening algorithm

One of the most important features of the bracket algebra is the *straightening algorithm* due to Alfred Young (1928). The general method of rewriting in terms of standard Young tableaux plays an important role in representation theory and has applications in many areas of mathematics. The specific straightening algorithm to be discussed here will be understood as the normal form reduction with respect to a Gröbner basis for the ideal of algebraic dependencies among the maximal minors of a matrix. Our presentation follows Hodge and Pedoe (1947) and Sturmfels and White (1987).

Let $X = (x_{ij})$ be an $n \times d$-matrix whose entries are indeterminates, and let $\mathbf{C}[x_{ij}]$ denote the corresponding polynomial ring in nd variables. Throughout this chapter we will think of X as a configuration of n vectors in the vector space $\mathbf{C}^d$. These vectors represent a configuration of n points in projective $(d-1)$-space P^{d-1}. It is our objective to study those polynomial functions in $\mathbf{C}[x_{ij}]$ which correspond to geometric properties of the projective point configuration X.

Consider the set $\Lambda(n, d) := \{[\lambda_1 \lambda_2 \ldots \lambda_d] \mid 1 \leq \lambda_1 < \lambda_2 < \ldots < \lambda_d \leq n\}$ of ordered d-tuples in $[n] := \{1, 2, \ldots, n\}$. The elements of $\Lambda(n, d)$ are called *brackets*. They will serve as indeterminates over $\mathbf{C}$. We define $\mathbf{C}[\Lambda(n, d)]$ to be the polynomial ring generated by the $\binom{n}{d}$-element set $\Lambda(n, d)$. Furthermore, we abbreviate $[\lambda] := [\lambda_1 \lambda_2 \ldots \lambda_d]$ and $[\lambda_{\pi_1} \lambda_{\pi_2} \ldots \lambda_{\pi_d}] := \text{sign}(\pi) \cdot [\lambda]$ for all permutations π of $\{1, \ldots, d\}$.

Consider the algebra homomorphism

$$\phi_{n,d} : \mathbf{C}[\Lambda(n,d)] \to \mathbf{C}[x_{ij}]$$

$$[\lambda] \mapsto \det \begin{pmatrix} x_{\lambda_1 1} & x_{\lambda_1 2} & \cdots & x_{\lambda_1 d} \\ x_{\lambda_2 1} & x_{\lambda_2 2} & \cdots & x_{\lambda_2 d} \\ \vdots & \vdots & \ddots & \vdots \\ x_{\lambda_d 1} & x_{\lambda_d 2} & \cdots & x_{\lambda_d d} \end{pmatrix}$$

which maps each bracket $[\lambda]$ to the $d \times d$-subdeterminant of X whose rows are indexed by λ. The map $\phi_{n,d}$ is called the *generic coordinatization*. At this point we stress the distinction between the "formal" bracket $[\lambda]$ and the associated determinant $\phi_{n,d}([\lambda])$. Later on we will follow the standard abuse of notation and identify these two objects.

Example 3.1.1. Let $d = 3$ and $n = 6$. The rows of the matrix

$$X = \begin{pmatrix} x_{11} & x_{12} & x_{13} \\ x_{21} & x_{22} & x_{23} \\ \vdots & \vdots & \vdots \\ x_{61} & x_{62} & x_{63} \end{pmatrix}$$

can be thought of as six points in the projective plane. Then the determinant

$$\phi_{6,3}([146]) = x_{11}x_{42}x_{63} - x_{11}x_{62}x_{43} - x_{41}x_{12}x_{63} + x_{41}x_{62}x_{13}$$
$$+ x_{61}x_{12}x_{43} - x_{61}x_{42}x_{13}$$

vanishes if and only if the points "1", "4" and "6" lie on a common line. In Example 3.4.3 it will be shown that the six points lie on a common quadratic curve in P^2 if and only if the polynomial $\phi_{3,6}([123][145][246][356]-[124][135][236][456])$ vanishes.

The image of the ring map $\phi_{n,d}$ coincides with the subring $\mathcal{B}_{n,d}$ of $\mathbf{C}[x_{ij}]$ which is generated by the $d \times d$-minors of X. We call $\mathcal{B}_{n,d}$ the *bracket ring*. Example 3.1.1 suggests that precisely the polynomials in the bracket ring $\mathcal{B}_{n,d}$ correspond to geometric properties. In the next section we will show that this is indeed the case. First, however, we analyze the structure of the bracket ring and we give an algorithm for computing in $\mathcal{B}_{n,d}$.

Example 3.1.2. Even for relatively small parameters, such as $d = 3$ and $n = 6$, it is rather cumbersome to compute in $\mathcal{B}_{n,d}$ using the variables x_{ij}. As an example, we consider the polynomial

$$F := x_{11}x_{22}x_{33}x_{41}x_{52}x_{63} - x_{11}x_{22}x_{33}x_{41}x_{53}x_{62} - x_{11}x_{22}x_{33}x_{51}x_{42}x_{63}$$

3.1. The straightening algorithm

$$+ x_{11}x_{22}x_{33}x_{61}x_{42}x_{53} - x_{11}x_{23}x_{32}x_{41}x_{52}x_{63} + x_{11}x_{23}x_{32}x_{41}x_{53}x_{62}$$
$$- x_{11}x_{23}x_{32}x_{51}x_{43}x_{62} + x_{11}x_{23}x_{32}x_{61}x_{43}x_{52} - x_{21}x_{12}x_{33}x_{41}x_{52}x_{63}$$
$$+ x_{21}x_{12}x_{33}x_{41}x_{53}x_{62} + x_{21}x_{12}x_{33}x_{51}x_{42}x_{63} - x_{21}x_{12}x_{33}x_{61}x_{42}x_{53}$$
$$+ x_{21}x_{13}x_{32}x_{41}x_{52}x_{63} - x_{21}x_{13}x_{32}x_{41}x_{53}x_{62} + x_{21}x_{13}x_{32}x_{51}x_{43}x_{62}$$
$$- x_{21}x_{13}x_{32}x_{61}x_{43}x_{52} - x_{31}x_{12}x_{23}x_{51}x_{42}x_{63} + x_{31}x_{12}x_{23}x_{51}x_{43}x_{62}$$
$$+ x_{31}x_{12}x_{23}x_{61}x_{42}x_{53} - x_{31}x_{12}x_{23}x_{61}x_{43}x_{52} + x_{31}x_{13}x_{22}x_{51}x_{42}x_{63}$$
$$- x_{31}x_{13}x_{22}x_{51}x_{43}x_{62} - x_{31}x_{13}x_{22}x_{61}x_{42}x_{53} + x_{31}x_{13}x_{22}x_{61}x_{43}x_{52}$$
$$- x_{11}x_{22}x_{43}x_{31}x_{52}x_{63} + x_{11}x_{22}x_{43}x_{31}x_{53}x_{62} + x_{11}x_{22}x_{43}x_{51}x_{32}x_{63}$$
$$- x_{11}x_{22}x_{43}x_{61}x_{32}x_{53} + x_{11}x_{23}x_{42}x_{31}x_{52}x_{63} - x_{11}x_{23}x_{42}x_{31}x_{53}x_{62}$$
$$+ x_{11}x_{23}x_{42}x_{51}x_{33}x_{62} - x_{11}x_{23}x_{42}x_{61}x_{33}x_{52} + x_{21}x_{12}x_{43}x_{31}x_{52}x_{63}$$
$$- x_{21}x_{12}x_{43}x_{31}x_{53}x_{62} - x_{21}x_{12}x_{43}x_{51}x_{32}x_{63} + x_{21}x_{12}x_{43}x_{61}x_{32}x_{53}$$
$$- x_{21}x_{13}x_{42}x_{31}x_{52}x_{63} + x_{21}x_{13}x_{42}x_{31}x_{53}x_{62} - x_{21}x_{13}x_{42}x_{51}x_{33}x_{62}$$
$$+ x_{21}x_{13}x_{42}x_{61}x_{33}x_{52} + x_{41}x_{12}x_{23}x_{51}x_{32}x_{63} - x_{41}x_{12}x_{23}x_{51}x_{33}x_{62}$$
$$- x_{41}x_{12}x_{23}x_{61}x_{32}x_{53} + x_{41}x_{12}x_{23}x_{61}x_{33}x_{52} - x_{41}x_{13}x_{22}x_{51}x_{32}x_{63}$$
$$+ x_{41}x_{13}x_{22}x_{51}x_{33}x_{62} + x_{41}x_{13}x_{22}x_{61}x_{32}x_{53} - x_{41}x_{13}x_{22}x_{61}x_{33}x_{52}.$$

The geometric meaning of this polynomial is as follows: The three lines "$\overline{12}$", "$\overline{34}$" and "$\overline{56}$" meet in a common point if and only if $F = 0$. The polynomial F lies in the subring $\mathcal{B}_{3,6}$. It has the two distinct representations (among others)

$$F = \phi_{6,3}([123][456] - [124][356]) = \phi_{6,3}(-[125][346] + [126][345]).$$

This example shows that the ring map $\phi_{n,d}$ is in general not injective. Let $I_{n,d} \subset \mathbf{C}[\Lambda(n,d)]$ denote the kernel of $\phi_{n,d}$. This is the ideal of algebraic dependencies or *syzygies* among the maximal minors of a generic $n \times d$-matrix. For instance, in Example 3.1.2 we saw that $[123][456] - [124][356] + [125][346] + [126][345] \in I_{6,3}$.

Remark 3.1.3. The bracket ring $\mathcal{B}_{n,d}$ is isomorphic to the quotient $\mathbf{C}[\Lambda(n,d)]/I_{n,d}$.

We now give an explicit Gröbner basis for the ideal $I_{n,d}$. The projective variety defined by the syzygy ideal $I_{n,d}$ is the $(n-d)d$-dimensional *Grassmann variety* whose points correspond to the d-dimensional vector subspaces of $\mathbf{C}^n$.

We shall need the following abbreviations. The *complement* of a d-tuple $\lambda \in \Lambda(n,d)$ is the unique $(n-d)$-tuple $\lambda^* \in \Lambda(n, n-d)$ with $\lambda \cup \lambda^* = \{1, 2, \ldots, n\}$. The sign of the pair (λ, λ^*) is defined as the sign of the permutation π which maps λ_i to i for $i = 1, 2, \ldots, d$ and λ_j^* to $d+j$ for $j = 1, 2, \ldots, n-d$.

Let $s \in \{1, 2, \ldots, d\}$, $\alpha \in \Lambda(n, s-1)$, $\beta \in \Lambda(n, d+1)$ and $\gamma \in \Lambda(n, d-s)$. We define the *van der Waerden syzygy* $[[\alpha\beta\gamma]]$ to be the following quadratic polynomial in $\mathbf{C}[\Lambda(n,d)]$:

$$[[\alpha\dot{\beta}\gamma]] := \sum_{\tau \in \Lambda(d+1,s)} \mathrm{sign}(\tau, \tau^*) \cdot [\alpha_1 \ldots \alpha_{s-1} \beta_{\tau_1^*} \ldots \beta_{\tau_{d+1-s}^*}] \cdot [\beta_{\tau_1} \ldots \beta_{\tau_s} \gamma_1 \ldots \gamma_{d-s}].$$

Example 3.1.4. Let $d = 3$, $n \geq 6$, $s = 3$ and consider the index tuples

$$\alpha = [\alpha_1, \alpha_2] \in \Lambda(n, 2), \quad \beta = [\beta_1, \beta_2, \beta_3, \beta_4] \in \Lambda(n, 4), \quad \gamma = [\,] \in \Lambda(n, 0).$$

The corresponding van der Waerden syzygy equals

$$\begin{aligned}[][[\alpha\dot{\beta}\gamma]] &= [[\alpha_1\alpha_2\,\dot{\beta}_1\dot{\beta}_2\dot{\beta}_3\dot{\beta}_4]] \\ &= [\alpha_1\alpha_2\beta_4][\beta_1\beta_2\beta_3] - [\alpha_1\alpha_2\beta_3][\beta_1\beta_2\beta_4] \\ &\quad + [\alpha_1\alpha_2\beta_2][\beta_1\beta_3\beta_4] - [\alpha_1\alpha_2\beta_1][\beta_2\beta_3\beta_4]. \end{aligned}$$

In Example 3.1.2 we encountered this syzygy with indices $\alpha = [1, 2]$ and $\beta = [3, 4, 5, 6]$.

Let us first verify that the van der Waerden syzygies are indeed algebraic dependencies among the maximal minors of X.

Lemma 3.1.5. *The polynomials $[[\alpha\dot{\beta}\gamma]]$ are contained in the ideal $I_{n,d}$.*

Proof. We need to show that the polynomial $\phi_{n,d}([[\alpha\dot{\beta}\gamma]]) \in \mathbf{C}[x_{ij}]$ evaluates to zero for every $n \times d$-matrix X. Consider the row vectors $x_{\alpha_1}, \ldots, x_{\alpha_{s-1}}, x_{\beta_1}, \ldots, x_{\beta_{d+1}}, x_{\gamma_1}, \ldots, x_{\gamma_{d-s}}$ of X which are indexed by the tuples α, β and γ respectively. We specialize the $d-1$ row vectors x_{α_i} and x_{γ_j} to arbitrary elements from $\mathbf{C}^d$, while the $d+1$ vectors x_{β_k} are left as indeterminates. After this specialization, the expression $\phi_{n,d}([[\alpha\dot{\beta}\gamma]])$ defines a multilinear $(d+1)$-form on $\mathbf{C}^d$. We see that this multilinear form is antisymmetric because the sum defining the van der Waerden syzygies is alternating. A well-known theorem from linear algebra states that there is no antisymmetric multilinear $(d+1)$-form on a d-dimensional vector space except the zero form. Since the above specialization was arbitrary, we conclude that $\phi_{n,d}([[\alpha\dot{\beta}\gamma]]) = 0$ in $\mathbf{C}[x_{ij}]$. ◁

Example 3.1.6. The ideal $I_{4,2}$ of algebraic relations among the six 2×2-minors of a generic 4×2-matrix is principal. It is generated by the *quadratic Plücker relation* $[[1\,\dot{2}\dot{3}\dot{4}]] = [12][34] - [13][24] + [14][23]$.

In order to perform computations in the bracket ring $\mathcal{B}_{n,d} = \mathbf{C}[\Lambda(n,d)]/I_{n,d}$, it is necessary to express every bracket polynomial F by a unique normal form modulo the syzygy ideal $I_{n,d}$. In particular, we need a method for deciding whether a given bracket polynomial $F \in \mathbf{C}[\Lambda(n,d)]$ is contained in $I_{n,d}$, i.e., whether F vanishes under the generic coordinatization $\phi_{n,d}$. Such a normal form procedure is the *straightening law* due to A. Young (1928). We will present this classical algorithm within the framework of Gröbner bases theory.

3.1. The straightening algorithm

We order the elements of $\Lambda(n,d)$ lexicographically, that is, $[\lambda] \prec [\mu]$ if there exists an m, $1 \leq m \leq d$, such that $\lambda_j = \mu_j$ for $1 \leq j \leq m-1$, and $\lambda_m < \mu_m$. This specifies a total order on the set of variables in $\mathbf{C}[\Lambda(n,d)]$. The induced degree reverse lexicographic monomial order on $\mathbf{C}[\Lambda(n,d)]$ will also be denoted by "$\prec$".

The monomial order "$\prec$" is called the *tableaux order*, as it is customary to write monomials in $\mathbf{C}[\Lambda(n,d)]$ as rectangular arrays or *tableaux*. Given $[\lambda^1], \ldots, [\lambda^k] \in \Lambda(n,d)$ with $[\lambda^1] \preceq \ldots \preceq [\lambda^k]$, then the monomial $T := [\lambda^1] \cdot [\lambda^2] \cdot \ldots \cdot [\lambda^k]$ is written as the tableau

$$T = \begin{bmatrix} \lambda_1^1 & \cdots & \lambda_d^1 \\ \lambda_1^2 & \cdots & \lambda_d^2 \\ \vdots & \ddots & \vdots \\ \lambda_1^k & \cdots & \lambda_d^k \end{bmatrix}.$$

A tableau T is said to be *standard* if its columns are sorted, that is, if $\lambda_s^1 \leq \lambda_s^2 \leq \ldots \leq \lambda_s^k$ for all $s = 1, 2, \ldots, d$; otherwise it is *non-standard*.

The van der Waerden syzygy $[[\alpha\dot{\beta}\gamma]]$ is called a *straightening syzygy* provided $\alpha_{s-1} < \beta_{s+1}$ and $\beta_s < \gamma_1$. Let $\mathcal{S}_{n,d}$ denote the set of all straightening syzygies.

Theorem 3.1.7. *The set $\mathcal{S}_{n,d}$ is a Gröbner basis for $I_{n,d}$ with respect to the tableaux order. A tableau T is standard if and only if T is not in the initial ideal* $\mathrm{init}_\prec(I_{n,d})$.

Proof. The set $\mathcal{S}_{n,d}$ is contained in $I_{n,d}$ by Lemma 3.1.5. Let $\mathcal{M} \subseteq \mathrm{init}_\prec(I_{n,d})$ denote the monomial ideal generated by the initial tableaux of the elements in $\mathcal{S}_{n,d}$. We need to prove the reverse inclusion $\mathrm{init}_\prec(I_{n,d}) \subseteq \mathcal{M}$. Our proof proceeds in two steps. We first show each non-standard tableau lies in $\mathcal{M}$, and we then prove that each monomial in $\mathrm{init}_\prec(I_{n,d})$ is a non-standard tableau.

Let $T = [\lambda^1][\lambda^2]\ldots[\lambda^k]$ be any non-standard tableau. There exist $i \in \{2, 3, \ldots, k\}$ and $s \in \{2, 3, \ldots, d\}$ such that $\lambda_s^{i-1} > \lambda_s^i$. We find that the factor $[\lambda^{i-1}][\lambda^i]$ is the initial tableau of the straightening syzygy $[[\alpha\dot{\beta}\gamma]]$ where $\alpha := [\lambda_1^{i-1}\lambda_2^{i-1}\ldots\lambda_{s-1}^{i-1}]$, $\beta := [\lambda_1^i\ldots\lambda_s^i\lambda_s^{i-1}\ldots\lambda_d^{i-1}]$, and $\gamma := [\lambda_{s+1}^i\ldots\lambda_d^i]$. Hence T lies in the ideal $\mathcal{M}$.

For our second step, we consider the polynomial ring $\mathbf{C}[x_{ij}]$ and we introduce the lexicographic monomial order "$<$" induced by the variable order $x_{11} > x_{12} > \ldots > x_{1d} > x_{21} > \ldots > x_{n1} > \ldots > x_{nd}$. We call "$<$" the *diagonal monomial order* on $\mathbf{C}[x_{ij}]$.

Given any tableau $T = [\lambda^1][\lambda^2]\ldots[\lambda^k] \in \mathbf{C}[\Lambda(n,d)]$, we consider its image $\phi_{n,d}(T) \in \mathbf{C}[x_{ij}]$ under the generic coordinatization. The initial monomial of this product of k maximal minors in the diagonal monomial order equals $\mathrm{init}\,\phi_{n,d}(T) = \prod_{i=1}^k x_{\lambda_1^i 1} x_{\lambda_2^i 2} \cdots x_{\lambda_d^i d}$.

The crucial step of our proof consists in the following combinatorial fact.

Lemma 3.1.8. Let $\{\lambda^1, \ldots, \lambda^k\}$ be any multisubset of $\Lambda(n, d)$. Then there exists a *unique* standard tableau $\tilde{T}$ such that

$$\text{init } \phi_{n,d}(\tilde{T}) = \prod_{i=1}^{k} x_{\lambda_1^i 1} x_{\lambda_2^i 2} \cdots x_{\lambda_d^i d}.$$

The standard tableau $\tilde{T}$ is obtained from the tableau $T = [\lambda^1][\lambda^2]\ldots[\lambda^k]$ by sorting each column.

Let us now suppose that some standard tableau T lies in the initial ideal $\text{init}_\prec(I_{n,d})$, say, $T = \text{init}(F)$ where $F \in I_{n,d}$. Without loss of generality we may assume that all tableaux occurring in F are standard. For, any non-standard tableau can be replaced by its normal form (which may not be unique) with respect to $\mathcal{S}_{n,d}$. Any such normal form is a linear combination of standard tableaux, by the first part of our proof.

Since F is non-zero but $\phi_{n,d}(F) = 0$, there exists a non-initial standard tableau T' in the expansion of F such that $\phi_{n,d}(T)$ and $\phi_{n,d}(T')$ have the same initial monomial in the diagonal monomial order on $\mathbf{C}[x_{ij}]$. This is a contradiction to Lemma 3.1.8, and our proof is complete. ◁

The Straightening Law for bracket polynomials is usually stated in the following form. Corollary 3.1.9 is an immediate consequence of Theorem 3.1.7 and Theorem 1.2.6.

Corollary 3.1.9 (Straightening law). The standard tableaux form a $\mathbf{C}$-vector space basis for the bracket ring $\mathcal{B}_{n,d}$.

The normal form reduction with respect to the Gröbner basis $\mathcal{S}_{n,d}$ is called the *straightening algorithm*. Let us see in an example how the straightening algorithm works.

Example 3.1.10. Let $n = 6$, $d = 3$, and consider the tableau $T = [\lambda^1][\lambda^2][\lambda^3] = [145][156][234]$. This tableau is non-standard because $\lambda_2^2 = 5 > \lambda_2^3 = 3$. We can reduce T modulo the straightening syzygy

$$[[1\ \dot{5}\dot{6}\dot{2}\dot{3}\ 4]] = [156][234] + [136][245] - [135][246]$$
$$- [126][345] + [125][346] + [123][456]$$

which results in the bracket polynomial

$$F = T - [145][[1\ \dot{5}\dot{6}\dot{2}\dot{3}\ 4]]$$
$$= [136][145][245] + [135][145][246] + [126][145][345]$$
$$- [125][145][346] - [123][145][456].$$

The second, fourth and fifth tableau in this expression are standard. The first and

3.1. The straightening algorithm

the third tableau are still non-standard, and we need to straighten them next. This process eventually terminates because all tableaux in the new expression F are smaller in the tableaux order than T. The unique representation of T as a linear combination of standard tableaux equals [123][145][456] − [124][145][356] + [134][145][256].

The Gröbner basis $\mathcal{S}_{n,d}$ of the syzygy ideal is by no means minimal. There are many proper subsets of $\mathcal{S}_{n,d}$ which are also Gröbner bases of $I_{n,d}$. For example, let

$$\mathcal{S}_{n,d}^* := \{[[\alpha \dot{\beta} \gamma]] \in \mathcal{S}_{n,d} : \alpha_i \leq \beta_i \text{ for all } i,\ 1 \leq i \leq s-1\}.$$

These syzygies straighten the leftmost violation in a given pair of rows. All our arguments in the proof of Theorem 3.1.7 are still valid, whence also $\mathcal{S}_{n,d}^*$ is a Gröbner basis. However, even $\mathcal{S}_{n,d}^*$ is not a reduced Gröbner basis.

We close this section with a description of the unique reduced Gröbner basis $\mathcal{R}_{n,d}$ for $I_{n,d}$ with respect to the tableaux order. Since each polynomial in $\mathcal{R}_{n,d}$ must be in reduced form modulo the other polynomials in $\mathcal{R}_{n,d}$, we see that each polynomial in $\mathcal{R}_{n,d}$ consists of a two-rowed non-standard tableau T minus the linear combination of standard tableaux obtained by applying the straightening algorithm to T.

Example 3.1.11 ($n = 6, d = 3$). The reduced Gröbner basis $\mathcal{R}_{6,3}$ for the syzygy ideal $I_{6,3}$ contains, among others, the following syzygies.

<u>[126][345]</u> − [123][456] + [124][356] − [125][346]

<u>[136][245]</u> + [123][456] + [134][256] − [135][246]

<u>[145][236]</u> + [125][346] − [135][246] − [124][356] + [134][256] + [123][456]

<u>[146][235]</u> + [125][346] − [135][246] + [123][456]

<u>[156][234]</u> + [124][356] − [134][256] − [123][456]

This list, which is ordered with respect to the tableaux order of their initial monomials, is sufficient to straighten all bracket polynomials with $n = 6, d = 3$ which are linear in each point, i.e., no number occurs twice in any tableau. Observe that only the underlined initial monomials are non-standard.

This description of the reduced Gröbner bases generalizes to an arbitrary polynomial ideal $I \subset \mathbf{C}[x_1, \ldots, x_n]$. Let $\mathcal{G}$ be any Gröbner basis for I and consider a set of monomials $\{u_1, \ldots, u_r\}$ which minimally generates the initial ideal init(I). Then the reduced Gröbner basis $\mathcal{G}_{\text{red}}$ of I equals $\mathcal{G}_{\text{red}} = \{u_1 - \text{normal form}_{\mathcal{G}}(u_1), \ldots, u_r - \text{normal form}_{\mathcal{G}}(u_r)\}$. Here the monomials u_i are called *minimally non-standard* (see also Exercises (1) and (2) of Sect. 1.2).

Exercises

(1) The bracket ring $\mathcal{B}_{n,d}$ is a graded subring of the polynomial ring $\mathbf{C}[x_{ij}]$. Show that the Hilbert series of $\mathcal{B}_{n,d}$ has the form $H(\mathcal{B}_{n,d}, z) = \sum_{k=0}^{\infty} a_{n,d,k} z^{dk}$. Compute the Hilbert series of $\mathcal{B}_{n,d}$ explicitly in the case $d = 2$.

(2) Apply the straightening algorithm to the polynomial

$$\det \begin{pmatrix} [124] & [143] & [423] \\ [125] & [153] & [523] \\ [126] & [163] & [623] \end{pmatrix} \in \mathbf{C}[\Lambda(6,3)].$$

Can you generalize your result?

(3) * A van der Waerden syzygy $[[\alpha\dot{\beta}\gamma]]$ with $s = 1$ is called a *Grassmann–Plücker syzygy*.
   (a) Show that the ideal $\mathcal{I}_{n,d}$ is generated by the Grassmann–Plücker syzygies. (This is sometimes called the *Second Fundamental Theorem of Invariant Theory*.)
   (b) The Grassmann–Plücker syzygies form a Gröbner basis whenever $d \leq 3$.
   (c) The Grassmann–Plücker syzygies are *not* a Gröbner basis for $d = 4$, $n = 8$.

(4) Consider the ideal $\mathcal{J}$ in $\mathbf{C}[x_{ij}]$ which is generated by $\phi_{n,d}(\mathcal{B}_{n,d})$. (What is its variety?) Compute the reduced Gröbner basis of $\mathcal{J}$ with respect to the diagonal monomial order on $\mathbf{C}[x_{ij}]$.

(5) * Here we assume familiarity with face rings of simplicial complexes (cf. Stanley (1983)).
   (a) The initial ideal $\text{init}_{\prec}(\mathcal{I}_{n,d})$ is square-free and hence corresponds to a simplicial complex $\Delta_{n,d}$ on $\Lambda(n,d)$. Describe the simplices of $\Delta_{n,d}$.
   (b) Show that $\Delta_{n,d}$ is the order complex of a partially ordered set on $\Lambda(n,d)$.
   (c) Show that $\Delta_{n,d}$ is a shellable ball, and conclude that its face ring $\mathbf{C}[\Delta_{n,d}] = \mathbf{C}[\Lambda(n,d)]/\text{init}_{\prec}(\mathcal{I}_{n,d})$ is Cohen–Macaulay. Find a Hironaka decomposition.
   (d) Show that the bracket ring $\mathcal{B}_{n,d}$ is Cohen–Macaulay. Give an explicit Hironaka decomposition for $\mathcal{B}_{n,d}$.

3.2. The first fundamental theorem

The group $SL(\mathbf{C}^d)$ of $d \times d$-matrices with determinant 1 acts by right multiplication on the ring $\mathbf{C}[x_{ij}]$ of polynomial functions on a generic $n \times d$-matrix $X = (x_{ij})$. The two fundamental theorems of classical invariant theory give an explicit description of the invariant ring $\mathbf{C}[x_{ij}]^{SL(\mathbf{C}^d)}$. It is clear that every $d \times d$-minor of X is invariant under $SL(\mathbf{C}^d)$. Therefore $\mathbf{C}[x_{ij}]^{SL(\mathbf{C}^d)}$ contains the

3.2. The first fundamental theorem

bracket ring $\mathcal{B}_{n,d}$ which is generated by all $d \times d$-minors. The main result of this section states that these two rings coincide.

Theorem 3.2.1 (First fundamental theorem of invariant theory). *The invariant ring $\mathbf{C}[x_{ij}]^{SL(\mathbf{C}^d)}$ is generated by the $d \times d$-minors of the matrix $X = (x_{ij})$.*

Together with the results of Sect. 3.1, this provides an explicit presentation

$$\mathbf{C}[x_{ij}]^{SL(\mathbf{C}^d)} = \mathbf{C}[\Lambda(n,d)]/I_{n,d} = \mathcal{B}_{n,d}.$$

The *Second Fundamental Theorem of Invariant Theory* states that the syzygy ideal $I_{n,d}$ is generated by certain sets of quadratic polynomials, such as the quadratic Grassmann–Plücker relations in Exercise 3.1. (3). An even stronger result was established in Theorem 3.1.7, where we exhibited an explicit Gröbner basis consisting of quadratic polynomials.

In order to prove Theorem 3.2.1, we introduce the following *multigrading* on the polynomial ring $\mathbf{C}[x_{ij}]$. Let $m \in \mathbf{C}[x_{ij}]$ be any monomial. For each column index $j \in \{1, 2, \ldots, d\}$ we define $\deg_j(m)$ to be the total degree of m in the subset of variables $\{x_{ij} : 1 \le i \le n\}$. The vector $\deg(m) := (\deg_1(m), \deg_2(m), \ldots, \deg_d(m))$ is called the *multi-degree* of m. Note that if a polynomial $f \in \mathbf{C}[x_{ij}]$ is *multi-homogeneous* of multi-degree $(\delta_1, \delta_2, \ldots, \delta_d)$, then f is homogeneous of total degree $\delta_1 + \delta_2 + \ldots + \delta_d$.

Here are a few examples. The monomial $x_{21}x_{31}^2 x_{22}x_{52}x_{13}x_{43}x_{53}x_{33} \in \mathbf{C}[x_{ij}]$ has multi-degree $(3, 2, 4)$. The polynomial $x_{21}x_{31}^2 x_{22}x_{52}x_{13}x_{33}x_{43}x_{53} - x_{11}x_{21}x_{51}x_{22}^2 x_{43}^3 x_{53}$ is multi-homogeneous of degree $(3, 2, 4)$. The polynomial in Example 3.1.2 is quadratic in each column of the 6×3-matrix, which means that it is multi-homogeneous of degree $(2, 2, 2)$.

Observation 3.2.2. Let $T := [\lambda^1][\lambda^2]\cdots[\lambda^k] \in \mathbf{C}[\Lambda(n,d)]$ be any tableau. Then its expansion $\phi_{n,d}(T)$ in $\mathbf{C}[x_{ij}]$ is multi-homogeneous of degree $(k, k, k, \ldots, k)$.

A polynomial $I \in \mathbf{C}[x_{ij}]$ is said to be a *relative invariant* of the general linear group $GL(\mathbf{C}^d)$ if there exists an integer $p \ge 0$ such that $I \circ A = \det(A)^p \cdot I$ for all $A \in GL(\mathbf{C}^d)$. The integer p is called the *index* of I. Clearly, every relative invariant of $GL(\mathbf{C}^d)$ is an (absolute) invariant of $SL(\mathbf{C}^d)$. But also the converse is essentially true.

Lemma 3.2.3. *Let $I \in \mathbf{C}[x_{ij}]$ be a homogeneous invariant of $SL(\mathbf{C}^d)$. Then there exists an integer $p \ge 0$ such that*

(i) *I has multi-degree $(p, p, \ldots, p)$, and*
(ii) *I is a relative invariant of $GL(\mathbf{C}^d)$ of index p.*

Proof. We fix two row indices $j_1, j_2 \in \{1, \ldots, d\}$. Let $D(j_1, j_2)$ denote the $d \times d$-diagonal matrix whose j_1-th diagonal entry equals 2, whose j_2-nd diagonal entry equals $\frac{1}{2}$, and all of whose other diagonal entries are equal to 1. Note that $D(j_1, j_2) \in SL(\mathbf{C}^d)$. This matrix transforms a monomial m into $m \circ D(j_1, j_2) = m \cdot 2^{\deg_{j_1}(m) - \deg_{j_2}(m)}$.

Since I was assumed to be an invariant, we have $I = I \circ D(j_1, j_2)$. This implies that each monomial m which occurs in the expansion of I must satisfy $\deg_{j_1}(m) = \deg_{j_2}(m)$. Since the indices j_1 and j_2 were chosen arbitrarily, the claim (i) follows.

Now let A be an arbitrary matrix in $GL(\mathbf{C}^d)$. We define the diagonal matrix $\tilde{A} := \operatorname{diag}(\det(A), 1, 1, \ldots, 1)$, and we observe that $A \cdot \tilde{A}^{-1} \in SL(\mathbf{C}^d)$. Now part (i) implies

$$I \circ A = I \circ (A \cdot \tilde{A}^{-1} \cdot \tilde{A}) = (I \circ (A \cdot \tilde{A}^{-1})) \circ \tilde{A} = I \circ \tilde{A} = \det(A)^p \cdot I.$$

This completes the proof of (ii). ◁

We now extend the $n \times d$-matrix X to a generic $(n + 2d) \times d$-matrix as follows:

$$\begin{pmatrix} A \\ X \\ B \end{pmatrix} := \begin{pmatrix} a_{11} & \cdots & a_{1d} \\ \vdots & \ddots & \vdots \\ a_{d1} & \cdots & a_{dd} \\ x_{11} & \cdots & x_{1d} \\ x_{21} & \cdots & x_{2d} \\ \vdots & \ddots & \vdots \\ x_{n1} & \cdots & x_{nd} \\ b_{11} & \cdots & b_{1d} \\ \vdots & \ddots & \vdots \\ b_{d1} & \cdots & b_{dd} \end{pmatrix}.$$

The polynomial ring $\mathbf{C}[a_{ij}, x_{ij}, b_{ij}]$ in the $(n+2d)d$ matrix entries is a superring of $\mathbf{C}[x_{ij}]$. Our strategy is the following: We will prove the first fundamental theorem (Theorem 3.2.1) for the matrix X by applying the straightening law (Theorem 3.1.7) to the larger matrix $\begin{pmatrix} A \\ X \\ B \end{pmatrix}$. The rows of this matrix are indexed by the ordered set $\{a_1 < \ldots < a_d < x_1 < x_2 < \ldots < x_n < b_1 < \ldots < b_d\}$. The corresponding bracket ring $\mathbf{C}[\Lambda(n+2d, d)]$ is generated by brackets of the form $[a_{i_1} \ldots a_{i_s} x_{j_1} \ldots x_{j_t} b_{k_1} \ldots x_{k_{d-s-t}}]$. The crucial idea is to study the effect of two suitable $\mathbf{C}$-algebra homomorphisms $\mathbf{C}[x_{ij}] \to \mathbf{C}[\Lambda(n + 2d, d)]$ on the subring of invariants. These homomorphisms are defined by

$$x_{ij} \mapsto [a_1 \ldots a_{j-1} x_i a_{j+1} \ldots a_d] \quad \text{and} \quad x_{ij} \mapsto [b_1 \ldots b_{j-1} x_i b_{j+1} \ldots b_d].$$

3.2. The first fundamental theorem

Example 3.2.4. Let $n = d = 4$ and consider the expansion of the determinant

$$\det \begin{pmatrix} x_{11} & x_{12} & x_{13} & x_{14} \\ x_{21} & x_{22} & x_{23} & x_{24} \\ x_{31} & x_{32} & x_{33} & x_{34} \\ x_{41} & x_{42} & x_{43} & x_{44} \end{pmatrix}.$$

This is a polynomial in $\mathbf{C}[x_{ij}]$ of multidegree $(1, 1, 1, 1)$, which is a relative $GL(\mathbf{C}^4)$-invariant of index 1. If we replace the variable x_{ij} by the bracket $[b_1 \ldots b_{j-1} \, x_i \, b_{j+1} \ldots b_4]$ then we obtain

$$\begin{aligned}
& [x_1 b_2 b_3 b_4][b_1 x_2 b_3 b_4][b_1 b_2 x_3 b_4][b_1 b_2 b_3 x_4] \\
& - [x_1 b_2 b_3 b_4][b_1 x_2 b_3 b_4][b_1 b_2 x_4 b_4][b_1 b_2 b_3 x_3] \\
& - [x_1 b_2 b_3 b_4][b_1 x_3 b_3 b_4][b_1 b_2 x_2 b_4][b_1 b_2 b_3 x_4] \\
& + [x_1 b_2 b_3 b_4][b_1 x_3 b_3 b_4][b_1 b_2 x_4 b_4][b_1 b_2 b_3 x_2] \\
& + [x_1 b_2 b_3 b_4][b_1 x_4 b_3 b_4][b_1 b_2 x_2 b_4][b_1 b_2 b_3 x_3] \\
& - [x_1 b_2 b_3 b_4][b_1 x_4 b_3 b_4][b_1 b_2 x_3 b_4][b_1 b_2 b_3 x_2] \\
& - [x_2 b_2 b_3 b_4][b_1 x_1 b_3 b_4][b_1 b_2 x_3 b_4][b_1 b_2 b_3 x_4] \\
& + [x_2 b_2 b_3 b_4][b_1 x_1 b_3 b_4][b_1 b_2 x_4 b_4][b_1 b_2 b_3 x_3] \\
& + [x_2 b_2 b_3 b_4][b_1 x_3 b_3 b_4][b_1 b_2 x_1 b_4][b_1 b_2 b_3 x_4] \\
& - [x_2 b_2 b_3 b_4][b_1 x_3 b_3 b_4][b_1 b_2 x_4 b_4][b_1 b_2 b_3 x_1] \\
& - [x_2 b_2 b_3 b_4][b_1 x_4 b_3 b_4][b_1 b_2 x_1 b_4][b_1 b_2 b_3 x_3] \\
& + [x_2 b_2 b_3 b_4][b_1 x_4 b_3 b_4][b_1 b_2 x_3 b_4][b_1 b_2 b_3 x_1] \\
& + [x_3 b_2 b_3 b_4][b_1 x_1 b_3 b_4][b_1 b_2 x_2 b_4][b_1 b_2 b_3 x_4] \\
& - [x_3 b_2 b_3 b_4][b_1 x_1 b_3 b_4][b_1 b_2 x_4 b_4][b_1 b_2 b_3 x_2] \\
& - [x_3 b_2 b_3 b_4][b_1 x_2 b_3 b_4][b_1 b_2 x_1 b_4][b_1 b_2 b_3 x_4] \\
& + [x_3 b_2 b_3 b_4][b_1 x_2 b_3 b_4][b_1 b_2 x_4 b_4][b_1 b_2 b_3 x_1] \\
& + [x_3 b_2 b_3 b_4][b_1 x_4 b_3 b_4][b_1 b_2 x_1 b_4][b_1 b_2 b_3 x_2] \\
& - [x_3 b_2 b_3 b_4][b_1 x_4 b_3 b_4][b_1 b_2 x_2 b_4][b_1 b_2 b_3 x_1] \\
& - [x_4 b_2 b_3 b_4][b_1 x_1 b_3 b_4][b_1 b_2 x_2 b_4][b_1 b_2 b_3 x_3] \\
& + [x_4 b_2 b_3 b_4][b_1 x_1 b_3 b_4][b_1 b_2 x_3 b_4][b_1 b_2 b_3 x_2] \\
& + [x_4 b_2 b_3 b_4][b_1 x_2 b_3 b_4][b_1 b_2 x_1 b_4][b_1 b_2 b_3 x_3] \\
& - [x_4 b_2 b_3 b_4][b_1 x_2 b_3 b_4][b_1 b_2 x_3 b_4][b_1 b_2 b_3 x_1] \\
& - [x_4 b_2 b_3 b_4][b_1 x_3 b_3 b_4][b_1 b_2 x_1 b_4][b_1 b_2 b_3 x_2] \\
& + [x_4 b_2 b_3 b_4][b_1 x_3 b_3 b_4][b_1 b_2 x_2 b_4][b_1 b_2 b_3 x_1].
\end{aligned}$$

The eight appearing index letters are sorted $x_1 \prec x_2 \prec x_3 \prec x_4 \prec b_1 \prec b_2 \prec b_3 \prec b_4$. The normal form of this large polynomial modulo the Gröbner basis for $I_{8,4}$ given in Theorem 3.1.7 equals the single standard tableau:

$$[x_1x_2x_3x_4][b_1b_2b_3b_4][b_1b_2b_3b_4][b_1b_2b_3b_4].$$

This example can be generalized as follows.

Lemma 3.2.5. Let $I = I(x_{ij}) \in \mathbf{C}[x_{ij}]$ be any relative $GL(\mathbf{C}^d)$-invariant of index p. Then the bracket polynomial

$$[b_1b_2\ldots b_d]^{p(d-1)} \cdot I\big([a_1\ldots a_{j-1}\, x_i\, a_{j+1}\ldots a_d]\big)$$
$$- [a_1a_2\ldots a_d]^{p(d-1)} \cdot I\big([b_1\ldots b_{j-1}\, x_i\, b_{j+1}\ldots b_d]\big)$$

is contained in the syzygy ideal $I_{n+2d,d} \subset \mathbf{C}[\Lambda(n+2d,d)]$.

Proof. We need to show that the image of the above bracket polynomial under the generic specialization $\phi_{n+2d,n}$ is zero. For simplicity of notation, we abbreviate the determinant $\phi_{n+2d,n}([\lambda])$ by the corresponding bracket $[\lambda]$.

Let $\mathrm{Adj}(A)$ denote the *adjoint matrix* of A. Its entry $\mathrm{Adj}(A)_{jk}$ is the correctly signed $(d-1) \times (d-1)$-minor of A which is obtained by deleting the j-th row and the k-th column of A. By Laplace expansion we obtain

$$[a_1\ldots a_{j-1}\, x_i\, a_{j+1}\ldots a_d] = \sum_{k=1}^{d} x_{ik}\, \mathrm{Adj}(A)_{jk},$$

and therefore

$$[a_1a_2\ldots a_d]^{p(d-1)} \cdot I(x_{ij}) = \det(A)^{p(d-1)} \cdot I(x_{ij})$$
$$= \det(\mathrm{Adj}(A))^p \cdot I(x_{ij})$$
$$= \big(I \circ \mathrm{Adj}(A)\big)(x_{ij})$$
$$= I\Big(\sum_{k=1}^{d} x_{ik}\, \mathrm{Adj}(A)_{jk}\Big)$$
$$= I\big([a_1\ldots a_{j-1}\, x_i\, a_{j+1}\ldots a_d]\big).$$

The same argument holds for the matrix B. This implies the desired identity in $\mathcal{B}_{n+2d,d}$:

$$[b_1b_2\ldots b_d]^{p(d-1)}[a_1a_2\ldots a_d]^{p(d-1)} \cdot I(x_{ij})$$
$$= [b_1b_2\ldots b_d]^{p(d-1)} I \cdot \big([a_1\ldots a_{j-1}\, x_i\, a_{j+1}\ldots a_d]\big)$$
$$= [a_1a_2\ldots a_d]^{p(d-1)} I \cdot \big([b_1\ldots b_{j-1}\, x_i\, b_{j+1}\ldots b_d]\big). \quad \triangleleft$$

We are now prepared to prove the First Fundamental Theorem, which states

3.2. The first fundamental theorem

that every invariant $I \in \mathbf{C}[x_{ij}]^{SL(\mathbf{C}^d)}$ can be written as a polynomial in the brackets $[x_{i_1} x_{i_2} \ldots x_{i_d}]$.

Proof of Theorem 3.2.1. By Lemma 3.2.3, we may assume that the given invariant I is a relative $GL(\mathbf{C}^d)$-invariant of index p. We apply the straightening algorithm to the polynomial

$$[b_1 b_2 \ldots b_d]^{p(d-1)} \cdot I\big([a_1 \ldots a_{j-1} x_i a_{j+1} \ldots a_d]\big) \in \mathbf{C}[\Lambda(n+2d,d)],$$

that is, we compute its normal form modulo the Gröbner basis given in Theorem 3.1.7. Since all row indices $b_1, \ldots, b_d$ are larger than the row indices $a_1, \ldots, a_d, x_1, x_2, \ldots, x_n$, the result is a linear combination of standard tableaux of the form

$$\sum_j T_j(a_1, \ldots, a_d, x_1, x_2, \ldots, x_n) \cdot [b_1 \ldots b_d]^{p(d-1)},$$

where the T_j are certain standard tableaux in the row indices $a_1, \ldots, a_d, x_1, x_2, \ldots, x_n$. Similarly, the polynomial

$$[a_1 a_2 \ldots a_d]^{p(d-1)} \cdot I\big([b_1 \ldots b_{j-1} x_i b_{j+1} \ldots b_d]\big)$$

is straightened to a polynomial

$$\sum_k [a_1 \ldots a_d]^{p(d-1)} \cdot T'_k(x_1, x_2, \ldots, x_n, b_1, \ldots, b_d),$$

where the T'_k are standard tableaux in the indices $x_1, x_2, \ldots, x_n, b_1, \ldots, b_d$.

By Lemma 3.2.5 and the straightening law (Corollary 3.1.9), these two standard tableaux expansions must be equal in $\mathbf{C}[\Lambda(n+2d,d)]$. But this is only possible if both sums are of the form

$$\sum_l [a_1 \ldots a_d]^{p(d-1)} \cdot T''_l(x_1, x_2, \ldots, x_n) \cdot [b_1 \ldots b_d]^{p(d-1)}$$

where the T''_l are certain standard tableaux only in the "old" indices $x_1, x_2, \ldots, x_n$. On the other hand, by the proof of Lemma 3.2.5, both polynomials in question are equal to

$$[a_1 \ldots a_d]^{p(d-1)} [b_1 \ldots b_d]^{p(d-1)} \cdot I(x_{ij}).$$

This implies the desired expansion

$$I(x_{ij}) = \sum_l T''_l(x_1, x_2, \ldots, x_n),$$

and the proof of Theorem 3.2.1 is complete. ◁

Our proof of the First Fundamental Theorem implies the following algorithm for rewriting a given $SL(\mathbf{C}^d)$-invariant polynomial $I \in \mathbf{C}[x_{ij}]$ in terms of brackets. We replace each variable x_{ij} in $I(x_{ij})$ by the corresponding bracket $[a_1 \ldots a_{j-1} x_i a_{j+1} \ldots a_d]$, and then we apply the straightening algorithm for the enlarged bracket ring $\mathbf{C}[\Lambda(n+d, d)]$ with respect to the order $a_1 < \ldots < a_d < x_1 < x_2 < \ldots < x_n$ on the row indices. If I is a relative invariant of index p, then $[a_1 a_2 \ldots a_d]^{p(d-1)}$ appears as a factor in the resulting standard representation. Dividing this factor out, we obtain the unique expansion of $I(x_{ij})$ in terms of standard tableaux in the row indices x_i.

This algorithm for the First Fundamental Theorem turns out to be rather slow for practical computations. In the remainder of this section we will discuss an alternative procedure which usually performs much better.

Let "$\prec$" denote the lexicographic monomial order on $\mathbf{C}[x_{ij}]$ induced from the variable order $x_{11} \prec x_{12} \prec \ldots \prec x_{1d} \prec x_{21} \prec x_{22} \prec \ldots \prec x_{2d} \prec \ldots \prec x_{n1} \prec x_{n2} \prec \ldots \prec x_{nd}$. This was called the *diagonal order* in Sect. 3.1. A monomial m in $\mathbf{C}[x_{ij}]$ is said to be *diagonal* if its degree is divisible by d and it can be written in the form

$$m = \prod_{i=1}^{k} \left(x_{\lambda_1^i 1} x_{\lambda_2^i 2} \cdots x_{\lambda_d^i d} \right) \qquad (*)$$

where $\lambda_1^i < \lambda_2^i < \ldots < \lambda_d^i$ for all $i = 1, 2, \ldots, k$. It is easy to see that the initial monomial of any (expanded) tableau is a diagonal monomial.

Lemma 3.2.6. Let T denote the tableau $[\lambda^1][\lambda^2]\ldots[\lambda^k] \in \mathbf{C}[\Lambda(n,d)]$. Then the initial monomial of its expansion $\phi_{n,d}(T) \in \mathbf{C}[x_{ij}]$ with respect to "$\prec$" equals the diagonal monomial m in $(*)$.

Conversely, every diagonal monomial is the initial monomial of some tableau. This tableau is unique if we require it to be standard.

Lemma 3.2.7. Let m be the diagonal monomial in $(*)$. Then there is a unique standard tableau T_m such that $\text{init}_\prec \left(\phi_{n,d}(T_m) \right) = m$.

Lemmas 3.2.6 and 3.2.7 are a reformulation of Lemma 3.1.8. The standard tableau T_m promised by Lemma 3.2.7 is constructed from the diagonal monomial m as follows. Consider the tableau

$$T := \begin{bmatrix} \lambda_1^1 & \lambda_2^1 & \cdots & \lambda_d^1 \\ \lambda_1^2 & \lambda_2^2 & \cdots & \lambda_d^2 \\ \vdots & \vdots & \ddots & \vdots \\ \lambda_1^k & \lambda_2^k & \cdots & \lambda_d^k \end{bmatrix},$$

and let T_m denote the unique standard tableau which is obtained from T by sorting all d columns.

3.2. The first fundamental theorem

Lemmas 3.2.6 and 3.2.7 imply the correctness of the following easy algorithm for the First Fundamental Theorem.

Algorithm 3.2.8.
Input: A polynomial $I \in \mathbf{C}[x_{ij}]$ which is an invariant of $SL(\mathbf{C}^d)$.
Output: A bracket polynomial $P \in \mathbf{C}[\Lambda(n,d)]$ whose expansion equals $I(x_{ij})$.

1. If $I = 0$ then output the bracket representation $P = 0$.
2. Let $m := \text{init}_\prec(I)$.
3. If m is not diagonal, then STOP and output "I is not an invariant".
4. Otherwise: let c be the coefficient of $\text{init}_\prec(I)$ in I, output the summand $c \cdot T_m$, replace I by $I - c \cdot \phi_{n,d}(T_m)$, and return to Step 1.

Both Algorithm 3.2.8 and the procedure used in our proof of Theorem 1.1.1 are instances of a general method for computing in subrings of polynomial rings. This is the method of SAGBI bases due to Robbiano and Sweedler (1990) and Kapur and Madlener (1989), which is the natural Subalgebra Analogue to Gröbner Bases for Ideals. Let R be any subalgebra of the polynomial ring $\mathbf{C}[x_1, \ldots, x_n]$, and let "$\prec$" be any monomial order. We define the *initial algebra* $\text{init}_\prec(R)$ to be the $\mathbf{C}$-algebra generated by $\text{init}_\prec(f)$, where f ranges over R. A finite subset $\{f_1, \ldots, f_m\}$ of R is called a SAGBI basis if the initial algebra $\text{init}_\prec(R)$ is generated by the initial monomials $\text{init}_\prec(f_1), \ldots, \text{init}_\prec(f_m)$.

The main difference to the theory of Gröbner bases lies in the fact that $\text{init}_\prec(R)$ need not be finitely generated even if R is finitely generated (Robbiano and Sweedler 1990: example 4.11). In such a case the subring R does not have a finite SAGBI basis with respect to "$\prec$".

On the other hand, in many nice situations a finite SAGBI basis $\{f_1, \ldots, f_m\}$ exists, in which case we can use the following easy *subduction algorithm* to test whether a given polynomial $f \in \mathbf{C}[x_1, \ldots, x_n]$ lies in the subring R:

While $f \neq 0$ and $\text{init}_\prec(f) \in \text{init}_\prec(R)$, find a representation $\text{init}_\prec(f) = \text{init}_\prec(f_1)^{v_1} \cdots \text{init}_\prec(f_m)^{v_m}$ and replace f by $f - f_1^{v_1} \cdots f_m^{v_m}$.

Our proof of Theorem 1.1.1 implies that the elementary symmetric polynomials form a SAGBI basis for the ring of symmetric polynomials $\mathbf{C}[x_1, \ldots, x_n]^{S_n}$. Similarly, Lemmas 3.2.6 and 3.2.7 imply the same result for the maximal minors and the bracket ring $\mathcal{B}_{n,d}$. Note that in this case the general subduction algorithm specializes to Algorithm 3.2.8.

Theorem 3.2.9. The set of $d \times d$-minors of the $n \times d$-matrix (x_{ij}) is a SAGBI basis for the bracket ring $\mathcal{B}_{n,d} = \mathbf{C}[x_{ij}]^{SL(\mathbf{C}^d)}$ with respect to the diagonal monomial order on $\mathbf{C}[x_{ij}]$.

We close this section with an example from projective geometry.

Example 3.2.10 (Projections of the quadrilateral set). Let $\mathcal{C}$ be an arbitrary configuration of six points on the projective line P^1. We write the homogeneous coordinates of the points in C as the columns of a generic matrix

$$\mathbf{X} = \begin{pmatrix} a_1 & b_1 & c_1 & d_1 & e_1 & f_1 \\ a_2 & b_2 & c_2 & d_2 & e_2 & f_2 \end{pmatrix}.$$

Consider the following polynomial function

$$\begin{aligned} I = & -a_1b_1c_1d_2e_2f_2 - a_1b_1c_2d_1e_2f_2 + a_1b_1c_2d_2e_1f_2 + a_1b_1c_2d_2e_2f_1 \\ & + a_1b_2c_1d_1e_2f_2 - a_1b_2c_2d_2e_1f_1 + a_2b_1c_1d_1e_2f_2 - a_2b_1c_2d_2e_1f_1 \\ & - a_2b_2c_1d_1e_1f_2 - a_2b_2c_1d_1e_2f_1 + a_2b_2c_1d_2e_1f_1 + a_2b_2c_2d_1e_1f_1. \end{aligned}$$

Is this polynomial invariant under the action of $SL(\mathbf{C}^2)$? If so, which geometric property of the configuration C is expressed by the vanishing of I? The answer to the first question is affirmative, and we can compute a bracket representation using either of the two given algorithms for the First Fundamental Theorem. For the second question see Exercise (1) below and Fig. 3.1.

In the straightening approach we replace $\mathbf{X}$ by the extended matrix

$$\mathbf{X}' = \begin{pmatrix} a_1 & b_1 & c_1 & d_1 & e_1 & f_1 & 0 & -1 \\ a_2 & b_2 & c_2 & d_2 & e_2 & f_2 & 1 & 0 \end{pmatrix}$$

whose last two columns are labeled 1 and 2. We express each matrix entry of $\mathbf{X}$ as a maximal minor of $\mathbf{X}'$ via $a_1 = [a1]$, $b_1 = [b1]$, ..., $f_2 = [f2]$. This transforms the invariant I into the bracket polynomial

$$\begin{aligned} & -[a1][b1][c1][d2][e2][f2] - [a1][b1][c2][d1][e2][f2] \\ & +[a1][b1][c2][d2][e1][f2] + [a1][b1][c2][d2][e2][f1] \\ & +[a1][b2][c1][d1][e2][f2] - [a1][b2][c2][d2][e1][f1] \\ & +[a2][b1][c1][d1][e2][f2] - [a2][b1][c2][d2][e1][f1] \\ & -[a2][b2][c1][d1][e1][f2] - [a2][b2][c1][d1][e2][f1] \\ & +[a2][b2][c1][d2][e1][f1] + [a2][b2][c2][d1][e1][f1]. \end{aligned}$$

We apply the straightening algorithm for $\mathbf{C}[\Lambda(8,2)]$ with respect to the order of column indices $a < b < c < d < e < f < 1 < 2$. In the specific strategy used in the author's implementation of the straightening algorithm this computation requires 58 steps. The output is the following linear combination of standard tableaux:

$$\begin{aligned} & -[ab][cd][ef][12][12][12] + [ab][ce][df][12][12][12] \\ & +[ac][bd][ef][12][12][12] - [ac][be][df][12][12][12] \\ & -[ad][be][cf][12][12][12]. \end{aligned}$$

3.2. The first fundamental theorem

Hence the invariant I has the bracket representation

$$I = -[ab][cd][ef]+[ab][ce][df]+[ac][bd][ef]-[ac][be][df]-[ad][be][cf].$$

For the same input polynomial I Algorithm 3.2.8 works as follows. The initial monomial of I with respect to the diagonal monomial order equals $m = -a_1 b_1 c_1 d_2 c_2 d_2$. The corresponding standard tableau T_m equals $-[ad][be][cf]$, and so we replace I by $I - \phi_{6,2}(T_m)$. Now the initial monomial equals $-a_1 b_1 \times c_2 d_1 e_2 f_2$, so we subtract (the expansion of) $-[ac][be][df]$. The new initial monomial equals $a_1 b_1 c_2 d_2 e_1 f_2$, so we subtract $[ac][bd][ef]$. The new initial monomial equals $a_1 b_2 c_1 d_1 e_2 f_2$, so we subtract $[ab][ce][df]$. The new initial monomial equals $-a_1 b_2 c_1 d_2 e_1 f_2$, so we subtract $-[ab][cd][ef]$. The result is zero and we are done. The number of steps needed in Algorithm 3.2.8 – here: five – is always equal to the size of the output.

Note that the bracket representation found by both methods is in general not minimal. In our example the minimal bracket representation of I has only two tableaux:

$$I = -[ad][cf][be] + [af][bc][ed].$$

For general $SL(\mathbf{C}^d)$-invariants, it remains an interesting research problem to find a good algorithm for computing a bracket representation having the minimal number of tableaux.

Exercises

(1) A configuration of six points $\mathbf{a} = (a_1 : a_2 : a_3), \ldots, \mathbf{f} = (f_1 : f_2 : f_3)$ in the projective plane P^2 is called a *quadrilateral set* if the triples **ace**, **adf**, **bcf** and **bde** are collinear (Fig. 3.1). The one-dimensional configuration $(a_1 : a_2), \ldots, (f_1 : f_2)$ in Example 3.2.10 is the projection of a quadrilateral set if and only if there exist complex numbers $a_3, b_3, c_3, d_3, e_3, f_3$ such that $(a_1 : a_2 : a_3), \ldots, (f_1 : f_2 : f_3)$ is a quadrilateral set. Prove that this geometric property is equivalent to the vanishing of the invariant I.

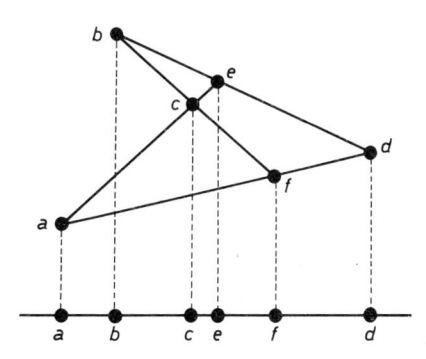

Fig. 3.1. The projection of a quadrilateral set

(2) Let $n = 4$, $d = 2$, and consider the polynomial

$$P(x_{ij}) = \begin{vmatrix} x_{11}^3 & x_{11}^2 x_{12} & x_{11} x_{12}^2 & x_{12}^3 \\ x_{21}^3 & x_{21}^2 x_{22} & x_{21} x_{22}^2 & x_{22}^3 \\ x_{31}^3 & x_{31}^2 x_{32} & x_{31} x_{32}^2 & x_{32}^3 \\ x_{41}^3 & x_{41}^2 x_{42} & x_{41} x_{42}^2 & x_{42}^3 \end{vmatrix} \in \mathbf{C}[x_{ij}].$$

Prove that P is an invariant of $SL(\mathbf{C}^2)$ and find a bracket representation. Interpreting the x_{ij} as homogeneous coordinates of four points on the projective line, what is the geometric meaning of the invariant P?

(3) Let $n = 4$, $d = 4$, and consider the rational function

$$Q(x_{ij}) = \frac{\begin{vmatrix} x_{11} & x_{12} & x_{13} \\ x_{21} & x_{22} & x_{23} \\ x_{31} & x_{32} & x_{33} \end{vmatrix} \begin{vmatrix} x_{11} & x_{12} & x_{14} \\ x_{21} & x_{22} & x_{24} \\ x_{41} & x_{42} & x_{44} \end{vmatrix} - \begin{vmatrix} x_{11} & x_{12} & x_{14} \\ x_{21} & x_{22} & x_{24} \\ x_{31} & x_{32} & x_{34} \end{vmatrix} \begin{vmatrix} x_{11} & x_{12} & x_{13} \\ x_{21} & x_{22} & x_{23} \\ x_{41} & x_{42} & x_{43} \end{vmatrix}}{\begin{vmatrix} x_{33} & x_{34} \\ x_{43} & x_{44} \end{vmatrix}}.$$

Show that Q is actually a polynomial in $\mathbf{C}[x_{ij}]^{SL(\mathbf{C}^4)}$, and find its bracket representation. (The resulting formula is called the *Bareiss expansion*.)

(4) * Compare the computational complexity of the subduction algorithm 3.2.8 with the straightening algorithm used in the proof of Theorem 3.2.1. Hint: Compare the number of standard tableaux with the number of non-standard tableaux.

3.3. The Grassmann–Cayley algebra

The Grassmann–Cayley algebra is an invariant algebraic formalism for expressing statements in synthetic projective geometry. The modern version to be presented here was developed in the 1970s by G.-C. Rota and his collaborators (Doubilet et al. 1974, Rota and Stein 1976). The main result in this section is an algorithm for expanding Grassmann–Cayley algebra expressions into bracket polynomials. As an illustration of these techniques, we give an "automated invariant-theoretic proof" of Desargues' theorem.

Let V be a $\mathbf{C}$-vector space of dimension d, and let $\Lambda(V)$ denote the *exterior algebra* over V. We refer to Greub (1967) for a detailed introduction to the exterior algebra. For geometric reasons we write the exterior product in $\Lambda(V)$ as "$\vee$" instead of the usual "$\wedge$", and refer to it as the *join* operation. The join (= exterior product) is multilinear, associative, and antisymmetric. The exterior algebra $\Lambda(V)$ is a graded $\mathbf{C}$-vector space of dimension 2^d, namely,

$$\Lambda(V) = \bigoplus_{k=0}^{d} \Lambda^k(V), \quad \text{where } \dim \Lambda^k(V) = \binom{d}{k}. \tag{3.3.1}$$

3.3. The Grassmann–Cayley algebra

Let $\{e_1, \ldots, e_d\}$ be any basis of V. Then a basis for $\Lambda^k(V)$ is given by

$$\{e_{j_1} \vee e_{j_2} \vee \ldots \vee e_{j_k} \mid 1 \leq j_1 < j_2 < \ldots < j_k \leq d\}. \tag{3.3.2}$$

Consider any k vectors $a_1, \ldots, a_k \in V$ and their expansions $a_i = \sum_{j=1}^{d} a_{ij} e_j$ in terms of the given basis. By multilinearity and antisymmetry, the expansion of their join equals

$$a_1 \vee a_2 \vee \ldots \vee a_k$$
$$= \sum_{1 \leq j_1 < \ldots < j_k \leq d} \begin{vmatrix} a_{1j_1} & a_{1j_2} & \cdots & a_{1j_k} \\ a_{2j_1} & a_{2j_2} & \cdots & a_{2j_k} \\ \vdots & \vdots & \ddots & \vdots \\ a_{kj_1} & a_{kj_2} & \cdots & a_{kj_k} \end{vmatrix} e_{j_1} \vee e_{j_2} \vee \ldots \vee e_{j_k}. \tag{3.3.3}$$

An element $A \in \Lambda^k(V)$ is said to be an *extensor (of step k)* if it has the form $A = a_1 \vee a_2 \vee \ldots \vee a_k$ for some $a_1, \ldots, a_k \in V$. In the following we abbreviate $A = a_1 a_2 \ldots a_k$.

We remark that our choice of basis identifies the ring of polynomial functions on $\Lambda^k(V)$ with the polynomial ring $\mathbf{C}[\Lambda(d,k)]$ defined in Sect. 3.1. The map $V^k \to \Lambda^k(V)$, $(a_1, a_2, \ldots, a_k) \mapsto a_1 a_2 \ldots a_k$ corresponds to the ring map $\phi_{d,k}$. Hence the set of extensors in $\Lambda^k(V)$ coincides with the affine algebraic variety defined by the ideal $I_{d,k} = \ker(\phi_{d,k})$. Since the ideal $I_{d,k}$ is homogeneous, we can also consider the projective variety defined by $I_{d,k}$. This projective variety is called the *Grassmann variety*.

The following argument shows that the points on the Grassmann variety are in bijection with the k-dimensional linear subspaces of V. Let $A = a_1 a_2 \ldots a_k$ be a non-zero extensor of step k. Then $a_1, a_2, \ldots, a_k$ is the basis of a k-dimensional linear subspace $\overline{A}$ of V. The subspace $\overline{A}$ is determined by the extensor A because $\overline{A} = \{v \in V : A \vee v = 0\}$. On the other hand, the extensor A is determined (up to scalar multiple) by $\overline{A}$, because the expansion (3.3.3) is invariant (up to scalar multiple) under a change of basis in $\overline{A}$.

Let $B = b_1 b_2 \cdots b_j$ be another non-zero extensor of step j. Then the join

$$A \vee B = a_1 \vee a_2 \vee \ldots \vee a_k \vee b_1 \vee \ldots \vee b_j = a_1 a_2 \cdots a_k b_1 \cdots b_j \tag{3.3.4}$$

is an extensor of step $j + k$. The following lemma explains why we use the term "join" for the exterior product in $\Lambda(V)$. Its proof is straightforward from the definitions.

Lemma 3.3.1. *The extensor $A \vee B$ is non-zero if and only if $a_1, a_2, \ldots, a_k, b_1, b_2, \ldots, b_j$ are distinct and linearly independent. In this case we have*

$$\overline{A} + \overline{B} = \overline{A \vee B} = \mathrm{span}\{a_1, a_2, \ldots, a_k, b_1, b_2, \ldots, b_j\}. \tag{3.3.5}$$

Lemma 3.3.1 states that the algebraic join of extensors corresponds to the geometric join of linear subpaces. This raises the question of how to define an algebraic operation which corresponds similarly to the geometric *meet* or intersection of linear subspaces.

We first identify the 1-dimensional vector space $\Lambda^d(V)$ with the ground field $\mathbf{C}$ via $e_1 e_2 \ldots e_d \mapsto 1$. Hence an extensor of step d can simply be expressed as the determinant of its d constituent vectors. We denote this determinant using the familiar bracket notation, say, $[a_1, a_2, \ldots, a_d] := a_1 a_2 \ldots a_d$. Let $A = a_1 a_2 \cdots a_j$ and $B = b_1 b_2 \cdots b_k$ be extensors with $j + k \geq d$. Then their *meet* is the element of $\Lambda^{j+k-d}(V)$ defined by

$$A \wedge B := \sum_\sigma \text{sign}(\sigma) [a_{\sigma(1)}, \ldots, a_{\sigma(d-k)}, b_1, \ldots, b_k] \cdot a_{\sigma(d-k+1)} \cdots a_{\sigma(j)}. \quad (3.3.6)$$

The sum is taken over all permutations σ of $\{1, 2, \ldots, j\}$ such that $\sigma(1) < \sigma(2) < \ldots < \sigma(d-k)$ and $\sigma(d-k+1) < \sigma(d-k+2) < \ldots < \sigma(j)$. Such permutations are called *shuffles* of the $(d-k, j-(d-k))$ split of A. We have given the definition for both join and meet only for pairs of extensors. These definitions are extended to arbitrary elements of $\Lambda(V)$ by distributivity. The *Grassmann–Cayley algebra* is the vector space $\Lambda(V)$ together with the operations $\vee$ and $\wedge$.

A useful notation for signed sums over shuffles such as (3.3.6) is the *dotted* notation, which we will frequently employ. We simply place dots over the shuffled vectors, with the summation and $\text{sign}(\sigma)$ implicit. Similarly, shuffles may be defined over splits into any number of parts. Here the brackets are always delimiters which define the parts of the split. If we wish to sum over several shuffles of disjoint sets, we will use separate symbols over each shuffled set. Thus (3.3.6) is equivalent to

$$A \wedge B = [\dot{a}_1, \ldots, \dot{a}_{d-k}, b_1, \ldots, b_k]\, \dot{a}_{d-k+1} \cdots \dot{a}_j. \quad (3.3.7)$$

If $j + k = d$ then we have $A \wedge B = [a_1, \ldots, a_j, b_1, \ldots, b_k]$. This is a scalar of step 0, and it needs to be distinguished from $A \vee B$, a scalar with the same numerical value but of step d. Thus $\Lambda^0(V)$ is a second copy of the ground field $\mathbf{C}$ in $\Lambda(V)$. We will now prove that the meet operation corresponds to our geometric intuition in the case when A and B themselves are non-degenerate and the dimension of $\overline{A} \cap \overline{B}$ is as small as possible.

Theorem 3.3.2.
(a) The meet is associative and anti-commutative in the following sense:

$$A \wedge B = (-1)^{(d-k)(d-j)} \cdot B \wedge A.$$

(b) The meet of two extensors is again an extensor.

3.3. The Grassmann–Cayley algebra

(c) We have $A \wedge B \neq 0$ if and only if $\overline{A} + \overline{B} = V$. In that case $\overline{A \wedge B} = \overline{A} \cap \overline{B}$, i.e., the meet corresponds to the intersection of linear subspaces.

Proof. Consider the following quadratic bracket polynomial in $C[\Lambda(2d,d)]$:

$$[\overset{\bullet}{a}_1, \ldots, \overset{\bullet}{a}_{d-k}, b_1, \ldots, b_k][\overset{\bullet}{a}_{d-k+1}, \ldots, \overset{\bullet}{a}_j, c_1, \ldots, c_{2d-j-k}]$$
$$- (-1)^{(d-k)(d-j)} [\overset{\circ}{b}_1, \ldots, \overset{\circ}{b}_{d-j}, a_1, \ldots, a_j][\overset{\circ}{b}_{d-k+1}, \ldots, \overset{\circ}{b}_j, c_1, \ldots, c_{2d-j-k}].$$

(3.3.8)

Using induction on $k + j$, it can be shown that this bracket polynomial lies in the syzygy ideal $I_{2d,d}$. The syzygy (3.3.8) is equivalent to $(A \wedge B) \vee C = (-1)^{(d-k)(d-j)} (B \wedge A) \vee C$ for all extensors C of step $2d - j - k$. This implies the statement (a).

It follows from the defining equation (3.3.6) that the meet $A \wedge B$ remains invariant (up to a scalar multiple) if we replace $b_1, \ldots, b_k$ by any other basis of $\overline{B}$. By the anti-commutativity of the meet operation, the same holds when $a_1, \ldots, a_j$ is replaced by any other basis of $\overline{A}$.

In view of our assumption $\overline{A} + \overline{B} = V$, we can choose a basis $v_1, v_2, \ldots, v_d$ of V such that $v_1, \ldots, v_{j+k-d}$ is a basis of $\overline{A} \cap \overline{B}$, $v_1, \ldots, v_j$ is a basis of $\overline{A}$, and $v_{j+1}, \ldots, v_d, v_1, \ldots, v_{j+k-d}$ is a basis for $\overline{B}$. There exist nonzero constants c_1 and c_2 such that $A = c_1 \cdot v_1 v_2 \ldots v_j$ and $B = c_2 \cdot v_{j+1} \ldots v_d v_1 \ldots v_{j+k-d}$. Substituting these representations into the defining equation (3.3.6), we find that

$$A \wedge B = \pm c_1 c_2 [v_1, v_2, \ldots, v_d] \cdot v_1 v_2 \ldots v_{j+k-d}. \quad (3.3.9)$$

This proves the statements (b) and (c) of Theorem 3.3.2. ◁

We now illustrate the translation of geometric incidences into the Grassmann–Cayley algebra. Each non-zero vector in V represents a point in the $(d-1)$-dimensional projective space P^{d-1}. This provides an identification of k-dimensional linear subspaces $\overline{A}$ of V with $(k-1)$-dimensional projective subspaces of P^{d-1}. Such subspaces are represented uniquely (up to scalar multiple) by the extensors of step k.

Example 3.3.3. Let a, b, c, d, e, f be any six points in the projective plane P^2 such that the lines $\overline{ad}, \overline{be}$ and $\overline{cf}$ are distinct (cf. Fig. 3.2). Under which algebraic condition are these three lines concurrent, i.e., have a point in common?

To answer this question we apply Theorem 3.3.2 (c). In the Grassmann–Cayley algebra the intersection point of the lines $\overline{ad}$ and $\overline{be}$ is represented as $(a \vee d) \wedge (b \vee e)$, an extensor of step 1. This point lies on the line $\overline{cf}$ if and only if $((a \vee d) \wedge (b \vee e)) \wedge (c \vee f) = 0$. Using the defining equation of the

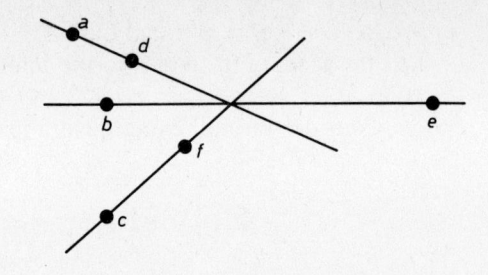

Fig. 3.2. The incidence relation expressed by (3.3.10)

meet (3.3.6), we obtain the bracket expansion

$$\big((a \vee d) \wedge (b \vee e)\big) \wedge (c \vee f) \\ = \big([abe]d - [dbe]a\big) \wedge cf = [abe][dcf] - [dbe][acf]. \quad (3.3.10)$$

Naturally, this quadratic bracket polynomial can be further expanded into a polynomial in the coordinates of the six points $a = (a_1 : a_2 : a_3), \ldots, f = (f_1 : f_2 : f_3)$. The resulting polynomial in the coordinates is homogeneous of degree 6 and has 48 monomials. It is listed completely in Example 3.1.2.

In similar fashion, any incidence relation or incidence theorem in projective geometry may be translated into a conjunction of simple Grassmann–Cayley algebra statements. Here *simple* means that the expression involves only join and meet, not addition. For instance, $(a \vee b) \wedge (c \vee d) \wedge (e \vee f)$ is a simple Grassmann–Cayley algebra expression. Conversely, every simple Grassmann–Cayley algebra statement may be translated back to projective geometry just as easily.

Now, generalizing the derivation in (3.3.10), every simple Grassmann–Cayley algebra expression of step 0 may be expanded into a bracket polynomial using only the definitions and basic properties of join and meet.

Algorithm 3.3.4 (Expanding Grassmann–Cayley algebra expressions into brackets).
Input: A Grassmann–Cayley algebra expression $C(a, b, \ldots)$ of step 0.
Output: A bracket polynomial equivalent to $C(a, b, \ldots)$.

1. Replace each occurrence of a subexpression $(a_1 \ldots a_j) \wedge (b_1 \ldots b_k)$ in C by $[\overset{\bullet}{a}_1, \ldots, \overset{\bullet}{a}_{d-k}, b_1, \ldots, b_k]\, \overset{\bullet}{a}_{d-k+1} \cdots \overset{\bullet}{a}_j$.
2. Erase unnecessary parentheses using the associativity of meet and join. Using distributivity, write $C(a, b, \ldots)$ as a linear combination of simple Grassmann–Cayley algebra expressions.
3. Extract bracket factors from each expression. For any remaining Grassmann–Cayley algebra factor $C'(a, b, \ldots)$ return to step 1.

When applying Algorithm 3.3.4, we will usually ignore global signs in the intermediate computations and in the output. This is justified by multilinearity and Theorem 3.3.2 (a).

3.3. The Grassmann–Cayley algebra

Most Grassmann–Cayley algebra statements resulting from geometric incidence relations have step 0 or d, in which case Algorithm 3.3.4 can be applied directly. If a Grassmann–Cayley algebra statement $C(a, b, \ldots)$ has step k with $0 < k < d$, then $C(a, b, \ldots) = 0$ is equivalent to the following universally quantified statement of step d:

$$\forall x_1, \ldots, x_{d-k} \in V : C(a, b, \ldots) \vee x_1 \vee \ldots \vee x_{d-k} = 0. \qquad (3.3.11)$$

In fact, here it suffices to take $x_1, \ldots, x_{d-k}$ from a basis of V, so (3.3.11) is equivalent to a finite conjunction of bracket statements.

In summary, Algorithm 3.3.4 gives an easy method for expanding simple Grassmann–Cayley algebra expressions into bracket polynomials. However, the converse problem, that of rewriting a bracket polynomial as a simple Grassmann–Cayley algebra expression, whenever possible, is not easy at all. This is the problem of *Cayley factorization*, which we will discuss in Sect. 3.5.

An often useful postprocessing to Algorithm 3.3.4 is the straightening algorithm (cf. Sect. 3.1). It rewrites the output bracket polynomial into a unique normal form, namely, as a linear combination of standard tableaux. In particular, we thus obtain an algorithm for testing whether two Grassmann–Cayley algebra expressions are equal. We illustrate these invariant-theoretic techniques by proving a classical incidence theorem of projective geometry.

Example 3.3.5 (Desargues' theorem). The corresponding sides of two triangles meet in collinear points if and only if the lines spanned by corresponding vertices are concurrent.

The vertices of the two triangles are labeled a, b, c and d, e, f as in Fig. 3.3. The corresponding sides of the triangles meet in collinear points if and only if the following Grassmann–Cayley algebra expression vanishes. We compute its bracket expansion using Algorithm 3.3.4.

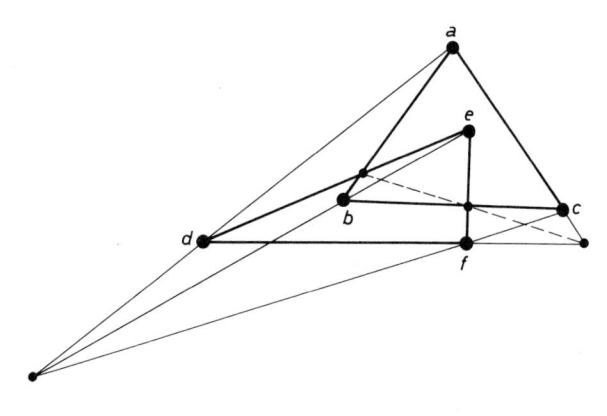

Fig. 3.3. Desargues' Theorem in the plane

$$(ab \wedge de) \vee (bc \wedge ef) \vee (ac \wedge df)$$
$$= ([ade]b - [bde]a) \vee ([bef]c - [cef]b) \vee ([adf]c - [cdf]a)$$
$$= -[ade][bef][cdf][bca] + [bde][cef][adf][abc].$$

Applying the straightening algorithm, we get the following standard bracket polynomial:

$$-[abc][abc][def][def] - [abc][abe][cdf][def]$$
$$-[abc][acd][bef][def] + [abc][ace][bdf][def].$$

This expression can be rewritten as

$$[abc][def] \cdot (-[abc][def] - [abe][cdf] - [acd][bef] + [ace][bdf])$$
$$= [abc][def] \cdot ([abe][dcf] - [dbe][acf]) = [abc][def] \cdot (ad \wedge be \wedge cf).$$

By Example 3.3.3, this expression vanishes if and only if a, b, c are collinear, or d, e, f are collinear, or the lines $\overline{ad}, \overline{be}$ and $\overline{cf}$ are concurrent. This proves Desargues' Theorem. ◁

Exercises

(1) Prove that the quadratic bracket polynomial in (3.3.8) is a syzygy.
(2) Expand the following Grassmann–Cayley algebra expressions into bracket polynomials of rank 3. In each case give a geometric interpretation:
  – $cd \wedge ((fg \wedge hk) \vee e) \wedge ab$
  – $(ab \wedge cd) \vee (ad \wedge bc) \vee (ac \wedge bd)$
  – $(ab \wedge cd) \vee (ad \wedge bc) \vee (ac \wedge de)$

3.4. Applications to projective geometry

In this section we discuss six applications which illustrate the use of Grassmann–Cayley algebra and bracket algebra as a computational tool in projective geometry.

Example 3.4.1 (Coordinatization of abstract configurations). A main problem in computational synthetic geometry (cf. Bokowski and Sturmfels 1989) is to find coordinates or non-realizability proofs for abstractly defined configurations; see also Sturmfels (1991). The methods of bracket and Grassmann–Cayley algebra are well suited for this problem. We illustrate this application with an example which is drawn from Sturmfels and White (1990). In that article it is shown that all 11_3 and all 12_3-configurations can be coordinatized over the field of rational numbers.

Consider a configuration $\mathcal{C}$ of eleven points in the projective plane, labeled

3.4. Applications to projective geometry

1, 2, ..., 9, 0 and A, such that precisely the following eleven triples are collinear:

$$124 \quad 235 \quad 346 \quad 457 \quad 568 \quad 679 \quad 780 \quad 89A \quad 901 \quad 0A2 \quad A13. \quad (3.4.1)$$

This configuration is called 11_3 because it consists of 11 points and 11 lines, with each point lying on 3 lines and each line containing 3 points.

In order to find coordinates for such a configuration we proceed as follows. Let us suppose that $x_1, x_3, x_4, x_5, x_8, x_9$ are arbitrary points in P^2. The configuration (3.4.1) translates into the following system of equations in the Grassmann–Cayley algebra:

$$
\begin{aligned}
& x_A \vee x_1 \vee x_3 = 0, \quad x_A = (x_8 \vee x_9) \wedge (x_0 \vee x_2), \\
& x_0 = (x_7 \vee x_8) \wedge (x_1 \vee x_9), \quad x_7 = (x_4 \vee x_5) \wedge (x_6 \vee x_9), \\
& x_6 = (x_3 \vee x_4) \wedge (x_5 \vee x_8), \quad x_2 = (x_1 \vee x_4) \wedge (x_3 \vee x_5).
\end{aligned}
\quad (3.4.2)
$$

We solve these equations by substituting $x_8 x_9 \wedge x_0 x_2$ for x_A in $x_A \vee x_1 \vee x_3$, then substituting $x_7 x_8 \wedge x_1 x_9$ for x_0 in the result, etc. Proceeding in this fashion, we find that (3.4.2) is equivalent to the vanishing of the following simple Grassmann–Cayley algebra expression:

$$
\Big(x_8 x_9 \wedge \big((x_4 x_5 \wedge (x_3 x_4 \wedge x_5 x_8) x_9) x_8 \wedge x_1 x_9\big)(x_1 x_4 \wedge x_3 x_5)\Big) \vee x_1 \vee x_3. \quad (3.4.3)
$$

We now apply Algorithm 3.3.4 to (3.4.3). This means we successively replace each subexpression $x_i x_j \wedge x_k x_l$ by $[kij]x_l - [lij]x_k$, using distributivity after each replacement. As the result we obtain the following bracket polynomial which is equivalent to (3.4.3):

$$
\begin{aligned}
& - [834][945][958][189][314][513] - [534][845][198][314][589][913] \\
& - [834][945][158][314][589][913] + [534][845][198][514][389][913] \\
& + [834][945][158][514][389][913] = 0
\end{aligned}
\quad (3.4.4)
$$

It remains to find six points in P^2 which satisfy the equation (3.4.4). We may assume

$$
\begin{aligned}
& x_1 = (1:0:0), \quad x_3 = (0:1:0), \quad x_4 = (0:0:1), \\
& x_5 = (1:1:1), \quad x_8 = (1:a:b), \quad x_9 = (1:c:d)
\end{aligned}
\quad (3.4.5)
$$

Under this choice of six vectors the bracket equation (3.4.4) specializes to

$$
\begin{aligned}
& a^3 d^2 - a^3 d^3 + 2a^2 bcd^2 - a^2 bcd - a^2 cd^2 + a^2 d^3 - a^2 d^2 - ab^2 c^2 d \\
& + abc^2 - abcd^2 + 2abcd - abc - abd^2 + ad^2 + b^2 c^3 - 2b^2 c^2 \\
& + b^2 cd + b^2 c - bc^3 + bc^2 d + bc^2 - 2bcd = 0.
\end{aligned}
$$

We choose the solution $a = \frac{3}{5}$, $b = \frac{57}{17}$, $c = 2$, $d = 5$ to this equation. Substituting the points (3.4.5) into the Grassmann–Cayley expressions (3.4.2), we obtain the following coordinates for the configuration (3.4.1):

$$(x_1, x_2, \ldots, x_0, x_A) = \begin{pmatrix} 1 & 1 & 0 & 0 & 1 & 0 & \frac{17}{185} & 1 & 1 & 1 \\ 0 & 0 & 1 & 0 & 1 & -\frac{17}{100} & \frac{17}{185} & \frac{3}{5} & 2 & \frac{18}{37} & -\frac{9}{4} \\ 0 & 1 & 0 & 1 & 1 & 1 & 1 & \frac{57}{17} & 5 & \frac{45}{37} & 0 \end{pmatrix}.$$

Example 3.4.2 (Final polynomials). Here is another coordinatization problem similar to Example 3.4.1. Consider a configuration $\mathcal{C}'$ of eight distinct points in projective 3-space P^3, labeled $1, 2, 3, 4, 5, 6, 7, 8$, such that precisely the following five quadruples are coplanar:

$$1256 \quad 1357 \quad 1458 \quad 2367 \quad 2468. \tag{3.4.6}$$

Does such a configuration exist?

We claim that the answer is "no", that is, $\mathcal{C}'$ cannot be coordinatized over **C**. Suppose that, on the contrary, there exist vectors $x_1, \ldots, x_8 \in \mathbf{C}^4$ such that, for $(i, j, k, l) \in \Lambda(8, 4)$, $\det(x_i, x_j, x_k, x_l)$ is zero if and only if $ijkl$ appears in (3.4.6). Then we get a **C**-algebra homomorphism

$$\phi : \mathcal{B}_{8,4} \to \mathbf{C}, \quad [i\,j\,k\,l] \mapsto \det(x_i, x_j, x_k, x_l)$$

having the property that $\phi([ijkl]) = 0$ if and only if $ijkl$ appears in (3.4.6).

We consider the following bracket polynomial

$$\begin{aligned} f := &\ [\underline{1256}][1734][1284][7234] + [\underline{1357}][1264][1284][7234] \\ &+ [\underline{1458}][1264][1237][7234] + [\underline{2367}][1734][1284][1254] \\ &+ [\underline{2468}][1734][1254][1237] + [3478][1264][1254][1237] \end{aligned}$$

in $\mathbf{C}[\Lambda(8, 4)]$. Using the straightening algorithm introduced in Sect. 3.1, it can be verified easily that f lies in the syzygy ideal $I_{8,4}$, that is, $f = 0$ in $\mathcal{B}_{8,4}$. On the other hand, the underlined brackets are mapped to zero by ϕ, and hence $\phi(f) = \phi([3478][1264][1254][1237]) \neq 0$. This is a contradiction, and the proof of our claim is complete.

The bracket polynomial $f \in \mathbf{C}[\Lambda(8, 4)]$ is said to be a *final polynomial* for the configuration (3.4.6). In general, final polynomials provide a systematic method of representing non-realizability proofs for abstract configurations. This method is originally due to Bokowski and Whiteley; it has been developed in detail in Bokowski and Sturmfels (1989).

We remark that the configuration in (3.4.6) is known as the *Vamos matroid* in matroid theory. Its non-realizability is equivalent to the following well-known incidence theorem in projective geometry. This incidence theorem is sometimes called the *bundle condition*.

3.4. Applications to projective geometry

Theorem. Given four lines $\ell_1, \ell_2, \ell_3, \ell_4$ in projective 3-space, no three in a plane, such that five of the six pairs (ℓ_i, ℓ_j) of lines intersect, then also the sixth pair of lines intersects.

Proof. Suppose that $\ell_1, \ell_2, \ell_3, \ell_4$ are lines in P^3, no three in a plane, such that $\ell_3 \cap \ell_4 = \emptyset$, but each of the pairs (ℓ_1, ℓ_2), (ℓ_1, ℓ_3), (ℓ_1, ℓ_4), (ℓ_2, ℓ_3), (ℓ_2, ℓ_4) intersects. Choose sufficiently generic points $x_1, x_5 \in \ell_1$, $x_2, x_6 \in \ell_2$, $x_3, x_7 \in \ell_3$, $x_4, x_8 \in \ell_4$. Then precisely the quadruples in (3.4.6) are coplanar. This is impossible by the final polynomial f. ◁

Example 3.4.3 (Pascal's theorem). In this example we outline an algorithm for both discovering and proving a certain class of geometry theorems in the plane. Suppose we are given the following problem:

Under which "geometric" condition do six points **a, b, c, d, e, f** in the projective plane lie on a common quadric? Find such a condition and prove that it is correct!!

Using homogeneous coordinates $\mathbf{a} = (a_1 : a_2 : a_3)$, $\mathbf{b} = (b_1 : b_2 : b_3)$, ..., $\mathbf{f} = (f_1 : f_2 : f_3)$, our problem can be rephrased as follows.

Find a synthetic interpretation or construction for the algebraic condition:

$$\exists (v_{200}, v_{020}, v_{002}, v_{110}, v_{101}, v_{011}) \in \mathbf{C}^6 \setminus \{0\}:$$

$$v_{200}a_1^2 + v_{020}a_2^2 + v_{002}a_3^2 + v_{110}a_1a_2 + v_{101}a_1a_3 + v_{011}a_2a_3 = 0$$

$$v_{200}b_1^2 + v_{020}b_2^2 + v_{002}b_3^2 + v_{110}b_1b_2 + v_{101}b_1b_3 + v_{011}b_2b_3 = 0 \quad (3.4.7)$$

$$\ldots \quad \ldots \quad \ldots \quad \ldots \quad \ldots$$

$$v_{200}f_1^2 + v_{020}f_2^2 + v_{002}f_3^2 + v_{110}f_1f_2 + v_{101}f_1f_3 + v_{011}f_2f_3 = 0$$

The existentially quantified variables in (3.4.7) are the coefficients $v_{200}, \ldots, v_{011}$ of the desired quadric $\sum_{i+j+k=2} v_{ijk} x_1^i x_2^j x_3^j$. It is our goal to compute an equivalent simple Grassmann–Cayley expression, which uses only the symbols **a, b, c, d, e, f**, $\wedge$ and $\vee$. In the first step we eliminate the v_{ijk}'s from (3.4.7), obtaining

$$\det \begin{pmatrix} a_1^2 & a_2^2 & a_3^2 & a_1a_2 & a_1a_3 & a_2a_3 \\ b_1^2 & b_2^2 & b_3^2 & b_1b_2 & b_1b_3 & b_2b_3 \\ c_1^2 & c_2^2 & c_3^2 & c_1c_2 & c_1c_3 & c_2c_3 \\ d_1^2 & d_2^2 & d_3^2 & d_1d_2 & d_1d_3 & d_2d_3 \\ e_1^2 & e_2^2 & e_3^2 & e_1e_2 & e_1e_3 & e_2e_3 \\ f_1^2 & f_2^2 & f_3^2 & f_1f_2 & f_1f_3 & f_2f_3 \end{pmatrix} = 0. \quad (3.4.8)$$

This elimination step involved only easiest linear algebra. In more general situations we would need Gröbner bases or resultants.

The degree 12 polynomial with 720 summands in (3.4.8) is invariant under projective transformations and can therefore be rewritten as a bracket polynomial in $\mathcal{B}_{6,3}$. Using Algorithm 3.2.8, we find that (3.4.8) equals the following sum of two standard tableaux:

$$-[\mathbf{abc}][\mathbf{ade}][\underline{\mathbf{bdf}}][\mathbf{cef}] + [\mathbf{abd}][\underline{\mathbf{ace}}][\mathbf{bcf}][\mathbf{def}] = 0. \qquad (3.4.9)$$

Our last step is now to find a simple Grassmann–Cayley algebra expression which is equivalent to (3.4.9). In general, translating bracket expressions into Grassmann–Cayley expressions is a very difficult problem. As we already remarked in Sect. 3.3, this problem is called the *Cayley factorization problem*. It will be discussed in Sect. 3.5. For our example Cayley factorization yields the following expression:

$$(\mathbf{ab} \wedge \mathbf{de}) \vee (\mathbf{bc} \wedge \mathbf{ef}) \vee (\mathbf{cd} \wedge \mathbf{fa}) = 0. \qquad (3.4.10)$$

It is easy to verify that (3.4.10) is equal to (3.4.9), using Algorithm 3.3.4 and subsequently the straightening algorithm. The expression (3.4.10) is equivalent to the synthetic statement,

"The intersection points $\overline{\mathbf{ab}} \cap \overline{\mathbf{de}}$, $\overline{\mathbf{bc}} \cap \overline{\mathbf{ef}}$ and $\overline{\mathbf{cd}} \cap \overline{\mathbf{fa}}$ are collinear." (3.4.11)

Thus we have "automatically" discovered Pascal's theorem.

Pascal's Theorem (see Fig. 3.4). Six points $\mathbf{a}, \mathbf{b}, \mathbf{c}, \mathbf{d}, \mathbf{e}$ and $\mathbf{f}$ in the projective plane lie on a common quadric if and only if the intersection points $\overline{\mathbf{ab}} \cap \overline{\mathbf{de}}$, $\overline{\mathbf{bc}} \cap \overline{\mathbf{ef}}$ and $\overline{\mathbf{cd}} \cap \overline{\mathbf{fa}}$ are collinear.

A special case of Pascal's theorem is Pappus' theorem. It is concerned with the case of degenerate quadrics, consisting of two lines. It states that *if* $\mathbf{a}, \mathbf{e}, \mathbf{c}$ *and* $\mathbf{d}, \mathbf{b}, \mathbf{f}$ *are collinear, then* $\overline{\mathbf{ab}} \cap \overline{\mathbf{de}}$, $\overline{\mathbf{bc}} \cap \overline{\mathbf{ef}}$ *and* $\overline{\mathbf{cd}} \cap \overline{\mathbf{fa}}$ *are collinear*. We obtain a proof of Pappus' theorem directly from the equality of (3.4.9) and (3.4.10). For, if $\mathbf{a}, \mathbf{e}, \mathbf{c}$ and $\mathbf{d}, \mathbf{b}, \mathbf{f}$ are collinear, then the two underlined brackets in (3.4.9) are zero, and hence (3.4.10) and (3.4.11) hold.

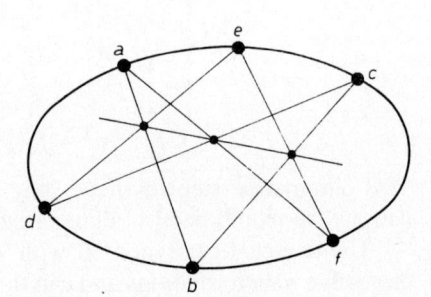

Fig. 3.4. Pascal's theorem

3.4. Applications to projective geometry

Example 3.4.4 (Parametric representation of quadrics in P^2). Let $\mathbf{a}, \mathbf{b}, \mathbf{c}, \mathbf{d}, \mathbf{e}$ be points in general position in P^2, and let C denote the unique quadratic curve through these five points. In this example we consider the problem of finding a parametric representation for C, directly in terms of $\mathbf{a}, \mathbf{b}, \mathbf{c}, \mathbf{d}, \mathbf{e}$. Such a parametrization must exist because quadrics are known to be rational curves. Let $f(\mathbf{x}) = f(x_1, x_2, x_3)$ denote the defining equation of C. From (3.4.9) we find

$$f(\mathbf{x}) = -[\mathbf{abc}][\mathbf{ade}][\mathbf{bd\,x}][\mathbf{ce\,x}] + [\mathbf{abd}][\mathbf{ace}][\mathbf{bc\,x}][\mathbf{de\,x}] = 0. \qquad (3.4.12)$$

In order to parametrize C, we take a general point $\lambda \mathbf{b} + \mathbf{c}$ on the secant line $\mathbf{b} \vee \mathbf{c}$, and we join it with $\mathbf{a}$ (see Fig. 3.5). A general point of the resulting line equals

$$\mathbf{x} = \mathbf{a} + \mu(\lambda \mathbf{b} + \mathbf{c}), \quad \text{where } \lambda, \mu \in \mathbf{C}. \qquad (3.4.13)$$

We now substitute this point into (3.4.12), and we expand the result as a polynomial in μ:

$$f(\mathbf{a} + \mu(\lambda \mathbf{b} + \mathbf{c})) = \mu^2 \cdot P(\lambda; \mathbf{a}, \mathbf{b}, \mathbf{c}, \mathbf{d}, \mathbf{e}) + \mu \cdot Q(\lambda; \mathbf{a}, \mathbf{b}, \mathbf{c}, \mathbf{d}, \mathbf{e}), \qquad (3.4.14)$$

where P, Q are homogeneous bracket polynomials of degree 4 with coefficients in $\mathbf{C}[\lambda]$. Note that (3.4.14) has no constant term with respect to μ because $f(\mathbf{a}) = 0$. Solving the right hand side of (3.4.14) for μ and clearing denominators (all points are given by homogeneous coordinates!!), we obtain the following parametrization of the quadratic curve C:

$$\mathbf{x}(\lambda) = -P(\lambda; \mathbf{a}, \mathbf{b}, \mathbf{c}, \mathbf{d}, \mathbf{e}) \cdot \mathbf{a} + Q(\lambda; \mathbf{a}, \mathbf{b}, \mathbf{c}, \mathbf{d}, \mathbf{e}) \cdot (\lambda \mathbf{b} + \mathbf{c}). \qquad (3.4.15)$$

Example 3.4.5 (Common transversals of four lines in P^3). This example concerns invariants of four lines $\ell_1, \ell_2, \ell_3, \ell_4$ in complex projective 3-space. By a *common transversal* of $\ell_1, \ell_2, \ell_3, \ell_4$ we mean a line $\ell \subset P^3$ such that

$$\ell \cap \ell_1 \neq \emptyset, \ \ell \cap \ell_2 \neq \emptyset, \ \ell \cap \ell_3 \neq \emptyset, \text{ and } \ell \cap \ell_4 \neq \emptyset. \qquad (3.4.16)$$

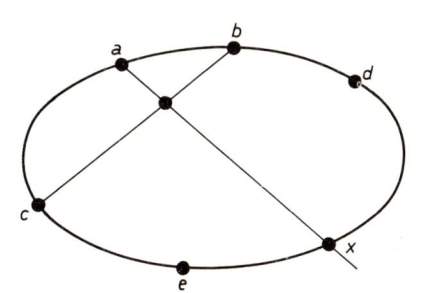

Fig. 3.5. Parametrization of the quadric through five given points

We will prove that a configuration of lines $\ell_1, \ell_2, \ell_3, \ell_4$ has either one transversal, or two transversals or infinitely many transversals. Moreover, we will compute the algebraic invariants which discriminate these three cases.

Each line ℓ_i can be written as the join of two points, say

$$\ell_1 = x_1 \vee x_2, \quad \ell_2 = x_3 \vee x_4, \quad \ell_3 = x_5 \vee x_6, \quad \ell_4 = x_7 \vee x_8. \tag{3.4.17}$$

As usual, the bracket $[ijkl]$ denotes the determinant $\det(x_i, x_j, x_k, x_l)$, and we have the Grassmann–Cayley algebra identities $\ell_1 \vee \ell_2 = [1234]$, $\ell_1 \vee \ell_3 = [1256]$, ..., $\ell_3 \vee \ell_4 = [5678]$.

Suppose that ℓ is a common transversal of $\ell_1, \ell_2, \ell_3, \ell_4$. The intersection point of ℓ and ℓ_1 equals $\lambda x_1 + \mu x_2$ for some $\lambda, \mu \in \mathbf{C}$. The plane spanned by this point and the line ℓ_2 is given by the Cayley expression $(\lambda x_1 + \mu x_2) \vee x_3 \vee x_4$, and the intersection point of this plane with the line ℓ_3 equals $\big((\lambda x_1 + \mu x_2) \vee x_3 \vee x_4\big) \wedge (x_5 \vee x_6)$. Hence the transversal equals

$$\ell = \big[((\lambda x_1 + \mu x_2) \vee x_3 \vee x_4) \wedge (x_5 \vee x_6)\big] \vee (\lambda x_1 + \mu x_2). \tag{3.4.18}$$

Since ℓ also meets the line $\ell_4 = x_7 \vee x_8$, we have the following Grassmann–Cayley algebra identities:

$$\big[((\lambda x_1 + \mu x_2) \vee x_3 \vee x_4) \wedge (x_5 \vee x_6)\big] \vee (\lambda x_1 + \mu x_2) \vee x_7 \vee x_8 = 0. \tag{3.4.19}$$

This is quadratic polynomial in λ, μ. We now apply Algorithm 3.3.4 to (3.4.19) to get

$$\big([1345][6178] - [1346][5178]\big)\lambda^2$$
$$+ \big([2345][6278] - [2346][5278]\big)\mu^2$$
$$+ \Big([1345][6278] - [1346][5278] \tag{3.4.20}$$
$$+ [2345][6178] - [2346][5178]\Big)\lambda\mu = 0.$$

The four lines have infinitely many common transversals if and only if all three bracket coefficients are zero. Generically, this is not the case and (3.4.20) has two distinct roots, corresponding to two distinct transversals. The discriminant of (3.4.20) is a polynomial of degree 4 in brackets; its unique expansion in terms of standard bracket monomials equals

$$- 2\,[1234][1256][3478][5678] - 2\,[1234][1256][3578][4678]$$
$$+ [1235][1235][4678][4678] - 2\,[1235][1236][4578][4678]$$
$$- 2\,[1235][1245][3678][4678] - 2\,[1235][1246][3478][5678]$$
$$+ 4\,[1235][1246][3578][4678] + [1236][1236][4578][4578]$$
$$- 2\,[1236][1246][3578][4578] + [1245][1245][3678][3678]$$
$$- 2\,[1245][1246][3578][3678] + [1246][1246][3578][3578].$$

3.4. Applications to projective geometry

We conclude that this invariant vanishes whenever there is exactly one transversal. It is instructive to verify that this expression is symmetric with respect to permuting the pairs of letters $\{1, 2\}$, $\{3, 4\}$, $\{5, 6\}$, $\{7, 8\}$, and it is antisymmetric with respect to permuting letters within each pair.

Here is a specific example of a configuration of four lines which has precisely one common transversal. Let

$$(x_1, x_2, x_3, x_4, x_5, x_6, x_7, x_8) = \begin{pmatrix} 1 & 0 & 0 & 0 & 0 & 1 & 0 & 1 \\ 0 & 1 & 0 & 0 & 1 & 0 & 2 & -1 \\ 0 & 0 & 1 & 0 & 1 & 0 & 1 & 0 \\ 0 & 0 & 0 & 1 & 0 & 1 & -1 & 2 \end{pmatrix}.$$

Then (3.4.20) specializes to $-(\lambda - \mu)^2$, which means the lines $x_1 \vee x_2$, $x_3 \vee x_4$, $x_5 \vee x_6$, $x_7 \vee x_8$ have precisely one common transversal.

Example 3.4.6 (Common transversals of five lines in P^3). We have seen in Example 3.4.5 that any four lines in P^3 have a common transversal. What is the situation for five lines? When do they have common transversals?

What is the algebraic condition for five lines $\ell_1, \ell_2, \ell_3, \ell_4, \ell_5$ in projective 3-space to be incident to a common sixth line ℓ?

Two equivalent answers to this question will be presented. For our first answer, we identify each line ℓ_i with the corresponding extensor $\ell_i = \sum_{1 \leq j < k \leq 4} \ell_i^{jk}$ $e_j \vee e_k$ in $\wedge_2 \mathbf{C}^4$. We call $(\ell_i^{12}, \ell_i^{13}, \ell_i^{14}, \ell_i^{23}, \ell_i^{24}, \ell_i^{34})$ the vector of *Plücker coordinates* for ℓ. Forming the meet of two lines defines an inner product on the space of lines:

$$[\ell_i, \ell_j] := \ell_i \vee \ell_j = \ell_i^{12} \ell_j^{34} - \ell_i^{13} \ell_j^{24} + \ell_i^{14} \ell_j^{23} + \ell_i^{23} \ell_j^{14} - \ell_i^{24} \ell_j^{13} + \ell_i^{34} \ell_j^{12}.$$

From our discussion above we derive the two basic facts about the inner product of lines:

1. Two lines ℓ_i and ℓ_j are incident if and only if $[\ell_i, \ell_j] = 0$.
2. An arbitrary six-dimensional vector $\ell = (\ell^{12}, \ell^{13}, \ell^{14}, \ell^{23}, \ell^{24}, \ell^{34})$ is the vector of Plücker coordinates of a line if and only if $[\ell, \ell] = 0$.

Here is the first solution to our problem. If ℓ is a common transversal, then it satisfies the following system of linear equations:

$$[\ell, \ell_1] = [\ell, \ell_2] = [\ell, \ell_3] = [\ell, \ell_4] = [\ell, \ell_5] = 0.$$

In the generic case this system has a unique solution vector ℓ which can be found by Cramer's rule:

$$\ell = \left(\det \begin{pmatrix} \ell_1^{12} & \ell_1^{13} & \ell_1^{14} & \ell_1^{23} & \ell_1^{24} \\ \vdots & \vdots & \vdots & \vdots & \vdots \\ \ell_5^{12} & \ell_5^{13} & \ell_5^{14} & \ell_5^{23} & \ell_5^{24} \end{pmatrix}, \det \begin{pmatrix} \ell_1^{12} & \ell_1^{13} & \ell_1^{14} & \ell_1^{23} & \ell_1^{34} \\ \vdots & \vdots & \vdots & \vdots & \vdots \\ \ell_5^{12} & \ell_5^{13} & \ell_5^{14} & \ell_5^{23} & \ell_5^{34} \end{pmatrix}, \right.$$

$$\det \begin{pmatrix} \ell_1^{12} & \ell_1^{13} & \ell_1^{14} & \ell_1^{24} & \ell_1^{34} \\ \vdots & \vdots & \vdots & \vdots & \vdots \\ \ell_5^{12} & \ell_5^{13} & \ell_5^{14} & \ell_5^{24} & \ell_5^{34} \end{pmatrix}, -\det \begin{pmatrix} \ell_1^{12} & \ell_1^{13} & \ell_1^{23} & \ell_1^{24} & \ell_1^{34} \\ \vdots & \vdots & \vdots & \vdots & \vdots \\ \ell_5^{12} & \ell_5^{13} & \ell_5^{23} & \ell_5^{24} & \ell_5^{34} \end{pmatrix},$$

$$\left. -\det \begin{pmatrix} \ell_1^{12} & \ell_1^{14} & \ell_1^{23} & \ell_1^{24} & \ell_1^{34} \\ \vdots & \vdots & \vdots & \vdots & \vdots \\ \ell_5^{12} & \ell_5^{14} & \ell_5^{23} & \ell_5^{24} & \ell_5^{34} \end{pmatrix}, -\det \begin{pmatrix} \ell_1^{13} & \ell_1^{14} & \ell_1^{23} & \ell_1^{24} & \ell_1^{34} \\ \vdots & \vdots & \vdots & \vdots & \vdots \\ \ell_5^{13} & \ell_5^{14} & \ell_5^{23} & \ell_5^{24} & \ell_5^{34} \end{pmatrix} \right).$$

This means that the lines $\ell_1, \ell_2, \ell_3, \ell_4, \ell_5$ are incident to common line if and only if the above vector satisfies

$$[\ell, \ell] = 2 \cdot \left(\ell^{12} \ell^{34} - \ell^{13} \ell^{24} + \ell^{14} \ell^{23} \right) = 0. \tag{3.4.21}$$

Notice that this expression is a polynomial condition of total degree 10 in the Plücker coordinates ℓ_i^{jk} of the five given lines. If it is satisfied, then the vector ℓ is the Plücker coordinate vector of the desired sixth line.

Let us now suppose that each line is given as the join of two distinct points $a_i = (a_i^1, a_i^2, a_i^3, a_i^4)$, $b_i = (b_i^1, b_i^2, b_i^3, b_i^4)$ in projective 3-space. We can easily express our condition as a polynomial in the coordinates of these ten points. If we replace $\ell_i^{jk} \to \det \begin{pmatrix} a_i^j & b_i^j \\ a_i^k & b_i^k \end{pmatrix}$ in Eq. (3.4.21), then we obtain the desired homogeneous polynomial $P(a_i^j, b_i^j)$ in 40 variables of total degree 20.

This situation is unsatisfactory both from a practical and a theoretical point of view. First, the polynomial P is so large that it cannot be written down in a nice way. Secondly, it would be desirable to rewrite the polynomial P as a polynomial function in the fundamental invariants of projective geometry, which are the brackets

$$[a_i b_i a_j b_j] := \det \begin{pmatrix} a_i^1 & b_i^1 & a_j^1 & b_j^1 \\ a_i^2 & b_i^2 & a_j^2 & b_j^2 \\ a_i^3 & b_i^3 & a_j^3 & b_j^3 \\ a_i^4 & b_i^4 & a_j^4 & b_j^4 \end{pmatrix}.$$

We wish to answer the following question.

3.4. Applications to projective geometry

Find an algebraic condition in terms of the brackets $[a_i b_i a_j b_j]$, $1 \leq i < j \leq 5$, which expresses the fact that the five lines $\overline{a_1 b_1}$, $\overline{a_2 b_2}$, $\overline{a_3 b_3}$, $\overline{a_4 b_4}$, $\overline{a_5 b_5}$ are incident to a common sixth line.

The following answer is due to Neil White (pers. comm.):

Theorem 3.4.7. Five lines $\overline{a_1 b_1}$, $\overline{a_2 b_2}$, $\overline{a_3 b_3}$, $\overline{a_4 b_4}$, $\overline{a_5 b_5}$ in projective 3-space have a common transversal if and only if

$$\det \begin{pmatrix} 0 & [a_1 b_1 a_2 b_2] & [a_1 b_1 a_3 b_3] & [a_1 b_1 a_4 b_4] & [a_1 b_1 a_5 b_5] \\ [a_2 b_2 a_1 b_1] & 0 & [a_2 b_2 a_3 b_3] & [a_2 b_2 a_4 b_4] & [a_2 b_2 a_5 b_5] \\ [a_3 b_3 a_1 b_1] & [a_3 b_3 a_2 b_2] & 0 & [a_3 b_3 a_4 b_4] & [a_3 b_3 a_5 b_5] \\ [a_4 b_4 a_1 b_1] & [a_4 b_4 a_2 b_2] & [a_4 b_4 a_3 b_3] & 0 & [a_4 b_4 a_5 b_5] \\ [a_5 b_5 a_1 b_1] & [a_5 b_5 a_2 b_2] & [a_5 b_5 a_3 b_3] & [a_5 b_5 a_4 b_4] & 0 \end{pmatrix} = 0.$$

Proof. If we use Plücker coordinates for the lines $\ell_i = \overline{a_i b_i}$, then we can use the identity $[a_i b_i a_j b_j] = [\ell_i, \ell_j]$ to rewrite all brackets as inner products of lines. The asserted condition translates to $\det([\ell_i, \ell_j]) = 0$ where $([\ell_i, \ell_j])$ denotes the 5×5-Gram matrix of the five lines. This 5×5-matrix can be written as the product of a 5×6-matrix with a 6×5-matrix:

$$\begin{pmatrix} \ell_1^{12} & \ell_1^{13} & \ell_1^{14} & \ell_1^{23} & \ell_1^{24} & \ell_1^{34} \\ \ell_2^{12} & \ell_2^{13} & \ell_2^{14} & \ell_2^{23} & \ell_2^{24} & \ell_2^{34} \\ \ell_3^{12} & \ell_3^{13} & \ell_3^{14} & \ell_3^{23} & \ell_3^{24} & \ell_3^{34} \\ \ell_4^{12} & \ell_4^{13} & \ell_4^{14} & \ell_4^{23} & \ell_4^{24} & \ell_4^{34} \\ \ell_5^{12} & \ell_5^{13} & \ell_5^{14} & \ell_5^{23} & \ell_5^{24} & \ell_5^{34} \end{pmatrix} \cdot \begin{pmatrix} \ell_1^{34} & \ell_2^{34} & \ell_3^{34} & \ell_4^{34} & \ell_5^{34} \\ -\ell_1^{24} & -\ell_2^{24} & -\ell_3^{24} & -\ell_4^{24} & -\ell_5^{24} \\ \ell_1^{14} & \ell_2^{14} & \ell_3^{14} & \ell_4^{14} & \ell_5^{14} \\ \ell_1^{23} & \ell_2^{23} & \ell_3^{23} & \ell_4^{23} & \ell_5^{23} \\ -\ell_1^{13} & -\ell_2^{13} & -\ell_3^{13} & -\ell_4^{13} & -\ell_5^{13} \\ \ell_1^{12} & \ell_2^{12} & \ell_3^{12} & \ell_4^{12} & \ell_5^{12} \end{pmatrix}.$$

Consider the vectors of 5×5-minors of these two matrices. Up to signs and reindexing, both vectors are equal to ℓ. By the Cauchy–Binet theorem, the dot product of these two vectors is equal to the determinant in question. Using the same notation as above, we can thus rewrite this determinant as

$$\ell^{12}\ell^{34} + \ell^{13}(-\ell^{24}) + \ell^{14}\ell^{23} + \ell^{23}\ell^{14} + \ell^{24}(-\ell^{13}) + \ell^{34}\ell^{12}.$$

This expression equals (3.4.21), and we are done. ◁

Exercises

(1) Derive (3.4.9) from (3.4.10).
(2) Compute the bracket polynomials P, Q in (3.4.12) and straighten them.
(3) * Consider the condition for ten points **a**, **b**, **c**, **d**, **e**, **f**, **g**, **h**, **i**, **j** in projective 3-space to lie on a quadric surface. In terms of coordinates this polynomial equals

$$\det \begin{pmatrix} a_1^2 & a_2^2 & a_3^2 & a_4^2 & a_1a_2 & a_1a_3 & a_1a_4 & a_2a_3 & a_2a_4 & a_3a_4 \\ b_1^2 & b_2^2 & b_3^2 & b_4^2 & b_1b_2 & b_1b_3 & b_1b_4 & b_2b_3 & b_2b_4 & b_3b_4 \\ c_1^2 & c_2^2 & c_3^2 & c_4^2 & c_1c_2 & c_1c_3 & c_1c_4 & c_2c_3 & c_2c_4 & c_3c_4 \\ d_1^2 & d_2^2 & d_3^2 & d_4^2 & d_1d_2 & d_1d_3 & d_1d_4 & d_2d_3 & d_2d_4 & d_3d_4 \\ e_1^2 & e_2^2 & e_3^2 & e_4^2 & e_1e_2 & e_1e_3 & e_1e_4 & e_2e_3 & e_2e_4 & e_3e_4 \\ f_1^2 & f_2^2 & f_3^2 & f_4^2 & f_1f_2 & f_1f_3 & f_1f_4 & f_2f_3 & f_2f_4 & f_3f_4 \\ g_1^2 & g_2^2 & g_3^2 & g_4^2 & g_1g_2 & g_1g_3 & g_1g_4 & g_2g_3 & g_2g_4 & g_3g_4 \\ h_1^2 & h_2^2 & h_3^2 & h_4^2 & h_1h_2 & h_1h_3 & h_1h_4 & h_2h_3 & h_2h_4 & h_3h_4 \\ i_1^2 & i_2^2 & i_3^2 & i_4^2 & i_1i_2 & i_1i_3 & i_1i_4 & i_2i_3 & i_2i_4 & i_3i_4 \\ j_1^2 & j_2^2 & j_3^2 & j_4^2 & j_1j_2 & j_1j_3 & j_1j_4 & j_2j_3 & j_2j_4 & j_3j_4 \end{pmatrix}.$$

(a) Rewrite this projective invariant as a bracket polynomial in $C[\Lambda(10, 4)]$ (Turnbull and Young 1926, White 1990).
(b) Does there exist a synthetic condition analogous to Pascal's theorem for 10 points in P^3 to lie on a quadric? (This is an unsolved geometry problem dating back to the 19th century; see Turnbull and Young (1926).)

(4) * Consider twelve points $x_1, x_2, \ldots, x_{12}$ in general position in the projective plane, let C_1 denote the quadric through x_1, x_2, x_3, x_4, x_5, let C_2 denote the quadric through $x_6, x_7, x_8, x_9, x_{10}$, and let ℓ denote the line through x_{11}, x_{12}. Find a bracket polynomial $R \in \mathcal{B}_{12,3}$ which vanishes if and only if $C_1 \cap C_2 \cap \ell \neq \emptyset$. (Hint: This is the *synthetic resultant* of two quadrics and a line. The degree of R in brackets equals 16.)

3.5. Cayley factorization

Cayley factorization stands for the problem of (re-)translating bracket polynomials into Grassmann–Cayley algebra expressions, or *Cayley expressions*, for short. This problem is much harder than its inverse, which is solved by Algorithm 3.3.4. As of today, no effective algorithm is known for the general Cayley factorization problem. An important partial result is Neil White's Cayley factorization algorithm for *multilinear* bracket polynomials (White 1991). In this section we give an exposition of this algorithm. Our first theorem, to be presented without proof, is the universal factorization result in Sturmfels and Whiteley (1991).

We have seen that there are four levels of description in projective geometry:

(1) Projective geometry
   $\updownarrow$
(2) Cayley algebra
   $\downarrow$ $\uparrow$ Cayley factorization
(3) Bracket algebra
   $\updownarrow$
(4) Coordinate algebra

3.5. Cayley factorization

We illustrate this diagram for the geometric statement in Examples 3.1.2 and 3.3.3:

(1) "The lines $\overline{ab}$, $\overline{cd}$ and $\overline{ef}$ are concurrent"
$\updownarrow$
(2) $(a \vee b) \wedge (c \vee d) \wedge (e \vee f) = 0$
$\downarrow$ $\uparrow$ Cayley factorization
(3) $[abe][dcf] - [abf][dce] = 0$
$\updownarrow$
(4) $a_1 b_2 e_3 c_1 d_2 f_3 + a_1 b_2 e_3 d_1 c_2 f_3 + \ldots$ (48 monomials) $\ldots = 0$

The translations (1) $\leftrightarrow$ (2) and (3) $\to$ (4) are straightforward. The arrow (4) $\to$ (3) is given by the First Fundamental Theorem (Algorithm 3.2.8), and the arrow (2) $\to$ (3) is the Cayley-bracket expansion (Algorithm 3.3.4). In what follows we will be concerned with the translation (3) $\to$ (2).

We define the *weight* of a tableaux $T \in \mathbf{C}[\Lambda(n, d)]$ as the vector $\omega(T) = (\omega_1, \ldots, \omega_n) \in \mathbf{N}^n$ where ω_i equals the number of occurrences of the letter i in T. A bracket polynomial $P \in \mathbf{C}[\Lambda(n, d)]$ is called *homogeneous* if each tableau in P has the same weight. Each syzygy in $I_{n,d}$ is homogeneous, and therefore the weight of a tableau and the property of being homogeneous depends only on the image in $\mathcal{B}_{n,d} = \mathbf{C}[\Lambda(n, d)]/I_{n,d}$.

Suppose that $C(a, b, c, \ldots)$ is a simple Cayley expression, i.e., it involves only join and meet, not addition. Let $P(a, b, c, \ldots)$ be its expansion in terms of brackets. Then $P(a, b, c, \ldots)$ is homogeneous, and the weight of $P(a, b, c, \ldots)$ counts the number of occurrences of $a, b, c, \ldots$ in $C(a, b, c, \ldots)$. A bracket polynomial or a simple Cayley expression is called *multilinear* if it is homogeneous of weight $(1, 1, \ldots, 1)$. We can now state our problem.

Cayley factorization problem
Input: A homogeneous bracket polynomial $P(a, b, c, \ldots)$.
Question: Does there exist a simple Cayley expression $C(a, b, c, \ldots)$ whose bracket expansion (modulo the syzygy ideal) is equal to $P(a, b, c, \ldots)$? If yes, output $C(a, b, c, \ldots)$; if no, output "NOT CAYLEY FACTORABLE".

A typical example of a successful Cayley factorization is the translation from (3.4.9) into (3.4.10), the synthetic condition for Pascal's theorem. Note that (3.4.9) has weight $(2, 2, 2, 2, 2, 2)$ and is therefore not multilinear. Before proceeding further, let us see that not all bracket polynomials – not even multilinear ones – are Cayley factorable.

Example 3.5.1. The multilinear bracket polynomial $P = [abc][def] + [abd][cef]$ does *not* factor in the Grassmann–Cayley algebra of rank 3. This can be seen by inspecting all possible multilinear simple Cayley expressions $C(a, b, c, d, e, f)$ of rank 3. None of these expressions is symmetric in two of its letters. However, the invariant P is antisymmetric in both (a, b) and (e, f) while it is symmetric in (c, d), and it is therefore not Cayley factorable. Another proof is given in Example 3.5.7.

This example has another more subtle aspect. If we multiply the given bracket polynomial by an appropriate tableau, then the resulting bracket polynomial becomes Cayley factorable. The expression $[acd][bcd] \cdot P$ does factor, as follows.

$$\{(ac \wedge bd) \vee (ad \wedge bc)\} \wedge ef \wedge cd$$
$$= \{([acb]d - [acd]b) \vee ([adb]c - [adc]b)\} \wedge \{[efc]d - [efd]c\}$$
$$= \{([acb][adb]dc - [acd][adb]bc - [acb][adc]db)\} \wedge \{[efc]d - [efd]c\}$$
$$= -[acd][adb][efc][bcd] + [acb][adc][efd][dbc]$$
$$= [acd][bcd]([abc][def] + [abd][cef]). \qquad (3.5.1)$$

This Cayley factorization would not be found by White's Algorithm 3.5.6 because the bracket polynomial in (3.5.1) is not multilinear. The multiplier tableau $[acd][bcd]$ corresponds to the subsidiary condition that both points a and b are not on the line through c and d. Under this non-degeneracy assumption we get a synthetic construction.

This example raises the question whether every homogeneous bracket polynomial with integer coefficients can be Cayley factored after a suitable multiplier has been chosen. Our first theorem states that this is true for rank $d \geq 3$.

Theorem 3.5.2 (Sturmfels and Whiteley 1991). Let $P(a, b, c, \ldots)$ be a homogeneous bracket polynomial with integer coefficients of rank $d \geq 3$. Then there exists a simple Cayley expression $C(a, b, c, \ldots)$ and a tableau $T(a, b, c, \ldots)$ such that the bracket expansion of C equals $T \cdot P$.

In this theorem we need the hypothesis $d \geq 3$ because there are no significant synthetic constructions on the projective line, except for coincidence of points. A rank 2 bracket polynomial $[ab][cd] - [ac][bd]$ will never factor to a synthetic construction unless the projective line is embedded into some higher-dimensional projective space. The proof of Theorem 3.5.2 is based on the classical construction of the arithmetic operations addition and multiplication in terms of synthetic projective geometry. We refer to Sturmfels and Whiteley (1991) and Bokowski and Sturmfels (1989: chapter 2) for details.

This factorization theorem is far too general to be of practical use. Our proof method generates a multiplier tableau T which has very high degree relative to P, and, the Cayley expression C does not tell the "true" synthetic interpretation of P, if such exists. Perhaps the main importance of Theorem 3.5.2 lies in the fact that it suggests the following problem.

Generalized Cayley factorization problem
Input: A homogeneous bracket polynomial $P(a, b, c, \ldots)$.
Question: Find a tableau $T(a, b, c, \ldots)$ of minimal degree such that $P \cdot T = C$ for some simple Cayley expression $C(a, b, c, \ldots)$.

3.5. Cayley factorization

We will now concentrate on the Cayley factorization problem for multilinear bracket polynomials. In order to state White's algorithm, we first derive a few structural properties of Cayley expressions. Suppose that $C(a, b, c, \ldots)$ is a simple and multilinear Cayley expression. We may represent C by a binary tree T whose leaves are labeled uniquely by $a, b, c, \ldots$ and whose inner nodes are labeled $\vee$ or $\wedge$. The step of a subtree T' of T is the step of the corresponding Cayley expression C'. We say that an operation $\vee$ or $\wedge$ in T is *trivial* if its operands have steps j and $d - j$ for some j, or, one of its operands has step 0 or d. We note that, under the numerical identification of steps 0 and d, a trivial $\vee$ may be replaced by a $\wedge$ and vice versa. Having a trivial operation amounts to having a subtree which evaluates to a bracket polynomial which factors out of C.

Lemma 3.5.3. Let $C(a, b, c, \ldots)$ be a non-zero multilinear simple Cayley expression. Then C is anti-symmetric in two arguments a and b if and only if a and b do not have a non-trivial $\wedge$ on the unique path joining them in the tree T which represents C.

Proof. Since C is multilinear, the tree T has two unique leaves labeled a and b respectively. If there are no non-trivial $\wedge$'s on the path from a to b in T, then, exchanging trivial $\wedge$'s for $\vee$'s, we may assume that there are only $\vee$'s on the path. Since $\vee$ is an associative, anti-commutative operation, we may rearrange C so that $(a \vee b)$ occurs explicitly. Thus C is anti-symmetric in a and b.

Conversely, suppose there is a non-trivial $\wedge$ on the path from a to b. Denote by $\wedge_1$ the first such. By modifying any trivial $\wedge$'s as above, we may assume that $C = ((a \vee S) \wedge_1 Z) \cdots$. We now specialize a few of the indeterminate points in S, so that S has the form $S = b \vee U$, with $a \notin \overline{U}$ and $b \notin \overline{U}$. This may be done inductively: if $S = X \vee Y$ then specialize either X or Y to have b as a join factor, while if $S = X \wedge Y$ then specialize both X and Y to have b as a join factor. This specialization has replaced some of the original points by b, and left all others indeterminate. Note that, by multilinearity, only points in S have been specialized to b. Denote by $\widehat{C} = \widehat{C}(a, b, c, \ldots)$ the image of C under this specialization. Observe that $\widehat{C}(b, a, c, \ldots) = 0$, because it has $b \vee b \vee U$ as a factor.

Now we claim that $\widehat{C}(a, b, c, \ldots) \neq 0$, which will prove that C is not anti-symmetric in a and b. First note that $(a \vee b \vee U) \wedge_1 Z)$ is a nonzero Cayley expression because b occurs at most once in Z and the operation $\wedge_1$ is nontrivial. Since the rest of $\widehat{C}$ contains entirely different letters, we conclude $\widehat{C} \neq 0$. ◁

Let $C(a, b, c, \ldots, z)$ be a simple Cayley expression. A subset of letters $\{a, b, \ldots, e\}$ is called an *atomic extensor* of C if its join $(a \vee b \vee \ldots \vee e)$ occurs explicitly in C. This is equivalent to saying that there is no non-trivial meet on the path (in the tree of C) between any two elements of $\{a, b, \ldots, e\}$. Note that each atomic extensor has cardinality at least 2. For instance, the Cayley expression $((a \vee b) \wedge (c \vee d)) \wedge (e \vee f)$ has the atomic extensors $\{a, b\}$, $\{c, d\}$ and $\{e, f\}$.

Let $P(a, b, c, \ldots, z)$ be a multilinear bracket polynomial. We define an equivalence relation $\sim$ on the letters occurring in P as follows. We set $a \sim b$ if $P(a, b, c, \ldots, z) = -P(b, a, c, \ldots, z)$, or, equivalently, if $P(a, a, c \ldots, z) = 0$. For instance, the bracket polynomial $[abc][def] - [abd][cef]$ has the equivalence classes $\{a, b\}$, $\{c, d\}$ and $\{e, f\}$. These are precisely the atomic extensors in its Cayley factorization. Note that the bracket polynomial $[abc][def] + [abd][cef]$ (discussed in Example 3.5.1) has the equivalence classes $\{a, b\}$, $\{e, f\}$, $\{c\}$ and $\{d\}$.

Corollary 3.5.4. Let P be a multilinear bracket polynomial which is Cayley factorable. Then each equivalence class of $\sim$ is an atomic extensor in some Cayley factorization of P.

Proof. If C is a simple Cayley expression which expands to P, and if A is an equivalence class of points under $\sim$, then by Lemma 3.5.3, there are no non-trivial meets between the points in A. Hence C may be rewritten so that the points of A are explicitly joined. Conversely, points in the same atomic extensor must be in the same equivalence class. ◁

Lemma 3.5.5. Let P be a multilinear bracket polynomial which is Cayley factorable, let A and B be atomic extensors of P, and suppose that $P = (A \wedge B) \vee Q$ for some linear combination Q of extensors with bracket coefficients. Then there exists a Cayley factorization $P = (A \wedge B) \vee U$ where U is a simple Cayley expression.

Proof. Let C be any Cayley factorization of P, and let T be its tree. We fix a generic hyperplane H in the ambient vector space V, and we specialize the points of A and B to be in generic position in H, with all other points remaining in generic position in V. Let $\widehat{P}, \widehat{C}$ and $\widehat{T}$ denote the images of P, C and T under this specialization. We identify A and B with an inner node in T (resp. $\widehat{T}$). Our hypothesis implies $\widehat{P} = 0$.

Suppose that A and B have a non-trivial join on the path between them in T, and therefore in $\widehat{T}$ also. Consider the node o_1 on the path from A to B which is nearest the root of $\widehat{T}$. If a non-trivial join not equal to o_1 occurs, then at most one of its operands is in H, and at least one of its operands is in generic position in V. The result of such a join is in generic position in V. Thus higher operations in $\widehat{T}$ never have more than one operand in H, and it follows that $\widehat{C} \neq 0$. The other case is that o_1 itself is a non-trivial join, both of whose operands are in generic position in H. The non-triviality of the join at o_1 in C implies that it is also non-trivial in $\widehat{C}$. Thus the result of the join has step at most $d - 1$, and is hence non-zero and in generic position in H. Again, higher operations in $\widehat{T}$ have at most one operand in H, and we again conclude that $\widehat{C} \neq 0$. In both cases we have contradicted $P = C$, hence there is no non-trivial join on the path from A to B. By associativity and anti-commutativity of the meet, we may rearrange so that $(A \wedge B)$ occurs explicitly in our Cayley factorization. ◁

3.5. Cayley factorization

Neither Lemma 3.5.3 nor Lemma 3.5.5 generalizes to non-multilinear Cayley expressions. These two lemmas are crucial for the correctness of the multilinear algorithm below, and their failure in the general case indicates that the general Cayley factorization problem is considerably harder than the multilinear problem. By a *primitive factor* of a simple Cayley expression C we mean an explicit subexpression $(E \wedge F)$ in C, where E and F are two atomic extensors. Note that any simple Cayley expression must have such a primitive factor, and Lemma 3.5.5 gives a criterion for detecting them in Cayley factorable multilinear bracket polynomials.

Algorithm 3.5.6 (Multilinear Cayley factorization) (White 1991).
Input: A multilinear bracket polynomial $P(a, b, c, \ldots)$ of rank d.
Output: A Cayley factorization C of P, if it exists; "NOT FACTORABLE" otherwise.

1. Find all atomic extensors of P. We defined $a \sim b$ by $P(a, a, c, \ldots) = 0$. This is checked using the straightening algorithm, where the transitivity of $\sim$ cuts down on the number of pairs of points we have to check. If there do not exist two atomic extensors whose sizes sum to at least d, then output "NOT FACTORABLE".
2. If there is an atomic extensor A of step d, then apply straightening with the d elements of A first in the linear order. The result has the bracket $[A]$ as an explicit factor, $P = [A] \cdot P'$. Remove A, store $P = A \wedge P'$, and proceed with the bracket polynomial P' replacing P. Repeat step 2 as appropriate. If $P = \pm 1$, then we are DONE, and the required Cayley factorization may be reconstructed.
3. Find two atomic extensors $E = \{e_1, e_2, \ldots, e_k\}$ and $F = \{f_1, f_2, \ldots, f_\ell\}$ with $k + \ell > d$ such that the criterion of Lemma 3.5.5 for $E \wedge F$ to be a primitive factor is satisfied: Apply the straightening algorithm to P, using an ordering in which $e_1 < \ldots < e_k < f_1 < \ldots < f_\ell$ comes first. Then the result has the form

$$\sum_{x_1,\ldots,y_1\ldots} \begin{bmatrix} e_1 & \ldots & \ldots & \ldots & \ldots & e_k & \overset{\bullet}{f}_1 & \ldots & \overset{\bullet}{f}_{d-k} \\ \overset{\bullet}{f}_{d-k+1} & \ldots & \overset{\bullet}{f}_\ell & x_1 & \ldots & \ldots & \ldots & \ldots & x_{2d-k-\ell} \\ y_1 & \ldots & \ldots & \ldots & \ldots & \ldots & \ldots & \ldots & y_d \\ \vdots & & \vdots & & & & & & \vdots \end{bmatrix}. \quad (3.5.2)$$

That is, every tableau has E in the first row, part of F filling up the rest of the first row, and the rest of F in the second row. The sum is over various terms with different choices for the x's and y's.

4. If such E and F do not exist, then return "NOT FACTORABLE". If they do exist, then choose new letters $g_1, g_2, \ldots, g_p$, where $p = k + \ell - d$, and store $G = E \wedge F$. Let G replace E and F in the collection of atomic extensors,

and proceed with

$$P' = \sum_{x_1,\ldots,y_1,\ldots} \begin{bmatrix} g_1 & \cdots & g_p & x_1 & \cdots & x_{2d-k-\ell} \\ y_1 & \cdots & \cdots & \cdots & \cdots & y_d \\ \vdots & & & \vdots & & \vdots \end{bmatrix} \quad (3.5.3)$$

where $x_1, \ldots, y_1, \ldots$ are the same as above.
5. Recompute the atomic extensors by trying to extend the current ones. Go to step 2.

The termination and correctness of this algorithm is implied by our two lemmas on the structure of multilinear simple Cayley expressions. Termination is guaranteed because the loops in steps 2 and 3 are over the finite set of atomic extensors, and both in step 2 and in step 4 the degree of the new bracket polynomial P' is one less than the degree of P. As for correctness, Lemma 3.5.3 implies that step 2 finds all atomic extensors of any potential Cayley factorization. Lemma 3.5.5 guarantees that no backtracking will be necessary in step 4. Once a tentative primitive factor $(E \wedge F)$ has been found in step 4, then $(E \wedge F)$ must be part of a Cayley factorization if one exists at all. We now illustrate how this algorithm works.

Example 3.5.7. We apply Algorithm 3.5.6 to the multilinear bracket polynomial $P = [abc][def] + [abd][cef]$ (cf. Example 3.5.1). Its atomic extensors are ab, ef, c and d. In step 2 we see that there is no atomic extensor of cardinality $3 = \text{rank}(P)$, and our unique choice in step 3 is the pair of extensors ab and ef. We apply the straightening algorithm in the ordering $a < b < e < f < c < d$ to P. The output desired in (3.5.2) would have each monomial equal to $[abe][f\ldots]$ or $[abf][e\ldots]$. This is not the case, because we get the output

$$[abe][fcd] - [abf][ecd] + 2[abc][efd].$$

Therefore we conclude in step 4 that P is not Cayley factorable. ◁

Example 3.5.8. We apply Algorithm 3.5.6 to the multilinear bracket polynomial

$$\begin{aligned} P = &[abc][def][ghk] - [abc][deg][fhk] \\ &- [abd][cef][ghk] + [abd][ceg][fhk]. \end{aligned} \quad (3.5.4)$$

The atomic extensors are ab, cd, e, fg and hk, so we skip step 2. In step 3 we choose the pair of atomic extensors ab and cd, and we apply the straightening algorithm in the usual lexicographic order. Here (3.5.4) is already standard, and it is of the form required in (3.5.2):

$$P = [ab \, \dot{c}][\dot{d} \, ef][ghk] - [ab \, \dot{c}][\dot{d} \, eg][fhk].$$

3.6. Invariants and covariants of binary forms

In step 4 we now choose a new letter u, we store $u = (ab \wedge cd)$, and we proceed with $P' = [uef][ghk] - [ueg][fhk]$. In step 5 we compute the atomic extensors ue, fg and hk, we skip step 2, and we then apply step 3 to ue and fg to get $P' = [ue \; \overset{\bullet}{f}][\overset{\bullet}{g} \; hk]$. In step 4 we now choose a new letter $\overset{\smile}{u}$, we store $v = (ue \wedge fg)$, and we proceed with $P'' = [vhk]$. Now the unique atomic extensor is vhk, and we are done after a single application of step 2. Our final result is the Cayley factorization

$$C = v \wedge hk = (ue \wedge fg) \wedge hk = (((ab \wedge cd) \vee e) \wedge fg) \wedge hk. \quad (3.5.5)$$

In this example it was relatively easy to guess the answer directly from the input. Note, however, that P could have been presented in a completely different form. For instance, if we switch a and k in the linear ordering of letters, then the straightened form of (3.5.4) has 19 tableaux, and the Cayley factorization (3.5.5) becomes far from obvious. ◁

The main bottleneck in Algorithm 3.5.6 is the application of the straightening algorithm in step 1 and step 3. McMillen and White (1991) have given a variant of the straightening algorithm which performs much better in step 3 – see White (1991) – and which is also of general invariant-theoretic interest. This variant is called the *dotted straightening algorithm*, and it can be understood as a special case of the straightening law in the superalgebra due to Grosshans et al. (1987).

Exercises

(1) What is the geometric interpretation of the Cayley expression in (3.5.1)?
(2) * Find a bracket polynomial P and a single bracket T such that P is not Cayley factorable but $P \cdot T$ is Cayley factorable.
(3) A Cayley factorable bracket polynomial generally has many distinct Cayley factorizations. For instance, another Cayley factorization of (3.5.5) is $(((fg \wedge hk) \vee e) \wedge ab) \wedge cd$.
   (a) Find all distinct Cayley factorizations of the Pascal condition (3.4.9).
   (b) Give an algorithm which finds all Cayley factorizations of a multilinear bracket polynomial.
(4) Compute the (rank 4) bracket expansion of $(((cdf \wedge agh) \vee bij) \wedge klm) \wedge eno$, and apply Algorithm 3.5.6 to the result.

3.6. Invariants and covariants of binary forms

The invariant theory of binary forms is a central chapter of classical invariant theory. It links the theory of projective invariants which was studied in the previous sections with the invariants and covariants of general polynomial systems. It is our principal objective to study binary forms from the perspective of computer algebra. The general path of our exposition follows Kung and Rota (1984), with the main difference that we avoid the use of the umbral calculus.

At this point we also refrain from employing techniques from the representation theory of $GL(\mathbf{C}^2)$, as these will be easier to understand (and appreciate) within the general context of Chap. 4. Instead we discuss many elementary examples, with a particular emphasis on projective geometry, and we illustrate the use of Gröbner bases as a tool for studying invariants of binary forms. In Sect. 3.7 we will prove Gordan's finiteness theorem, derive Kempe's circular straightening algorithm, and determine fundamental sets of covariants for binary forms of low degree.

Our investigations will concern those properties of polynomial functions on the complex projective line P^1 which are independent of the choice of coordinates. A *binary form* of *degree n* is a homogeneous polynomial

$$f(x, y) = \sum_{k=0}^{n} \binom{n}{k} a_k x^k y^{n-k} \tag{3.6.1}$$

in two complex variables x and y. The numbers a_k are the *coefficients* of $f(x, y)$. For mainly technical reasons the coefficients are scaled by binomial factors $\binom{n}{k}$ in (3.6.1). A *linear change of variables* (c_{ij}) is a transformation of the variables x and y given by

$$x = c_{11}\bar{x} + c_{12}\bar{y}, \quad y = c_{21}\bar{x} + c_{22}\bar{y} \tag{3.6.2}$$

such that the determinant of the transformation matrix, $c_{11}c_{22} - c_{12}c_{21}$, is nonzero. Under a linear change of variables (3.6.2), the binary form $f(x, y)$ is transformed into another binary form $\bar{f}(\bar{x}, \bar{y})$ in the new variables $\bar{x}$ and $\bar{y}$. It is defined by

$$\bar{f}(\bar{x}, \bar{y}) = \sum_{k=0}^{n} \binom{n}{k} a_k (c_{11}\bar{x} + c_{12}\bar{y})^k (c_{21}\bar{x} + c_{22}\bar{y})^{n-k}. \tag{3.6.3}$$

After expanding and regrouping terms, we obtain a binary form

$$\bar{f}(\bar{x}, \bar{y}) = \sum_{k=0}^{n} \binom{n}{k} \bar{a}_k \bar{x}^k \bar{y}^{n-k} \tag{3.6.4}$$

in the new variables $\bar{x}$ and $\bar{y}$. The new coefficients $\bar{a}_k$ are linear combinations of the a_i whose coefficients are polynomials in the c_{ij}. This representation is described explicitly in the following proposition, whose proof we omit.

Proposition 3.6.1. The coefficients $\bar{a}_k$ of the transformed binary form $\bar{f}(\bar{x}, \bar{y})$ satisfy

$$\bar{a}_k = \sum_{i=0}^{n} \left(\sum_{j=\max(0,i-n+k)}^{\min(i,k)} \binom{k}{j} \binom{n-k}{i-j} c_{11}^j c_{12}^{i-j} c_{21}^{k-j} c_{22}^{n-k-i+j} \right) a_i \tag{3.6.5}$$

for $k = 0, 1, 2, \ldots, n$.

3.6. Invariants and covariants of binary forms

Changing our point of view slightly, we now assume that the coefficients $a_0, a_1, \ldots, a_n$ in (3.6.1) are algebraically independent variables over the ground field **C**. Consider the polynomial ring $\mathbf{C}[a_0, a_1, \ldots, a_n, x, y]$ in both the old variables x and y and the new variables a_i. The group $GL(\mathbf{C}^2)$ of invertible 2×2-matrices (c_{ij}) acts linearly on the $(n+3)$-dimensional vector space spanned by $a_0, \ldots, a_n, x, y$. This action $x \mapsto \bar{x}, y \mapsto \bar{y}, a_i \mapsto \bar{a}_i$ is described by the formulas (3.6.2) and (3.6.5). It extends to an action of $GL(\mathbf{C}^2)$ on the polynomial ring $\mathbf{C}[a_0, \ldots, a_n, x, y]$.

Example 3.6.2. Consider the case $n = 2$ of a binary quadric $f(x, y) = a_2 x^2 + 2a_1 xy + a_0 y^2$. The linear transformation group $GL(\mathbf{C}^2)$ acts on $\mathbf{C}[a_0, a_1, a_2, x, y]$ via

$$(c_{ij}): \begin{pmatrix} x \\ y \end{pmatrix} \mapsto \begin{pmatrix} \bar{x} \\ \bar{y} \end{pmatrix} = \begin{pmatrix} c_{11} & c_{12} \\ c_{21} & c_{22} \end{pmatrix}^{-1} \begin{pmatrix} x \\ y \end{pmatrix},$$

$$\begin{pmatrix} a_0 \\ a_1 \\ a_2 \end{pmatrix} \mapsto \begin{pmatrix} \bar{a}_0 \\ \bar{a}_1 \\ \bar{a}_2 \end{pmatrix} = \begin{pmatrix} c_{22}^2 & 2c_{12}c_{22} & c_{12}^2 \\ c_{21}c_{22} & c_{11}c_{22} + c_{12}c_{21} & c_{11}c_{12} \\ c_{21}^2 & 2c_{11}c_{21} & c_{11}^2 \end{pmatrix} \begin{pmatrix} a_0 \\ a_1 \\ a_2 \end{pmatrix}.$$

Here the coordinate vector (x, y) is transformed by the *inverse* matrix of (c_{ij}), while the coefficient vector (a_0, a_1, a_2) is transformed by a 3×3-matrix which is the second symmetric power of (c_{ij}).

It is our objective to study and characterize the subring $\mathbf{C}[a_0, a_1, \ldots, a_n, x, y]^{GL(\mathbf{C}^2)}$ of relative invariants with respect to the linear transformation group $GL(\mathbf{C}^2)$. A polynomial $I \in \mathbf{C}[a_0, a_1, \ldots, a_n, x, y]$ is said to be a *covariant* of index g if

$$I(\bar{a}_0, \bar{a}_1, \ldots, \bar{a}_n, \bar{x}, \bar{y}) = (c_{11}c_{22} - c_{21}c_{22})^g \cdot I(a_0, a_1, \ldots, a_n, x, y).$$

A covariant $I \in \mathbf{C}[a_0, a_1, \ldots, a_n, x, y]$ is *homogeneous* if it is homogeneous both as a polynomial in the variables $a_0, a_1, \ldots, a_n$ and in the variables x, y. In that case the total degree of I in $a_0, a_1, \ldots, a_n$ is called the *degree* of the covariant I, and its total degree in x, y is called the *order* of I. A covariant of order 0, that is, a covariant $I \in \mathbf{C}[a_0, a_1, \ldots, a_n]$ with no occurrences of the variables x and y, is said to be an *invariant*.

In many situations it becomes necessary to consider a collection of binary forms

$$f_i(x, y) = \sum_{k=0}^{n_i} \binom{n_i}{k} a_{ik} x^k y^{n_i - k}, \quad i = 1, 2, \ldots, r.$$

As before we assume that the a_{ik} are algebraically independent over **C**, and we consider the natural action of the group $GL(\mathbf{C}^2)$ of invertible 2×2-matrices on the polynomial ring in x, y and the a_{ik}. A polynomial $I \in \mathbf{C}[a_{10}, a_{11}, \ldots, a_{1n_1}, \ldots, a_{r0}, a_{r1}, \ldots, a_{rn_r}, x, y]$ is called a *joint covariant* of the forms

$f_1, f_2, \ldots, f_r$ if it is a relative invariant of the $GL(\mathbf{C}^2)$-action. We say that I is a *joint invariant* of $f_1, f_2, \ldots, f_r$ if I does not depend on x and y at all.

Whenever a (joint) covariant $I(a_{ik}, x, y)$ vanishes identically under some specialization of the a_{ik}, this means that a certain *geometric condition* is satisfied by the specialized forms $f_1, f_2, \ldots, f_r$. Conversely, every geometric (i.e., invariant) condition in the coefficients of $f_1, f_2, \ldots, f_r$ is expressible as a boolean combination of a finite set of covariants.

Let us take a look at some examples of covariants of binary forms. We will be particularly interested in the geometric meaning of these covariants. Consider a binary quadric

$$f_2 = a_2 x^2 + 2a_1 xy + a_0 y^2,$$

a binary cubic

$$f_3 = b_3 x^3 + 3b_2 x^2 y + 3b_1 xy^2 + b_0 y^3,$$

and a binary quartic

$$f_4 = c_4 x^4 + 4c_3 x^3 y + 6c_2 x^2 y^2 + 4c_1 xy^3 + c_0 y^4.$$

(a) We consider the polynomial $D_2(a_0, a_1, a_2) := a_0 a_2 - a_1^2$ in the coefficients of the binary quadric. In order to determine whether D_2 is an invariant, we replace a_0 by $\bar{a}_0 = c_{22}^2 a_0 + 2c_{12}c_{22}a_1 + c_{12}^2 a_2$, we replace a_1 by $\bar{a}_1 = c_{21}c_{22}a_0 + c_{11}c_{22}a_1 + c_{12}c_{21}a_1 + c_{11}c_{12}a_2$, and we replace a_2 by $\bar{a}_2 = c_{21}^2 a_0 + 2c_{11}c_{21}a_1 + c_{11}^2 a_2$. As the result we obtain

$$\begin{aligned}
D_2(\bar{a}_0, \bar{a}_1, \bar{a}_2) &= (c_{22}^2 a_0 + 2c_{12}c_{22}a_1 + c_{12}^2 a_2)(c_{21}^2 a_0 + 2c_{11}c_{21}a_1 + c_{11}^2 a_2) \\
&\quad - (c_{21}c_{22}a_0 + c_{11}c_{22}a_1 + c_{12}c_{21}a_1 + c_{11}c_{12}a_2)^2 \\
&= (c_{22}c_{11} - c_{21}c_{12})^2 (a_0 a_2 - a_1)^2 \\
&= (c_{22}c_{11} - c_{21}c_{12})^2 D_2(a_0, a_1, a_2).
\end{aligned}$$

This identity shows that D_2 is a covariant of degree 2, of order 0, and of index 2. The invariant D_2 is called the *discriminant* of the binary quadric f_2. It vanishes for a binary quadric f_2 if and only if f_2 has a double root on the projective line P^1.

(b) The *discriminant*

$$D_3(b_0, b_1, b_2, b_3) := -9b_1^2 b_2^2 - 54b_1 b_2 b_3 b_0 + 27b_3^2 b_0^2 + 108b_2^3 b_0 + 4b_1^3 b_3$$

of the binary cubic f_3 is an invariant of degree 4 and index 6. It vanishes if and only if f_3 has a double root.

3.6. Invariants and covariants of binary forms

(c) Our next example is the *Hessian*

$$H_3(b_0, b_1, b_2, b_3, x, y) := \det \begin{pmatrix} \frac{\partial^2 f_3}{\partial x^2} & \frac{\partial^2 f_3}{\partial x \partial y} \\ \frac{\partial^2 f_3}{\partial x \partial y} & \frac{\partial^2 f_3}{\partial y^2} \end{pmatrix}$$

of the binary cubic f_3. It is a covariant of degree 2, index 2, and order 2. By expansion we find

$$\frac{1}{36} H_3 = (b_3 b_1 - b_2^2) x^2 + (-b_2 b_1 + b_3 b_0) xy + (-b_1^2 + b_2 b_0) y^2.$$

The Hessian vanishes identically (i.e., all three coefficient polynomials are zero) if and only if f_3 has a triple root.

(d) Next consider the polynomial

$$R_{23}(a_0, a_1, a_2, b_0, b_1, b_2, b_3) =$$
$$b_0^2 a_2^3 - 6 b_0 a_2^2 b_2 a_0 + 6 b_0 a_2 b_3 a_0 a_1 - 6 b_2 a_0^2 b_3 a_1 - 6 a_1 b_1 a_2^2 b_0$$
$$- 18 a_1 b_1 a_2 b_2 a_0 + 9 b_2^2 a_0^2 a_2 + 12 a_1^2 b_1 b_3 a_0 + 12 a_1^2 b_2 a_2 b_0 - 8 a_1^3 b_3 b_0$$
$$+ 9 a_0 b_1^2 a_2^2 - 6 a_0^2 b_1 a_2 b_3 + a_0^3 b_3^2.$$

This is the *Sylvester resultant* of the quadric f_2 and the cubic f_3. The resultant R_{23} is our first example of a joint invariant. It vanishes if and only if f_2 and f_3 have a root in common.

(e) The *Hessian*

$$H_4(c_0, c_1, c_2, c_3, c_4, x, y) := \det \begin{pmatrix} \frac{\partial^2 f_4}{\partial x^2} & \frac{\partial^2 f_4}{\partial x \partial y} \\ \frac{\partial^2 f_4}{\partial x \partial y} & \frac{\partial^2 f_4}{\partial y^2} \end{pmatrix}$$

of the quartic f_4 expands to

$$(c_4 c_2 - c_3^2) x^4 + (2 c_4 c_1 - 2 c_3 c_2) x^3 y + (2 c_3 c_1 + c_4 c_0 - 3 c_2^2) x^2 y^2$$
$$+ (2 c_3 c_0 - 2 c_2 c_1) x y^3 + (c_2 c_0 - c_1^2) y^4$$

after division by the constant 144. The Hessian H_4 is a covariant of degree 2, order 4 and index 6 which vanishes (identically) if and only if all four roots of f_4 coincide.

(f) Our next invariant is the *catalecticant*

$$C_4(c_0, c_1, c_2, c_3, c_4) := \det \begin{pmatrix} c_0 & c_1 & c_2 \\ c_1 & c_2 & c_3 \\ c_2 & c_3 & c_4 \end{pmatrix}$$
$$= c_0 c_2 c_4 - c_0 c_3^2 - c_1^2 c_4 + 2 c_1 c_2 c_3 - c_2^3.$$

This invariant cannot be interpreted easily in terms of the roots. For an interpretation we need the general fact that every binary quartic can be written as a sum

$$f_4(x, y) = (\mu_1 x - \nu_1 y)^4 + (\mu_2 x - \nu_2 y)^4 + (\mu_3 x - \nu_3 y)^4$$

of three perfect powers. Now the catalecticant C_4 of a quartic $f_4(x, y)$ vanishes if and only if $f_4(x, y)$ can be expressed as the sum of only *two* perfect powers. A similar invariant C_{2m} exists for every binary form of even degree.

(g) Finally, we consider the remainder obtained by dividing $f_4(x, y)$ into $f_3(x, y)$ with respect to the variable x. After factoring out y^2, this remainder equals

$$F_{34}(b_0, b_1, b_2, b_3, c_0, c_1, c_2, c_3, c_4) =$$
$$(6b_3^2 c_2 - 3b_3 b_1 c_4 - 12 b_2 b_3 c_3 + 9 b_2^2 c_4) x^2$$
$$+ (-b_3 b_0 c_4 - 12 b_1 b_3 c_3 + 9 b_1 b_2 c_4 + 4 b_3^2 c_1) xy$$
$$+ (b_3^2 c_0 + 3 b_0 b_2 c_4 - 4 b_0 b_3 c_3) y^2.$$

This is a joint covariant of the cubic f_3 and the quartic f_4. It vanishes if and only if f_3 divides f_4, or, geometrically, if all three roots of f_3 are also roots of f_4. In computer algebra (cf. Loos 1982) the polynomial F_{34} is known as the *first principal subresultant* in the remainder sequence of f_4 and f_3.

Since the ground field **C** is algebraically closed, every binary form of degree n can be factored into n linear factors. The coefficients $a_0, a_1, \ldots, a_n$ of the binary form

$$f(x, y) = \sum_{k=0}^{n} \binom{n}{k} a_k x^k y^{n-k} \qquad (3.6.6)$$
$$= (\mu_1 x - \nu_1 y)(\mu_2 x - \nu_2 y) \cdots (\mu_n x - \nu_n y)$$

can thus be expressed as polynomial functions in the roots (ν_1, μ_1), (ν_2, μ_2), $\ldots$, (ν_n, μ_n) of $f(x, y)$. Here (ν_i, μ_i) is the homogeneous coordinate vector of the i-th root of f on the projective line P^1. In the examples (a)–(e) and (f) discussed above we gave a geometric interpretation of the given covariant in terms of the roots of the binary forms. Here we go one step further: Using the root representation derived from (3.6.6), we shall characterize covariants as those bracket polynomials in the roots which satisfy an easy regularity and symmetry condition.

We treat the coordinates $\mu_1, \nu_1, \mu_2, \nu_2, \ldots, \mu_n, \nu_n$ as algebraically independent variables over **C**, and we denote with $\mathbf{C}[\mu_1, \nu_1, \ldots, \mu_n, \nu_n, x, y]$ the ring of polynomials in both the homogenized roots and the original variables x and y. The expansion in terms of roots defines a **C**-algebra homomorphism

3.6. Invariants and covariants of binary forms

$$\Psi : \mathbf{C}[a_0, a_1, \ldots, a_n, x, y] \to \mathbf{C}[\mu_1, \nu_1, \ldots, \mu_n, \nu_n, x, y]$$

$$a_{n-k} \mapsto \frac{(-1)^k}{n!} \sum_{\pi \in S_n} \nu_{\pi(1)} \cdots \nu_{\pi(k)} \mu_{\pi(k+1)} \cdots \mu_{\pi(n)}$$

$$= \frac{(-1)^k}{n!} \mu_1 \cdots \mu_n \cdot \sigma_k\left(\frac{\nu_1}{\mu_1}, \ldots, \frac{\nu_n}{\mu_n}\right).$$

(3.6.7)

Here σ_k denotes the k-th elementary symmetric function in n variables. The image $\Psi(I)$ of a polynomial $I \in \mathbf{C}[a_0, a_1, \ldots, a_n, x, y]$ under the expansion map Ψ is called the *representation* of I *in terms of homogenized roots*. The following lemma states that each covariant can be recovered uniquely from this representation.

Lemma 3.6.3. *The expansion homomorphism Ψ is injective.*

Proof. Let P be any polynomial in the coefficients and suppose that $\Psi(P) = 0$ as a polynomial in the homogenized roots. Let now $a_0, a_1, \ldots, a_n \in \mathbf{C}$ be arbitrary field elements, and factor the corresponding binary form into linear factors $(\mu_i x - \nu_i y)$ as in (3.6.6). Here μ_i and ν_i are elements of the ground field $\mathbf{C}$. From the definition of Ψ we get the following identity in the bivariate polynomial ring $\mathbf{C}[x, y]$:

$$P(\ldots, a_k, \ldots, x, y)$$
$$= P\left(\ldots, \frac{(-1)^{n-k}}{n!} \sum_{\pi \in S_n} \mu_{\pi(1)} \cdots \mu_{\pi(k)} \nu_{\pi(k+1)} \cdots \nu_{\pi(n)}, \ldots, x, y\right)$$
$$= [\Psi(P)](\mu_1, \nu_1, \ldots, \mu_n, \nu_n, x, y) = 0.$$

This proves that the polynomial P is equal to zero in the ring $\mathbf{C}[a_0, a_1, \ldots, a_n, x, y]$. ◁

This embedding raises the question under which conditions a polynomial in the homogenized roots can be expressed in terms of the coefficients $a_0, a_1, \ldots, a_n$. The answer is given in our next result, which is a homogenized version of the main theorem of symmetric functions (Theorem 1.1.1). We first need the following definitions. A monomial

$$M = \mu_1^{u_1} \mu_2^{u_2} \cdots \mu_n^{u_n} \nu_1^{v_1} \nu_2^{v_2} \cdots \nu_n^{v_n} x^{w_1} y^{w_2}$$

in the homogenized roots is said to be *regular of degree d* provided

$$u_1 + v_1 = u_2 + v_2 = \ldots = u_n + v_n = d.$$

A polynomial $R \in \mathbf{C}[\mu_1, \nu_1, \ldots, \mu_n, \nu_n, x, y]$ is *regular (of degree d)* if every monomial in the expansion of R is regular (of degree d). A polynomial $R(\mu_1, \nu_1, \ldots, \mu_n, \nu_n, x, y)$ is said to be *symmetric* provided

$$R(\mu_1, \nu_1, \ldots, \mu_n, \nu_n, x, y) = R(\mu_{\pi(1)}, \nu_{\pi(1)}, \ldots, \mu_{\pi(n)}, \nu_{\pi(n)}, x, y)$$

for all permutations $\pi \in S_n$.

Proposition 3.6.4. A polynomial $R \in \mathbf{C}[\mu_1, \nu_1, \ldots, \mu_n, \nu_n, x, y]$ is contained in the image of the expansion map Ψ if and only if R is regular and symmetric.

Proof. We see from (3.6.7) that the images of the variables $a_0, a_1, \ldots, a_n, x, y$ under Ψ are both regular and symmetric. Since the property to be regular and symmetric is preserved under addition and multiplication, it follows that every element of image(Ψ) is regular and symmetric.

To prove the converse, let R be symmetric and regular of degree d. Then R can be rewritten as

$$R(\mu_1, \nu_1, \ldots, \mu_n, \nu_n, x, y) = (\mu_1 \ldots \mu_n)^d \cdot \widehat{R}\Big(\frac{\nu_1}{\mu_1}, \ldots, \frac{\nu_n}{\mu_n}, x, y\Big),$$

where $\widehat{R}$ is a symmetric function (in the usual sense) in the ratios $\frac{\nu_i}{\mu_i}$. By Theorem 1.1.1, we can write $\widehat{R}$ as a polynomial Q with coefficients in $\mathbf{C}[x, y]$ in the elementary symmetric functions $\sigma_k(\frac{\nu_1}{\mu_1}, \ldots, \frac{\nu_n}{\mu_n})$. Multiplying Q by $(\mu_1 \ldots \mu_n)^d$ and distributing factors of $\mu_1 \ldots \mu_n$, we obtain a representation of R as a polynomial function in the magnitudes $\mu_1 \cdots \mu_n \sigma_k(\frac{\nu_1}{\mu_1}, \ldots, \frac{\nu_n}{\mu_n})$. By (3.6.7) this completes the proof of Proposition 3.6.4. ◁

In the following we give Gröbner bases illustrations of Lemma 3.6.3 and Proposition 3.6.4 for the special case of a binary quadric

$$f(x, y) = a_2 x^2 + 2a_1 xy + a_0 y^2 = \mu_1 \mu_2 x^2 - (\mu_1 \nu_2 + \nu_1 \mu_2) xy + \nu_1 \nu_2 y^2.$$

In order to invert the expansion homomorphism

$$\Psi : \mathbf{C}[a_0, a_1, a_2, x, y] \to \mathbf{C}[\mu_1, \nu_1, \mu_2, \nu_2, x, y]$$

we consider the ideal

$$I := \langle a_2 - \mu_1 \mu_2, \ 2a_1 + \mu_1 \nu_2 + \nu_1 \mu_2, \ a_0 - \nu_1 \nu_2 \rangle$$

in the polynomial ring $\mathbf{C}[a_0, a_1, a_2, \mu_1, \nu_1, \mu_2, \nu_2]$. This ideal is the vanishing ideal of the graph of the map Ψ. We compute the Gröbner basis

$$\mathcal{G} = \{\underline{\mu_1 \mu_2} - a_2, \quad \underline{\mu_1 \nu_2} + \nu_1 \mu_2 + 2a_1, \quad \underline{\mu_1 a_0} + \nu_1^2 \mu_2 + 2\nu_1 a_1,$$
$$\underline{\mu_2^2 \nu_1} + 2\mu_2 a_1 + \nu_2 a_2, \quad \underline{\mu_2^2 a_0} + 2\mu_2 \nu_2 a_1 + \nu_2^2 a_2, \quad \underline{\nu_1 \nu_2} - a_0\}$$

3.6. Invariants and covariants of binary forms

of I with respect to the lexicographic order induced from $\mu_1 > \mu_2 > \nu_1 > \nu_2 > a_0 > a_1 > a_2$. The injectivity of the expansion map Ψ (Lemma 3.6.3) is equivalent to the fact that no polynomial from $\mathbf{C}[a_0, a_1, a_2]$ appears in the Gröbner basis $\mathcal{G}$. As an example for inverting Ψ we consider the polynomial

$$R = \mu_1^2 \nu_2^2 - \nu_1 \mu_1 \mu_2 \nu_2 + \nu_1^2 \mu_2^2 \in \mathbf{C}[\mu_1, \nu_1, \mu_2, \nu_2]$$

which is symmetric and regular of degree 2. Taking the normal form of R modulo $\mathcal{G}$ we obtain its unique preimage $\Psi^{-1}(R) = 4a_1^2 - 3a_0 a_2$.

We now return to the general case, and we consider the action by linear substitution of the group $GL(\mathbf{C}^2) = \{(c_{ij})\}$ on the polynomial ring $\mathbf{C}[\mu_1, \nu_1, \ldots, \mu_n, \nu_n, x, y]$ in the homogenized roots. For any invertible 2×2-matrix (c_{ij}) the resulting transformation is described by the formulas

$$\begin{pmatrix} \nu_i \\ \mu_i \end{pmatrix} \mapsto \begin{pmatrix} \bar{\nu}_i \\ \bar{\mu}_i \end{pmatrix} = \begin{pmatrix} c_{22} & -c_{12} \\ -c_{21} & c_{11} \end{pmatrix} \begin{pmatrix} \nu_i \\ \mu_i \end{pmatrix} \quad \text{and}$$

$$\begin{pmatrix} x \\ y \end{pmatrix} \mapsto \begin{pmatrix} \bar{x} \\ \bar{y} \end{pmatrix} = \frac{1}{c_{11}c_{22} - c_{12}c_{21}} \begin{pmatrix} c_{22} & -c_{12} \\ -c_{21} & c_{11} \end{pmatrix} \begin{pmatrix} x \\ y \end{pmatrix}. \quad (3.6.8)$$

In order to characterize the relative invariants of this $GL(\mathbf{C}^2)$-action, we use the bracket notation for all vectors in question. As in the previous sections we set

$$[i\ j] := \mu_i \nu_j - \nu_i \mu_j \quad \text{and} \quad [i\ \mathbf{u}] := \mu_i y - \nu_i x$$

for $i, j \in \{1, \ldots, n\}$. The subring generated by these brackets in $\mathbf{C}[\mu_1, \nu_1, \ldots, \mu_n, \nu_n, x, y]$ is called the *bracket ring*. Every homogeneous bracket polynomial R is a relative $GL(\mathbf{C}^2)$-invariant, which means that there exists an integer $g \in \mathbf{N}$ such that $(c_{ij}) \circ R = \det(c_{ij})^g \cdot R$ for all linear transformations $(c_{ij}) \in GL(\mathbf{C}^2)$. Also the converse is true.

Lemma 3.6.5. A homogeneous polynomial R in $\mathbf{C}[\mu_1, \nu_1, \ldots, \mu_n, \nu_n, x, y]$ is a relative $GL(\mathbf{C}^2)$-invariant if and only if R is contained in the bracket ring.

Proof. If we restrict the linear action to the subgroup $SL(\mathbf{C}^2)$ of unimodular matrices, then we obtain the familiar action by right multiplication on the generic $(n+1) \times 2$-matrix

$$\begin{pmatrix} \nu_1 & \mu_1 \\ \vdots & \vdots \\ \nu_n & \mu_n \\ x & y \end{pmatrix}.$$

By the First Fundamental Theorem (Theorem 3.2.1), every invariant $R(\mu_1, \nu_1, \ldots, \nu_n, x, y)$ can be written as a polynomial in the brackets

$$[1\ 2], [1\ 3], \ldots [1\ n], [2\ 3], \ldots [n-1\ n], [1\ \mathbf{u}], [2\ \mathbf{u}], \ldots [n\ \mathbf{u}]. \quad \triangleleft$$

We are now ready to prove the main result of this section.

Theorem 3.6.6. The expansion map Ψ defines an isomorphism between the subring of covariants in $\mathbf{C}[a_0, a_1, \ldots, a_n, x, y]$ and the subring of symmetric regular bracket polynomials in $\mathbf{C}[\mu_1, \nu_1, \ldots, \mu_n, \nu_n, x, y]$. If $I(a_0, a_1, \ldots, a_n, x, y)$ is a covariant of degree d and order t, then $\Psi(I)$ is a symmetric bracket polynomial with each of the indices $1, 2, \ldots, n$ occurring d times and the letter $\mathbf{u}$ occurring t times.

Proof. Let I be any polynomial in $\mathbf{C}[a_0, a_1, \ldots, a_n, x, y]$, and let $R = \Psi(I)$ be its image in $\mathbf{C}[\mu_1, \nu_1, \ldots, \mu_n, \nu_n, x, y]$ under the expansion map. The expansion map Ψ commutes with the $GL(\mathbf{C}^2)$-action on both rings, and hence I is a covariant if and only if R is a bracket polynomial (cf. Lemma 3.6.5). By Proposition 3.6.4, R is regular and symmetric. Conversely, every regular symmetric bracket polynomial R is the representation $R = \Psi(I)$ in terms of homogenized roots of some covariant I. The fact that Ψ restricts to an isomorphism follows from Lemma 3.6.3.

Suppose now that the above I is a (homogeneous) covariant of degree d and order t. Since $\Psi(x) = x$ and $\Psi(y) = y$, also the bracket polynomial R is homogeneous of order t in (x, y). From formula (3.6.7) we see that, for all i and k, the total degree of R in (μ_i, ν_i) equals the degree of I in a_k. This completes the proof of Theorem 3.6.6. ◁

Let us illustrate this representation theorem by some examples in the case $n = 3$. We consider the seven bracket polynomials

$R_1 := [1\,2][2\,\mathbf{u}][3\,\mathbf{u}]$,

$R_2 := [1\,2][1\,3][2\,\mathbf{u}][3\,\mathbf{u}]$,

$R_3 := [1\,\mathbf{u}] + [2\,\mathbf{u}] + [3\,\mathbf{u}]$,

$R_4 := [1\,2][1\,2][1\,3][1\,3][2\,3][2\,3]$,

$R_5 := [1\,\mathbf{u}][2\,\mathbf{u}][3\,\mathbf{u}]$,

$R_6 := - [1\,2][1\,2][3\,\mathbf{u}][3\,\mathbf{u}] - [1\,3][1\,3][2\,\mathbf{u}][2\,\mathbf{u}] - [2\,3][2\,3][1\,\mathbf{u}][1\,\mathbf{u}]$
$\quad\quad + 2[1\,2][3\,1][2\,\mathbf{u}][3\,\mathbf{u}] + 2[2\,3][1\,2][1\,\mathbf{u}][3\,\mathbf{u}] + 2[3\,1][2\,3][1\,\mathbf{u}][2\,\mathbf{u}]$,

$R_7 := - 4[1\,2][1\,2][3\,\mathbf{u}][3\,\mathbf{u}] + 4[1\,2][1\,3][2\,\mathbf{u}][3\,\mathbf{u}] - 4[1\,3][1\,3][2\,\mathbf{u}][2\,\mathbf{u}]$.

The bracket polynomial R_1 is neither symmetric nor regular, R_2 is regular of degree 2 but not symmetric, and R_3 is symmetric but not regular. The bracket polynomial R_4 is symmetric and regular of degree 4: it is the root representation of the discriminant D_3 of the binary cubic $f_3(x, y)$. R_5 is symmetric and regular of degree 1: it equals the binary cubic $f_3(x, y)$, here viewed as a covariant of itself. R_6 is symmetric and regular of degree 2: it equals (up to a multiplicative constant) the root representation of the Hessian of H_3 of the binary cubic $f_3(x, y)$. The last bracket polynomial R_7 is regular but it appears to be not

3.6. Invariants and covariants of binary forms

symmetric. However, we can see that R_7 is symmetric if we use some syzygies. Indeed, R_7 equals the expansion of R_6 in terms of standard tableaux (for the order $1 \leq 2 \leq 3 \leq u$) and thus also represents the Hessian.

By suitably extending the notion of regularity, all results of this section generalize to several binary forms. Let $f_1(x, y), \ldots, f_r(x, y)$ be binary forms of degrees $n_1, \ldots, n_r$, and let $v_i^{(k)}, \mu_i^{(k)}$, $i = 1, 2, \ldots, n_k$, be the homogenized roots of the k-th binary form $f_k(x, y)$. The *brackets* are the polynomials in the algebra $\mathbf{C}[v_i^{(k)}, \mu_i^{(k)}, x, y]$ of homogenized roots of the form

$$[i^{(k)}\, j^{(l)}] := \mu_i^{(k)} v_j^{(l)} - \mu_j^{(l)} v_i^{(k)} \quad \text{or} \quad [i^{(k)}\, \mathbf{u}] := \mu_i^{(k)} x - \mu_j^{(l)} y.$$

Here a bracket polynomial R is *symmetric* if it is invariant under permuting the roots of each form $f_k(x, y)$ individually, and R is *regular of degree* $(d_1, \ldots, d_r)$ if each root of $f_k(x, y)$ occurs with the same multiplicity d_k. With this definition of symmetric regular bracket polynomials, the assertion of Theorem 3.6.7 extends to several binary forms.

It was stated in Sect. 1.3 that we wish to provide algorithmic tools for going back and forth between geometric statements and invariants/covariants. In the case of binary forms this means that we would like to compute the covariants for a given geometric property of binary forms, and vice versa. If such a geometric property is presented as a bracket polynomial in the roots, then we can use the Gröbner basis inversion of Ψ to compute the corresponding covariant. Conversely, whenever we are given a covariant, then we can expand it into homogenized roots and use Algorithm 3.2.8 for computing the corresponding bracket polynomial. We close this section by giving examples for these transformations.

Consider two binary quadrics

$$f(x, y) = a_2 x^2 + 2 a_1 x y + a_0^2 y = (\mu_1 x - v_1 y)(\mu_2 x - v_2 y) \quad \text{and}$$
$$g(x, y) = b_2 x^2 + 2 b_1 x y + b_0^2 y = (\mu_3 x - v_3 y)(\mu_4 x - v_4 y).$$

The polynomial

$$R := [1\,3][1\,4][2\,3][2\,4]$$
$$= (v_1 \mu_3 - \mu_1 v_3)(v_1 \mu_4 - \mu_1 v_4)(v_2 \mu_3 - \mu_2 v_3)(v_2 \mu_4 - \mu_2 v_4)$$

is symmetric and regular of degree $(2, 2)$. Clearly, R vanishes if and only if f and g have a root in common. In order to find the coefficient representation $\Psi(R)$ of the invariant R, we consider the ideal

$$\langle a_0 - v_1 v_2,\ b_0 - v_3 v_4,\ a_2 - \mu_1 \mu_2,\ b_2 - \mu_3 \mu_4,\ 2 b_1 + \mu_3 v_4 + v_3 \mu_4,\ 2 a_1 + \mu_1 v_2 + v_1 \mu_2 \rangle$$

of algebraic relations between the coefficients and the roots. We compute its

Gröbner basis $\mathcal{G}$ with respect to the purely lexicographic order induced from

$$v_1 > \mu_1 > v_2 > \mu_2 > v_3 > \mu_3 > v_4 > \mu_4 > a_0 > a_1 > a_2 > b_0 > b_1 > b_2.$$

The normal form of R modulo the Gröbner basis $\mathcal{G}$ equals

$$\begin{aligned}\Psi^{-1}(R) = \quad & 4\,a_0 a_2 b_1^2 + 4\,a_1^2 b_0 b_2 - 4\,a_0 a_1 b_1 b_2 - 4\,a_1 a_2 b_0 b_1 \\ & - 2\,a_0 a_2 b_0 b_2 + a_0^2 b_2^2 + a_2^2 b_0^2.\end{aligned}$$

This is the expansion of the Sylvester resultant of the two quadrics $f(x, y)$ and $g(x, y)$ in terms of their coefficients.

Next we illustrate the reverse direction for the bilinear invariant $L_2 = a_0 b_2 - 2 a_1 b_1 + a_2 b_0$ of $f(x, y)$ and $g(x, y)$. In order to "automatically" derive the geometric interpretation of L_2, we first compute its expansion

$$\begin{aligned}\Psi(L_2) = \quad & 2\,\mu_1 \mu_2 v_3 v_4 + 2\,\mu_3 \mu_4 v_1 v_2 - \mu_1 v_2 \mu_3 v_4 - \mu_1 v_2 v_3 \mu_4 \\ & - v_1 \mu_3 v_4 \mu_2 - v_1 \mu_2 v_3 \mu_4\end{aligned}$$

in terms of homogenized roots. We apply Algorithm 3.2.8 to find the standard tableaux expansion

$$\begin{aligned}\Psi(L_2) &= -[1\,2][3\,4] + 2\,[1\,3][2\,4] \\ &= -(\mu_1 v_2 - v_1 \mu_2)(\mu_3 v_4 - v_3 \mu_4) + 2\,(\mu_1 v_3 - v_1 \mu_3)(\mu_2 v_4 - v_2 \mu_4).\end{aligned}$$

Note also the alternative bracket representation

$$\begin{aligned}\Psi(L_2) &= [1\,3][2\,4] + [1\,4][2\,3] \\ &= (\mu_1 v_3 - v_1 \mu_3)(\mu_2 v_4 - v_2 \mu_4) + (\mu_1 v_4 - v_1 \mu_4)(\mu_2 v_3 - v_2 \mu_3)\end{aligned}$$

which makes the symmetries ("1" $\leftrightarrow$ "2" and "3" $\leftrightarrow$ "4") more transparent. As the result we conclude that the joint invariant L_2 of two binary quadrics vanishes if and only if the projective cross-ratio

$$(1, 2; 3, 4) = \frac{[1\,3][2\,4]}{[1\,4][2\,3]}$$

of the two pairs of roots is equal to -1.

Exercises

(1) Prove Proposition 3.6.1.
(2) Find a joint covariant of the binary cubic f_3 and the binary quartic f_4 which vanishes if and only if f_3 and f_4 have two or more roots in common.
(3) Define the catalecticant C_{2m} of a binary form $f_{2m}(x, y)$ of degree $2m$, and

show that C_{2m} vanishes if and only if $f_{2m}(x, y)$ can be written as a sum of m or less perfect powers.

(4) Let I be a covariant of degree d and order t of the binary form $f_n(x, y)$ of degree n. Prove that the index of I equals $g = \frac{1}{2}(dn - t)$.

(5) The bracket monomial $[1\,2][1\,3][1\,4][2\,3][2\,4][3\,4]$ represents the discriminant of the binary quartic

$$f_4 = c_4 x^4 + 4c_3 x^3 y + 6c_2 x^2 y^2 + 4c_1 x y^3 + c_0 y^4$$

in terms of its homogenized roots. Write this invariant as a polynomial in the coefficients c_0, c_1, c_2, c_3, c_4.

(6) Express the catalecticant C_4 of the binary quartic $f_4(x, y)$ as a bracket polynomial in the roots of f_4.

(7) Express the first principal subresultant F_{34} of the binary quartic $f_4(x, y)$ and the binary cubic $f_3(x, y)$ as a bracket polynomial in the seven roots and **u**.

3.7. Gordan's finiteness theorem

We now come to the central problem of classical invariant theory: *Does there exist a finite set of generators for the ring of covariants?* For binary forms the affirmative answer to this question was given by Gordan (1863). Gordan's finiteness result is undoubtedly one of the most important theorems of 19th century constructive algebra. Both Gordan's original proof and the proof presented here are algorithmic. They can in principle be used to compute fundamental systems of (joint) covariants for binary forms of any degree.

Theorem 3.7.1 (Gordan's finiteness theorem). *There exists a finite set of covariants $\{I_1, \ldots, I_k\}$ of a binary form of degree n such that every covariant $I \in \mathbf{C}[a_0, a_1, \ldots, a_n, x, y]$ can be expressed as a polynomial function $I = p(I_1, \ldots, I_k)$.*

In order to prove Theorem 3.7.1, we need one lemma about finite group actions on quotients of polynomial rings. Let $\Gamma \subset GL(\mathbf{C}^r)$ be any *finite* matrix group, and let I be an ideal in the polynomial ring $\mathbf{C}[x_1, \ldots, x_r]$ which is fixed under the action of Γ on I. In this situation, we get an induced action of Γ on the quotient ring $\mathbf{C}[x_1, \ldots, x_r]/I$.

Lemma 3.7.2.
(a) Suppose that $p_1, \ldots, p_m$ generate the invariant ring $\mathbf{C}[x_1, \ldots, x_r]^\Gamma$. Then $(\mathbf{C}[x_1, \ldots, x_r]/I)^\Gamma$ is generated by the images of $p_1, \ldots, p_m$ under the canonical surjection $\mathbf{C}[x_1, \ldots, x_r] \to \mathbf{C}[x_1, \ldots, x_r]/I$.
(b) The invariant ring $(\mathbf{C}[x_1, \ldots, x_r]/I)^\Gamma$ is generated by the Reynolds images $(x_1^{i_1} x_2^{i_2} \ldots x_r^{i_r})^*$ of all monomials $x_1^{i_1} x_2^{i_2} \ldots x_r^{i_r}$ with degree $i_1 + i_2 + \ldots + i_r \leq |\Gamma|$.

Proof. We need to show that every element $p \in (\mathbf{C}[x_1, \ldots, x_r]/I)^\Gamma$ can be written as a polynomial in (the images of) $p_1, \ldots, p_m$. The fact that p is invariant means that $p - p^* \in I$. But the canonical preimage of p^* is an invariant in $\mathbf{C}[x_1, \ldots, x_r]$ and thus is expressible as a polynomial in $p_1, \ldots, p_m$. Hence $p = p^* + (p - p^*) \in p^* + I$ can be written as in terms of the residues $p_1 + I, \ldots, p_m + I$, and (a) is proved. Statement (b) follows directly from (a) and Noether's degree bound (Theorem 2.1.4). ◁

As before in Sect. 2.5, we can apply Gröbner basis methods to compute an optimal generating set, which is usually much smaller than the one given in Lemma 3.7.2 (b). Note, however, that in general we cannot obtain a Hironaka decomposition because the invariant ring $(\mathbf{C}[x_1, \ldots, x_r]/I)^\Gamma$ need not be Cohen–Macaulay.

Proof of Theorem 3.7.1. The bracket subring in $\mathbf{C}[\mu_1, \nu_1, \ldots, \mu_n, \nu_n, x, y]$ is denoted by

$$\mathcal{B} := \mathbf{C}\big[[1\,2], [1\,3], \ldots [1\,n], [2\,3], \ldots [n-1\,n], [1\,\mathbf{u}], [2\,\mathbf{u}], \ldots [n\,\mathbf{u}]\big].$$

We write $\mathcal{B}_{\text{reg}}$ for the subring of regular bracket polynomials in $\mathcal{B}$, and we write $\mathcal{B}_{\text{reg,sym}}$ for the subring of symmetric regular bracket polynomials in $\mathcal{B}_{\text{reg}}$. By Theorem 3.6.6, it suffices to show that $\mathcal{B}_{\text{reg,sym}}$ is finitely generated as a C-algebra.

A bracket monomial $M \in \mathcal{B}$ is said to be *minimally regular* if it is regular and none of its proper factors are regular. Let $\mathcal{M}$ denote the set of minimally regular bracket monomials in $\mathcal{B}$. By minimality, no two elements of $\mathcal{M}$ are comparable in the divisibility order. Gordan's lemma (Lemma 1.2.2) implies that the set $\mathcal{M}$ is finite, say $\mathcal{M} = \{M_1, M_2, \ldots, M_r\}$. In other words, the M_i are polynomials in the homogenized roots which generate the subring of regular bracket polynomials:

$$\mathcal{B}_{\text{reg}} = \mathbf{C}[M_1, M_2, \ldots, M_r].$$

The symmetric group S_n acts on the bracket ring $\mathcal{B}$ by permuting the letters "1", "2", ..., "n". Since minimally regular monomials remain minimally regular after permuting letters, the symmetric group S_n acts on the subring $\mathcal{B}_{\text{reg}}$ of regular bracket polynomials by permuting its generators $M_1, M_2, \ldots, M_r$.

In this situation we can apply Lemma 3.7.2 (a) in order to conclude that the subring

$$\mathcal{B}_{\text{reg,sym}} = \mathcal{B}_{\text{reg}}^{S_n} = \mathbf{C}[M_1, M_2, \ldots, M_r]^{S_n} \tag{3.7.1}$$

of symmetric regular bracket polynomials is finitely generated. Lemma 3.7.2 (b) implies that an explicit finite set of generators for this invariant ring is given by the Reynolds images $(M_1^{i_1} M_2^{i_2} \ldots M_r^{i_r})^*$ of all monomials $M_1^{i_1} M_2^{i_2} \ldots M_r^{i_r}$ with degree $i_1 + i_2 + \ldots + i_r \leq n!$. This completes the proof of Gordan's finiteness theorem. ◁

The above proof is not entirely satisfactory because the invocation of Gor-

3.7. Gordan's finiteness theorem

dan's lemma appears to be a non-constructive step. Kung and Rota (1984) suggest two possible algorithms for computing a generating set $\mathcal{M} = \{M_1, M_2, \ldots, M_r\}$ of the subring $\mathcal{B}_{\text{reg}}$ of regular bracket polynomials. The first approach consists in expressing the exponent vectors of all regular bracket monomials as the non-negative solutions of a linear diophantine system. Using algorithms from *integer programming*, we can compute a Hilbert basis for the monoid of all solutions; see Sect. 1.4. The bracket monomials corresponding to these exponent vectors then generate the ring $\mathcal{B}_{\text{reg}}$.

The disadvantage of the general integer programming approach lies in the fact that it ignores the specific combinatorial structure of our problem. The second and more efficient algorithm is based on the following result of Kempe.

Theorem 3.7.3 (Kempe's Lemma). *The ring $\mathcal{B}_{\text{reg}}$ of regular bracket polynomials is generated by all bracket monomials which are regular of degree 1 or 2.*

The proof of Kempe's lemma, given in sections 6.2 and 6.3 of Kung and Rota (1984), is based on the *circular straightening algorithm*. In the spirit of Sect. 3.1 we will interpret circular straightening in terms of Gröbner bases. This leads to a new proof of Theorem 3.7.3, based on Proposition 3.7.4 and Lemma 3.7.5 below.

Consider the set $\mathcal{S}_{n,2}$ of straightening syzygies in $C[\Lambda(n, 2)]$,

$$P_{i_1 i_2 i_3 i_4} := \underline{[i_1 i_3][i_2 i_4]} - [i_1 i_2][i_3 i_4] - [i_1 i_4][i_2 i_3] \quad (1 \leq i_1 < i_2 < i_3 < i_4 \leq n).$$

Let $\prec_{\text{circ}}$ be any monomial order on $C[\Lambda(n, 2)]$ which selects the underlined initial monomial for each syzygy $P_{i_1 i_2 i_3 i_4}$. Establishing the existence of such a monomial order is a non-trivial exercise in solving linear systems of inequalities. It can also be deduced from the proof of Kung and Rota (1984: lemma 6.1) in conjunction with the main theorem in Reeves and Sturmfels (1992).

Proposition 3.7.4. *The set $\mathcal{S}_{n,2}$ is a Gröbner basis of the syzygy ideal with respect to $\prec_{\text{circ}}$.*

First proof. Suppose that n is the smallest positive integer for which Proposition 3.7.4 is false. By the Buchberger criterion (Buchberger 1985: theorem 6.2), there exist two polynomials $P_{i_1 i_2 i_3 i_4}$ and $P_{j_1 j_2 j_3 j_4}$ whose $\mathcal{S}$-polynomial with respect to $\prec_{\text{circ}}$ does not reduce to zero. By Exercise 1.2. (6), the initial monomials of $P_{i_1 i_2 i_3 i_4}$ and $P_{j_1 j_2 j_3 j_4}$ cannot be relatively prime, and therefore the set of indices $\{i_1, i_2, i_3, i_4, j_1, j_2, j_3, j_4\}$ has cardinality at most seven. After relabeling we assume that $\{i_1, i_2, i_3, i_4, j_1, j_2, j_3, j_4\} \subseteq \{1, 2, \ldots, 7\}$. Our minimality assumption implies $n \leq 7$. It is clear that $n \geq 5$ because there are no syzygies for $n \leq 3$ and for $n = 4$ the syzygy ideal $I_{4,2} = \langle [13][24] - [12][34] + [14][23] \rangle$ is principal. It therefore suffices to verify the Gröbner basis property in the three cases $n = 5, 6, 7$. This can be done easily by computer. ◁

Here is an alternative and more conceptual proof, based on enumerative combinatorics.

Second proof. Let $\text{init}_\prec(I_{n,2})$ denote the initial ideal of $I_{n,2}$ with respect to the standard tableaux order in Sect. 3.1, and let $\text{init}_{\prec_{\text{circ}}}(I_{n,2})$ denote the initial ideal with respect to the circular order $\prec_{\text{circ}}$. Let J denote the ideal in $\mathbf{C}[\Lambda(n,2)]$ generated by all initial monomials $\text{init}_{\prec_{\text{circ}}}(P_{i_1 i_2 i_3 i_4}) = [i_1 i_3][i_2 i_4]$, where $1 \leq i_1 < i_2 < i_3 < i_4 \leq n$. We need to show that the inclusion $J \subseteq \text{init}_{\prec_{\text{circ}}}(I_{n,2})$ is an equality. This will be done using proposition 7.3 of Sturmfels and Zelevinsky (1992). In order to apply this result, we need to verify that the monomial order $\prec_{\text{circ}}$ is compatible with the set $\mathcal{S}_{n,2}$.

Every square-free monomial ideal can be identified with a simplicial complex (Stanley 1983: section II.1). Let Δ_n and Δ'_n be the simplicial complexes on $\Lambda(n,2)$ such that

$$J = I_{\Delta_n} \quad \text{and} \quad \text{init}_\prec(I_{n,2}) = I_{\Delta'_n}. \tag{3.7.2}$$

Now, by Exercise 3.1.5, the complex Δ'_n is isomorphic to the chain complex of a poset on $\Lambda(n,2)$. This poset is isomorphic to the graded poset $J([n-2] \times [2])$, consisting of the order ideals in the product of chains $[n-2] \times [2]$. It follows from Stanley (1986: example 3.5.5) that Δ_n is a pure $(2n-4)$-dimensional complex with $\frac{1}{n-1}\binom{2n-4}{n-2}$ maximal faces. By Theorem 3.1.7, this number equals the degree of the syzygy ideal $I_{n,2}$.

Every maximal face of the complex Δ_n contains the set $\{[1\,2], [2\,3], \ldots, [n-1\,n], [n\,1]\}$. Hence Δ_n is the free join of an $(n-1)$-simplex with a certain simplicial complex Σ_n of dimension $n-4$ on the set $\Lambda \setminus \{[1\,2], [2\,3], \ldots, [n-1\,n], [n\,1]\}$. Identify this set with the set of diagonals in a regular planar n-gon. The faces of the complex Σ_n are precisely the sets of non-crossing diagonals. It is known (Lee 1989) that Σ_n is the boundary complex of a simplicial $(n-3)$-polytope Q_n. The polar to the polytope Q_n is the famous *associahedron* or *Stasheff polytope*. We conclude that Δ_n is a pure $(2n-4)$-dimensional complex. By Lee (1989: theorem 3), the number of maximal faces of Δ_n equals the Catalan numbers

$$f_{2n-4}(\Delta_n) = \frac{1}{n-1}\binom{2n-4}{n-2} = f_{2n-4}(\Delta'_n). \tag{3.7.3}$$

We now apply proposition 7.3 of Sturmfels and Zelevinsky (1992) to complete our proof. ◁

As in Sect. 3.6, we work in the polynomial ring $\mathbf{C}[\Lambda(n+1,2)]$, where we identify the index $n+1$ with the letter $\mathbf{x}$. A bracket monomial in $\mathbf{C}[\Lambda(n+1,2)]$ is called *cyclically standard* if it is standard with respect to the Gröbner basis in Proposition 3.7.4. We call a bracket monomial *elemental* if it is minimally regular and cyclically standard. For instance, if $n = 5$ then the bracket monomials $[1\,2][3\,4][5\,\mathbf{x}]$, $[1\,2][3\,\mathbf{x}][4\,\mathbf{x}][5\,\mathbf{x}]$, and $[1\,2][2\,3][3\,4][4\,5][5\,1]$ are elemental. It follows from Proposition 3.7.4 that the elemental bracket monomials generate $\mathcal{B}_{\text{reg}}$ as a C-algebra. Therefore Theorem 3.7.3 is implied by the following lemma.

3.7. Gordan's finiteness theorem

Lemma 3.7.5. Every elemental bracket monomial in B_{reg} is regular of degree at most 2.

Proof. Suppose M is elemental and regular of degree d. By the second proof of Proposition 3.7.4, the brackets occurring in M can be extended to a triangulation of the $(n+1)$-gon having vertices $1, 2, \ldots, n, \mathbf{x}$. We now use the elementary fact that (the edge graph of) any triangulation of a regular $(n+1)$-gon has at least two 2-valent vertices. This implies that at least one of the indices $1, 2, \ldots, n$ occurs in at most two distinct brackets of M. Therefore M is regular of degree at most 2. ◁

Our results suggest the following algorithm for computing a fundamental system of covariants $\{I_1, \ldots, I_k\}$ of a binary n-form. First compile a list of all elemental bracket monomials, and then apply Lemma 3.7.2 (b) to this list to get a presentation as in (3.7.1).

Example 3.7.6 (Fundamental system of covariants for the binary cubic). For $n = 3$ there are precisely four elemental bracket monomials:

$$A := [1\,2][2\,3][3\,1], \quad B := [1\,2][3\,\mathbf{x}], \quad C := [2\,3][1\,\mathbf{x}], \quad D := [1\,\mathbf{x}][2\,\mathbf{x}][3\,\mathbf{x}].$$

By Lemma 3.7.1, the ring of covariants is generated by the finite set

$$\{(A^i B^j C^k D^l)^* : i, j, k, l \in \mathbf{N}, i + j + k + l \leq 6\}, \tag{3.7.4}$$

where "$*$" denotes the Reynolds operator for the action of the symmetric group S_3 on the bracket algebra. For instance,

$$(B^2 C^4)^* = \tfrac{1}{3}\Big([1\,2]^2[1\,3]^4[2\,\mathbf{x}]^4[3\,\mathbf{x}]^2 + [2\,1]^2[2\,3]^4[1\,\mathbf{x}]^4[3\,\mathbf{x}]^2$$
$$+ [3\,1]^2[3\,2]^4[1\,\mathbf{x}]^4[2\,\mathbf{x}]^2\Big).$$

The set (3.7.4) is by no means a minimal generating set for the covariant ring. To reduce its size, we first observe that both $A^2 = (A^2)^*$ and $D = D^*$ are covariants already. For, the bracket monomial $\tfrac{1}{27}A^2$ equals the discriminant, and D is the form itself. By Proposition 2.1.2 (c), we have $(A^i B^j C^k D^l)^* = A^{i-i'} D^l \cdot (A^{i'} B^j C^k)^*$ where $i' = 1$ if i is odd and $i' = 0$ if i is even. We conclude that the set

$$\{A^2, D\} \cup \{(A^i B^j C^k)^* : i, j, k \in \mathbf{N}, \ i \in \{0, 1\}, \ i + j + k \leq 6\} \tag{3.7.5}$$

suffices to generate all covariants. By explicitly evaluating all covariants in (3.7.5), we can show that the following four covariants are sufficient.

Proposition 3.7.7. A minimal generating set for the covariants of the binary cubic consists of the following four covariants:

$f = a_3 x^3 + 3a_2 x^2 y + 3a_1 xy^2 + a_0 y^3$, the form itself;
$D = a_3^2 a_0^2 - 6a_3 a_0 a_1 a_2 - 3a_1^2 a_2^2 + 4a_2^3 a_0 + 4a_1^3 a_3$, the discriminant;
$H = (a_3 a_1 - a_2^2) x^2 + (a_3 a_0 - a_2 a_1) yx + (a_2 a_0 - a_1^2) y^2$, the Hessian; and
$T = (a_3^2 a_0 - 3a_3 a_2 a_1 + 2a_2^3) x^3 + (3a_3^2 a_1 + 3a_3 a_2 a_0 - 6a_3 a_1^2) yx^2 + (-3a_2 a_1^2 + 6a_2^2 a_0 - 3a_1 a_3 a_0) y^2 x + (3a_1 a_2 a_0 - 2a_1^3 - a_0^2 a_3) y^3$, the Jacobian of the form and the Hessian.

The unique minimal syzygy among these covariants equals $f^2 D - T^2 - 4H^3 = 0$.

Proof. We need to show that each covariant in (3.7.5) can be expressed as a polynomial in f, D, H and T. Using computer algebra, we find the following presentations:

$(B^2)^* = (C^2)^* = -6H, \ (BC)^* = 3H, \ (BC^2)^* = \frac{9}{2}T, \ (B^2 C)^* = -\frac{9}{2}T,$

$(B^4)^* = (C^4)^* = 54H^2, \ (BC^3)^* = (B^3 C) = -27H^2, \ (B^2 C^2)^* = 27H^2,$

$(BC^4)^* = (B^3 C^2)^* = -\frac{81}{2} HT, \ (B^2 C^3)^* = (B^4 C)^* = \frac{81}{2} HT,$

$(BC^5)^* = (B^5 C)^* = \frac{27}{2} J^2 + 297 H^3, \ (B^2 C^4)^* = (B^4 C^2)^* = \frac{27}{2} J^2 - 189 H^3,$

$(B^3 C^3)^* = -27 J^2 + 135 H^3, \quad (AB^3)^* = (AC^3)^* = 27 Df,$

$(AB^2 C)^* = (ABC^2)^* = -\frac{27}{2} Df, \ (AB^5)^* = (AC^5)^* = -405 DHf,$

$(AB^4 C)^* = (ABC^4)^* = -\frac{405}{2} DHf, \ (AB^2 C^3)^* = (AB^3 C^2) = \frac{-81}{2} DHf.$

The Reynolds images of all other bracket monomials $A^i B^j C^k$ in (3.7.5) are zero.

It remains to be shown that the ideal of algebraic relations among f, D, H and T is the principal ideal generated by $f^2 D - T^2 - 4H^3$. The validity of this syzygy is easily checked, and since $f^2 D - T^2 - 4H^3$ is an irreducible polynomial, it suffices to show that f, D and H are algebraically independent. But this follows from the fact that their lexicographic initial monomials $a_3 x^3$, $a_3^2 a_0^2$ and $a_3 a_1 x^2$ are algebraically independent. ◁

In closing, we briefly mention the known results for binary forms of higher degree n. Explicit fundamental systems of invariants and covariants are known only for $n \leq 8$. For details see Meyer (1892), Springer (1977: section 3.4),

3.7. Gordan's finiteness theorem

and Dixmier and Lazard (1988). However, as in the case of finite groups, it is possible for general n to determine a priori the Hilbert series of the invariant and covariant ring of a binary n-form. This result is based on the representation theory of $GL(\mathbf{C}^2)$. It will be presented in Sect. 4.2.

Exercises

(1) Prove that every invariant of the binary quadric $f_2(x, y)$ is a polynomial function in the discriminant $D_2 = a_0 a_2 - a_1^2$. Prove the same result for the binary cubic.

(2) * Let $\{I_1, I_2, \ldots, I_k\}$ be a minimal generating set for the joint covariants of a binary cubic $f_3(x, y)$ and a binary quartic $f_4(x, y)$. Find an upper bound for the degrees of the fundamental covariants I_j.

(3) Determine the Hilbert function of the ring of covariants of a binary cubic.

(4) * Find finite generating set for the invariants and covariants of a binary quartic $f_4(x, y)$. Write each element in your generating set as a bracket polynomial in the roots. (Hint: Use the list 3.4.4 in Springer (1977).)

4 Invariants of the general linear group

This chapter deals with methods for computing the invariants of an arbitrary polynomial representation of the general linear group $GL(\mathbf{C}^n)$. The main algorithm, to be presented in Sect. 4.6, is derived from Hilbert (1893). We will discuss Hilbert's algorithm from the point of view of Gröbner bases theory. This chapter is less elementary than the previous three. While most of the presentation is self-contained, familiarity with basic notions of commutative algebra and representation theory will be assumed.

4.1. Representation theory of the general linear group

Throughout this chapter we let $\Gamma = GL(\mathbf{C}^n)$ denote the group of invertible complex $n \times n$-matrices. This section provides a crash course in the representation theory of Γ. All stated results are well known, and we will omit most of the proofs. This theory is essentially due to I. Schur, with extensions by H. Weyl and A. Young. A comprehensive introduction with many geometric applications can be found in Fulton and Harris (1991).

A *representation* of Γ (or Γ-*module*) is a pair (V, ρ) where V is a $\mathbf{C}$-vector space and

$$\rho : \Gamma \to GL(V),$$
$$A = (a_{ij})_{1 \leq i, j \leq n} \mapsto \rho(A) = \bigl(\rho_{kl}(A)\bigr)_{1 \leq k, l \leq N} \quad (4.1.1)$$

is a group homomorphism. The *dimension* N of the representation (V, ρ) is the dimension of the vector space V. We say that (V, ρ) is a *polynomial representation* (of *degree* d) if the matrix entries $\rho_{kl}(A) = \rho_{kl}(a_{11}, a_{12}, \ldots, a_{nn})$ are polynomial functions (homogeneous of degree d). If the action ρ is understood, then we sometimes write $A \circ v$ instead of $\rho(A) \cdot v$, where $v \in V$ and $A \in \Gamma$.

Examples of representations 4.1.1.

(a) The trivial representation: $V = \mathbf{C}^n$, $(N = n, d = 1)$.
(b) The determinant: $V = \mathbf{C}^1$, $\rho = \det$, $(N = 1, d = n)$.
(c) The action ρ by *left multiplication* on the space of $n \times s$-matrices

$$V = \mathbf{C}^{n \times s} = \underbrace{\mathbf{C}^n \oplus \mathbf{C}^n \oplus \ldots \oplus \mathbf{C}^n}_{s \text{ times}}. \qquad (N = s \cdot n, d = 1)$$

(d) The adjoint representation: $V = \mathbf{C}^{n \times n}$, $\rho = $ the action by conjugation:

$A \circ M = AMA^{-1}$. This is not a polynomial representation but a rational representation, i.e., the $\rho_{kl}(A)$ are rational functions. Note that $\det(A)^{n-1}$ is a common denominator for the rational functions $\rho_{kl}(A)$.

The following result explains why we restrict ourselves to polynomial representation.

Proposition 4.1.2. Given any rational representation $\rho : \Gamma \to GL(V)$, there exists an integer k and polynomial representation $\rho' : \Gamma \to GL(V)$ such that $\rho = \rho' \cdot \det^{-k}$.

From now on all representations are assumed to be polynomial representations of Γ.

(e) The d-th symmetric power representation: $V = S_d \mathbf{C}^n =$ the space of homogeneous polynomials of degree d in $x_1, x_2, \ldots, x_n$, $\rho = $ action by linear substitution, $N = \binom{n+d-1}{d}$. For instance, for $d = 3$, $n = 2$ we have $S_3 \mathbf{C}^2 = $ binary cubics $= \text{span}\{x^3, x^2 y, xy^2, y^3\} \simeq \mathbf{C}^4$, and ρ is the group homomorphism

$$\begin{pmatrix} a_{11} & a_{12} \\ a_{21} & a_{22} \end{pmatrix}$$

$$\mapsto \begin{pmatrix} a_{11}^3 & a_{11}^2 a_{12} & a_{11} a_{12}^2 & a_{12}^3 \\ 3a_{11}^2 a_{21} & a_{11}^2 a_{22} + 2a_{11} a_{12} a_{21} & 2a_{11} a_{12} a_{22} + a_{12}^2 a_{21} & 3a_{12}^2 a_{22} \\ 3a_{11} a_{21}^2 & 2a_{11} a_{21} a_{22} + a_{12} a_{21}^2 & a_{11} a_{22}^2 + 2a_{12} a_{21} a_{22} & 3a_{12} a_{22}^2 \\ a_{21}^3 & a_{21}^2 a_{22} & a_{21} a_{22}^2 & a_{22}^3 \end{pmatrix}.$$

(f) The d-th exterior power representation: $V = \wedge_d \mathbf{C}^n = $ the space of alternating d-forms on $\mathbf{C}^n$. A basis of V is $\{e_{i_1} \wedge \ldots \wedge e_{i_d} : 1 \le i_1 < \ldots < i_d \le n\}$. Here $N = \binom{n}{d}$ and $\rho(A) = \wedge_d A$, the d-th compound matrix whose entries are the $d \times d$-minors of A.

(g) The d-th tensor power representation:

$$V = \otimes_d \mathbf{C}^n = \underbrace{\mathbf{C}^n \otimes \mathbf{C}^n \otimes \ldots \otimes \mathbf{C}^n}_{d \text{ times}}. \qquad (N = n^d)$$

(h) Building new representations from old ones: For any two Γ-representations (V, ρ) and (W, σ), we can form their *direct sum* $(V \oplus W, \rho \oplus \sigma)$ and their *tensor product* $(V \otimes W, \rho \otimes \sigma)$. If $\{v_1, \ldots, v_N\}$ and $\{w_1, \ldots, w_M\}$ are bases of V and W respectively, then $\{v_1, \ldots, v_N, w_1, \ldots, w_M\}$ is a basis of $V \oplus W$, and $\{v_i \otimes w_j : 1 \le i \le N, 1 \le j \le M\}$ is a basis of $V \otimes W$. The two new representations are defined by

$$(\rho \oplus \sigma)(A) : \begin{cases} v_i \mapsto \rho(A) v_i \\ w_j \mapsto \sigma(A) w_j \end{cases} \qquad (4.1.2)$$

$$(\rho \otimes \sigma)(A) : v_i \otimes w_j \mapsto (\rho(A) v_i) \otimes (\sigma(A) w_j).$$

4.1. Representation theory of the general linear group

A main tool for studying representations of the general linear group is the theory of symmetric polynomials. We write $\mathbf{C}[A] = \mathbf{C}[a_{11}, a_{12}, \ldots, a_{nn}]$ for the ring of polynomial functions on $n \times n$-matrices. We consider the action of Γ on $\mathbf{C}[A]$ via the adjoint representation (d). Let $\mathbf{C}[A]^\Gamma$ denote the invariant ring. The *character* of a polynomial representation (V, ρ) is the polynomial $\text{tr}_\rho(A) := \text{trace}(\rho(A))$ in $\mathbf{C}[A]$. Since the trace is invariant under conjugation, we have $\text{tr}_\rho(B^{-1}AB) = \text{tr}_\rho(A)$.

Remark 4.1.3. For every polynomial representation (V, ρ), the character tr_ρ lies in $\mathbf{C}[A]^\Gamma$.

Let $t_1, t_2, \ldots, t_n$ be new variables representing the eigenvalues of a generic $n \times n$-matrix A. We write $\text{diag}(t_1, t_2, \ldots, t_n)$ for the corresponding diagonal matrix. Every symmetric polynomial in $t_1, t_2, \ldots, t_n$ can be written as an invariant polynomial in the entries of A, and vice versa, by Exercise 1.3. (4).

Lemma 4.1.4. The map $f \mapsto f\big(\text{diag}(t_1, t_2, \ldots, t_n)\big)$ defines an isomorphism between the invariant ring $\mathbf{C}[A]^\Gamma$ and the ring of symmetric polynomials $\mathbf{C}[t_1, t_2, \ldots, t_n]^{S_n}$. The image of the character $\text{tr}_\rho(A)$ under this isomorphism,

$$f_\rho(t_1, t_2, \ldots, t_n) := \text{trace}\big(\rho(\text{diag}(t_1, t_2, \ldots, t_n))\big), \qquad (4.1.3)$$

is called the *formal character* of the representation (V, ρ). Note that the dimension of V can easily be read off from the formal character: $\dim V = N = f_\rho(1, 1, \ldots, 1)$.

Examples 4.1.5. The formal characters of the representations in Examples 4.1.1 are:

(a) $V = \mathbf{C}^n$, $f_\rho = t_1 + t_2 + \ldots + t_n$.
(b) $V = \mathbf{C}^1$, $f_\rho = t_1 t_2 \ldots t_n$.
(c) $V = \mathbf{C}^{n \times s}$, $f_\rho = s(t_1 + t_2 + \ldots + t_n)$.
(e) $V = S_d \mathbf{C}^n$, $f_\rho = $ the sum of all degree d monomials in $t_1, \ldots, t_n$ (the d-th complete symmetric polynomial); for instance, for $V = S_2 \mathbf{C}^3$ we have $f_\rho = t_1^3 + t_1^2 t_2 + t_1 t_2^2 + t_2^3$.
(f) $V = \wedge_d \mathbf{C}^n$, $f_\rho = $ the d-th elementary symmetric polynomial in $t_1, t_2, \ldots, t_n$.
(g) $V = \otimes_d \mathbf{C}^n$, $f_\rho = (t_1 + t_2 + \ldots + t_n)^d$.

Lemma 4.1.6. The formal characters of the direct sum and the tensor product satisfy the relations $f_{\rho \oplus \sigma} = f_\rho + f_\sigma$ and $f_{\rho \otimes \sigma} = f_\rho \cdot f_\sigma$.

Proof. This follows immediately from the rules for traces of matrices: $\text{trace}(A \oplus B) = \text{trace}(A) + \text{trace}(B)$ and $\text{trace}(A \otimes B) = \text{trace}(A) \cdot \text{trace}(B)$. ◁

A *submodule* of a representation (V, ρ) is a pair $(W, \rho|_W)$, where W is a Γ-invariant subspace of V and $\rho|_W$ is the restriction of ρ to W. We say that

(V, ρ) is *irreducible* if V contains no proper Γ-invariant subspace. Otherwise it is *reducible*. For instance, the representations $S_d \mathbf{C}^n$ and $\wedge_d \mathbf{C}^n$ are irreducible, while the representation $\otimes_d \mathbf{C}^n$ is reducible for $d, n \geq 2$.

Theorem 4.1.7 (Schur). Every Γ-representation is a direct sum of irreducible representation, and it is uniquely determined by its formal character (up to isomorphism).

A standard problem is to decompose the tensor product of two irreducible representations into irreducibles. Only in certain extreme cases it can happen that such a tensor product is again irreducible. For instance, for $n = 2$ the representation $W_{(2,1)} \mathbf{C}^2 := \wedge_2 \mathbf{C}^2 \otimes \mathbf{C}^2$ is irreducible. It has the formal character $f_\rho = t_1 t_2 (t_1 + t_2)$.

Example 4.1.8 (Decomposition into irreducible representations).
(a) Every tensor in $\otimes_2 \mathbf{C}^n$ can be written uniquely as a sum of a symmetric tensor and an antisymmetric tensor. For instance, we have the decomposition

$$\otimes_2 \mathbf{C}^3 = S_2 \mathbf{C}^3 \oplus \wedge_2 \mathbf{C}^3$$
$$(t_1 + t_2 + t_2)^3 = (t_1^2 + t_2^2 + t_3^2 + t_1 t_2 + t_1 t_3 + t_2 t_3) + (t_1 t_2 + t_1 t_3 + t_2 t_3)$$

(b) The statement in (a) is false for tensors in $\otimes_d \mathbf{C}^n$ with $d \geq 3$. For instance,

$$\otimes_3 \mathbf{C}^2 = S_3 \mathbf{C}^2 \oplus W_{(2,1)} \mathbf{C}^2 \oplus W_{(2,1)} \mathbf{C}^2$$
$$(t_1 + t_2)^3 = (t_1^3 + t_1^2 t_2 + t_1 t_2^2 + t_2^3) + (t_1^2 t_2 + t_1 t_2^2) + (t_1^2 t_2 + t_1 t_2^2)$$

The *Grothendieck ring* $M(\Gamma)$ is the $\mathbf{Z}$-algebra generated by all (isomorphism classes of) Γ-representations, having addition $\oplus$ and multiplication $\otimes$. Theorem 4.1.7 states in other words that the formal character defines a monomorphism from the Grothendieck ring $M(\Gamma)$ into $\mathbf{Z}[t_1, \ldots, t_n]^{S_n}$, the ring of symmetric polynomials with integer coefficients. By Theorem 1.1.1 and Example 4.1.5 (f), this map is in fact an isomorphism.

Corollary 4.1.9. The Grothendieck ring $M(\Gamma)$ is isomorphic to $\mathbf{Z}[t_1, \ldots, t_n]^{S_n}$.

We will now describe the irreducible Γ-representations. First note that every irreducible polynomial Γ-representation is homogeneous of some degree d (Exercise 1 below). The following basic fact is also due to Schur.

Proposition 4.1.10. Every irreducible Γ-representation of degree d is a submodule of $\otimes_d \mathbf{C}^n$.

Let $\lambda = (\lambda_1, \lambda_2, \ldots, \lambda_n)$ be a partition of the integer d. The *Ferrers diagram* of λ is the set $\{(i, j) \in \mathbf{Z}^2 : 1 \leq i \leq n, 1 \leq j \leq \lambda_i\}$. A *standard Young tableau of shape λ* (for short: SYT λ) is a filling T of the Ferrers diagram of λ with the

4.1. Representation theory of the general linear group

integers $1, 2, \ldots, d$ (without repetitions) such that the rows and the columns are increasing.

For instance, the partition $\lambda = (3, 3, 1)$ of $d = 7$ has the Ferrers diagram

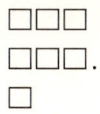

```
□ □ □
□ □ □ .
□
```

Examples of SYTλ's corresponding to this partition are

```
1 2 3     1 2 3     1 2 3     1 2 4            1 4 6
4 5 6,    4 5 7,    4 6 7,    3 5 6,   ... ,   2 5 7.
7         6         5         7                3
```

The tableau
```
1 3 6
2 4 5
7
```
is not a SYTλ because its last column is not increasing.

With each SYTλ T we associate an idempotent linear map $c_T : \otimes_d \mathbf{C}^n \to \otimes_d \mathbf{C}^n$. This map is called the *Young symmetrizer* of T, and it is defined as follows. Let rowstb(T) denote the subgroup of permutations of $\{1, 2, \ldots, d\}$ which preserve the set of entries in each row of T. Similarly, let colstb(T) denote the subgroup of permutations of $\{1, 2, \ldots, d\}$ which preserve the entries in each column of T. We define c_T by giving its image for decomposable tensors:

$$c_T : v_1 \otimes v_2 \otimes \cdots \otimes v_d \mapsto \sum_{\sigma \in \text{colstb}(T)} \sum_{\tau \in \text{rowstb}(T)} (\text{sign } \sigma) \cdot v_{\sigma\tau(1)} \otimes v_{\sigma\tau(2)} \otimes \ldots \otimes v_{\sigma\tau(d)}.$$

Let $W_T \mathbf{C}^n$ denote the image of the Young symmetrizer c_T. Thus $W_T \mathbf{C}^n$ is the subspace of all tensors in $\otimes_d \mathbf{C}^n$ which are symmetric with respect to the rows of T and antisymmetric with respect to the columns of T. The Γ-representation $W_T \mathbf{C}^n$ is called the *Weyl module* associated with the SYTλ T. If T and T' are SYT of the same shape λ, then W_T and $W_{T'}$ are isomorphic Γ-modules, and we sometimes write $W_\lambda := W_T \simeq W_{T'}$.

Theorem 4.1.11. *The Weyl modules $W_T \mathbf{C}^n$ are precisely the irreducible Γ-modules of degree d. The Young symmetrizers c_T define an isomorphism of Γ-modules*

$$\otimes_d \mathbf{C}^n \simeq \bigoplus_{\lambda \vdash d} \bigoplus_{T \text{ SYT}\lambda} W_T \mathbf{C}^n. \tag{4.1.4}$$

The first direct sum in (4.1.4) is over all partitions λ of d, and the second direct sum is over all standard Young tableaux T of shape λ. The Young symmetrizer c_T is the projection from $\otimes_d \mathbf{C}^n$ onto $W_T \mathbf{C}^n$. For $d = 2$ the decomposition (4.1.4) looks like

$$c_{\substack{1\\2}} \oplus c_{12} : \otimes_2 \mathbf{C}^n \to W_{12}\mathbf{C}^n \oplus W_{\substack{1\\2}}\mathbf{C}^n \simeq S_2\mathbf{C}^n \oplus \wedge_2\mathbf{C}^n \qquad (4.1.5)$$
$$v_1 \otimes v_2 \to (v_1 \otimes v_2 + v_2 \otimes v_1) + (v_1 \otimes v_2 - v_2 \otimes v_1).$$

For $d = 3$ there are three different partitions and four different SYT's. We have

$$\otimes_3 \mathbf{C}^n \simeq W_{\substack{1\\2\\3}}\mathbf{C}^n \oplus W_{\substack{12\\3}}\mathbf{C}^n \oplus W_{\substack{13\\2}}\mathbf{C}^n \oplus W_{123}\mathbf{C}^n. \qquad (4.1.6)$$

As is seen in Example 4.1.8 (b), the last summand in (4.1.6) is zero if $n = 2$.

We now construct an explicit basis for the Weyl module $W_T \mathbf{C}^n$. As a consequence we will obtain a description of the formal character of $W_T \mathbf{C}^n$. A *semistandard Young tableau* of shape $\lambda \vdash d$ (for short: SSYT λ) is a filling U of the Ferrers diagram of λ with the integers $1, 2, \ldots, n$ (repetition is allowed!) such that the rows of U are weakly increasing and the columns of U are strictly increasing. A *standard bitableau* of shape $\lambda \vdash d$ is a pair (T, U) where T is a SYT λ and U is a SSYT λ. With each standard bitableau (T, U) we associate a basis vector of $\otimes_d \mathbf{C}^n$ as follows. Set $e_{(T,U)} := e_{i_1} \otimes e_{i_2} \otimes \ldots \otimes e_{i_d}$, where $i_j \in [n]$ is in the cell of U which is occupied by $j \in [d]$ in T.

Theorem 4.1.12. *The set $\{c_T(e_{(T,U)}) : (T, U) \text{ standard bitableau}\}$ is a basis for $\otimes_d \mathbf{C}^n$.*

The proof of Theorem 4.1.12 is based on two important techniques in algebraic combinatorics. To show that the set in question is spanning, one uses the *straightening law for bitableaux* in Désarménien et al. (1978). This is a generalization of the straightening law for bracket monomials which was discussed in Sect. 3.1. To show linear independence, it suffices to show that the number of standard bitableaux equals n^d. This can be done using the *Knuth–Robinson–Schensted correspondence* (Knuth 1970).

Example 4.1.13 ($d = 3$, $n = 2$). From Theorem 4.1.12 we get the following explicit formulas for the isomorphism in Example 4.1.8 (b) or (4.1.6):

$$c_{123}(e_{123,111}) = c_{123}(e_1 \otimes e_1 \otimes e_1) = 6 \cdot e_1 \otimes e_1 \otimes e_1$$
$$c_{123}(e_{123,112}) = c_{123}(e_1 \otimes e_1 \otimes e_2)$$
$$= 2(e_1 \otimes e_1 \otimes e_2 + e_1 \otimes e_2 \otimes e_1 + e_2 \otimes e_1 \otimes e_1)$$
$$c_{123}(e_{123,122}) = c_{123}(e_1 \otimes e_2 \otimes e_2)$$
$$= 2(e_1 \otimes e_2 \otimes e_2 + e_2 \otimes e_1 \otimes e_2 + e_2 \otimes e_2 \otimes e_1)$$
$$c_{123}(e_{123,222}) = c_{123}(e_2 \otimes e_2 \otimes e_2) = 6 e_2 \otimes e_2 \otimes e_2$$
$$c_{\substack{13\\2}}(e_{\substack{13\\2}\substack{11\\2}}) = c_{\substack{13\\2}}(e_1 \otimes e_2 \otimes e_1) = 2e_1 \otimes e_2 \otimes e_1 - 2e_2 \otimes e_1 \otimes e_1$$
$$c_{\substack{13\\2}}(e_{\substack{13\\2}\substack{12\\2}}) = c_{\substack{13\\2}}(e_1 \otimes e_2 \otimes e_2) = e_1 \otimes e_2 \otimes e_2 - e_2 \otimes e_1 \otimes e_2$$

4.1. Representation theory of the general linear group

$$c_{12\atop 3}(e_{12\atop 3\ 2}{}_{11}) = c_{12\atop 3}(e_1 \otimes e_1 \otimes e_2) = 2e_1 \otimes e_1 \otimes e_2 - 2e_2 \otimes e_1 \otimes e_1$$

$$c_{12\atop 3}(e_{12\atop 3\ 2}{}_{12}) = c_{12\atop 3}(e_1 \otimes e_2 \otimes e_2) = e_1 \otimes e_2 \otimes e_2 - e_2 \otimes e_2 \otimes e_1$$

Theorems 4.1.11 and 4.1.12 have a number of important consequences.

Corollaries 4.1.14. Let d, n be positive integers and λ a partition of d.

(1) The set $\{c_T(e_{(T,U)}) : U \text{ SSYT}\lambda\}$ is a basis for the Weyl module $W_T \mathbf{C}^n$.
(2) The formal character of the Weyl module $W_\lambda \mathbf{C}^n \simeq W_T \mathbf{C}^n$ equals

$$s_\lambda(t_1, t_2, \ldots, t_n) = \sum_{U \text{ SSYT}\lambda} \prod_{i=1}^n t_i^{\#i\text{'s in } U}. \tag{4.1.7}$$

The monomial $\prod_{i=1}^n t_i^{\#i\text{'s in } U}$ is called the *weight* of the SSYTλ U. It turns out that the formal character in (4.1.7) is equal to the *Schur polynomial* $s_\lambda = s_\lambda(t_1, \ldots, t_n)$ as defined in Sect. 1.1. This result is a non-trivial identity in the theory of symmetric polynomials; for the proof and many details we refer to Macdonald (1979).

Example 4.1.15. We consider the case $d = 6$, $n = 3$, $\lambda = (4, 2)$. There are precisely 27 SSYTλ. Here is a complete list of all of them:

```
1 1 1 1    1 1 1 1    1 1 1 1    1 1 1 2    1 1 1 2    1 1 1 2
2 2        2 3        3 3        2 2        2 3        3 3

1 1 1 3    1 1 1 3    1 1 1 3    1 1 2 2    1 1 2 2    1 1 2 2
2 2        2 3        3 3        2 2        2 3        3 3

1 1 2 3    1 1 2 3    1 1 2 3    1 1 3 3    1 1 3 3    1 1 3 3
2 2        2 3        3 3        2 2        2 3        3 3

1 2 2 2    1 2 2 2    1 2 2 3    1 2 2 3    1 2 3 3    1 2 3 3
2 3        3 3        2 3        3 3        2 3        3 3

2 2 2 2    2 2 2 3    2 2 3 3
3 3        3 3        3 3
```

We form the sum of the weights of all 27 SSYTλ (in the above order):

$$\begin{aligned}
s_{(4,2)} &= t_1^4 t_2^2 + t_1^4 t_2 t_3 + t_1^4 t_3^2 + t_1^3 t_2^3 + t_1^3 t_2^2 t_3 + t_1^3 t_2 t_3^2 + t_1^3 t_2^2 t_3 + t_1^3 t_2 t_3^2 + t_1^3 t_3^3 \\
&\quad + t_1^2 t_2^4 + t_1^2 t_2^3 t_3 + t_1^2 t_2^2 t_3^2 + t_1^2 t_2^3 t_3 + t_1^2 t_2^2 t_3^2 + t_1^2 t_2 t_3^3 + t_1^2 t_2^2 t_3^2 + t_1^2 t_2 t_3^3 \\
&\quad + t_1^2 t_3^4 + t_1 t_2^4 t_3 + t_1 t_2^3 t_3^2 + t_1 t_2^3 t_3^2 + t_1 t_2^2 t_3^3 + t_1 t_2^2 t_3^3 + t_1 t_2 t_3^4 + t_2^4 t_3^2 \\
&\quad + t_2^3 t_3^3 + t_2^2 t_3^4 \\
&= (t_1^2 + t_1 t_2 + t_2^2)(t_1^2 + t_1 t_3 + t_3^2)(t_2^2 + t_2 t_3 + t_3^2).
\end{aligned}$$

This Schur polynomial is the formal character of the 27-dimensional Weyl module $W_{(4,2)}\mathbf{C}^3$. By the above remarks, it satisfies the identity

$$s_{(4,2)} = \frac{\begin{vmatrix} t_1^{4+2} & t_2^{4+2} & t_3^{4+2} \\ t_1^{2+1} & t_2^{2+1} & t_3^{2+1} \\ t_1^{0+0} & t_2^{0+0} & t_3^{0+0} \end{vmatrix}}{\begin{vmatrix} t_1^2 & t_2^2 & t_3^2 \\ t_1^1 & t_2^1 & t_3^1 \\ 1 & 1 & 1 \end{vmatrix}}.$$

Any partition $\lambda \vdash d$ can be encoded into a monomial $\omega(\lambda) := t_1^{\nu_1} t_2^{\nu_2} \cdots t_n^{\nu_n}$ as follows: the exponent ν_i is the cardinality of the i-th column in the Ferrers diagram of λ. Equivalently, $\nu_i = \#\{j : \lambda_j \geq i\}$. It is easy to see that $\omega(\lambda)$ is the lexicographically leading monomial of the Schur polynomial $s_\lambda(t_1, t_2, \ldots, t_n)$. This monomial uniquely characterizes the partition λ and hence the Weyl module $W_\lambda \mathbf{C}^n$. We call $\omega(\lambda)$ the *highest weight* of $W_\lambda \mathbf{C}^n$. For instance, $t_1^4 t_2^2$ is the highest weight of the Weyl module in Example 4.1.15.

A main problem in representation theory is to decompose a given representation (ρ, V) into a direct sum of irreducible representations

$$V \simeq \bigoplus_\lambda c_\lambda \, W_\lambda \mathbf{C}^n. \tag{4.1.8}$$

A generally satisfactory solution to this problem is the list of all nonnegative integers c_λ, which are called *multiplicities*. Once the multiplicities are known, then one may (or may not) ask for a more explicit description of the isomorphism in (4.1.8).

Our discussion shows that the problem of determining the *multiplicities* c_λ is a problem in the theory of symmetric polynomials. The Schur polynomials s_λ are a $\mathbf{Z}$-basis for the Grothendieck ring $M(\Gamma) \simeq \mathbf{Z}[t_1, \ldots, t_n]^{S_n}$. We need to find the coefficients in the expansion

$$f_\rho = \sum c_\lambda s_\lambda, \tag{4.1.9}$$

where f_ρ is the formal character of the given representation (ρ, V). This can be done using the following subduction algorithm (cf. Algorithm 3.2.8):

Algorithm 4.1.16 (Expanding a Symmetric Polynomial into Schur Polynomials). Input: A symmetric polynomial $f \in \mathbf{Z}[t_1, \ldots, t_n]^{S_n}$ homogeneous of degree d. Output: The unique representation $f = \sum_{\lambda \vdash d} c_\lambda s_\lambda$ in terms of Schur polynomials.

1. If $f = 0$ then output the zero polynomial.

4.1. Representation theory of the general linear group

2. Let $t_1^{v_1} t_2^{v_2} \cdots t_n^{v_n}$ be the lexicographically leading monomial of f, let $\lambda \vdash d$ be the unique partition with $\omega(\lambda) = t_1^{v_1} t_2^{v_2} \cdots t_n^{v_n}$, and let c be the coefficient of $\omega(\lambda)$ in f.
3. Compute the Schur polynomial s_λ (e.g., using the formula in Sect. 1.1).
4. Output the summand $c \cdot s_\lambda$, replace f by $f - c \cdot s_\lambda$, and return to step 1.

A much more efficient version of this algorithm and other conversion algorithms for symmetric polynomials have been implemented in J. Stembridge's MAPLE package "SF" (available at no cost from J. Stembridge, University of Michigan, Ann Arbor).

Example 4.1.17. Let $S_3 S_2 \mathbf{C}^3$ denote the space of polynomial functions of degree 3 in the coefficients of a ternary quadric

$$a_{200} x^2 + a_{020} y^2 + a_{002} z^2 + a_{110} xy + a_{101} xz + a_{011} yz. \quad (4.1.10)$$

A basis is given by the set of monomials $a_{200}^{i_1} a_{020}^{i_2} a_{002}^{i_3} a_{110}^{j_3} a_{101}^{j_2} a_{011}^{j_1}$, where $i_1 + i_2 + i_3 + j_1 + j_2 + j_3 = 3$ in nonnegative integers. The action of $\Gamma = GL(\mathbf{C}^3)$ by linear substitution on (x, y, z) gives rise to a linear action ρ on $S_3 S_2 \mathbf{C}^3$. Thus $S_3 S_2 \mathbf{C}^3$ is a Γ-module of dimension $56 = \binom{\binom{2+3-1}{2}+3-1}{3}$. Its formal character equals

$$f_\rho(t_1, t_2, t_3) = \sum_{i_1+i_2+i_3+j_1+j_2+j_3=3} t_1^{2i_1+j_2+j_3} t_2^{j_1+2i_2+j_3} t_3^{j_1+j_2+2i_3}. \quad (4.1.11)$$

Using Algorithm 4.1.6 we obtain the following decomposition into Schur polynomials

$$\begin{aligned} f_\rho(t_1, t_2, t_3) &= (t_1^6 + \text{lower terms}) + (t_1^4 t_2^2 + \text{lower terms}) + t_1^2 t_2^2 t_3^2 \\ &= s_{(6)}(t_1, t_2, t_3) + s_{(4,2)}(t_1, t_2, t_3) + s_{(3,3,3)}(t_1, t_2, t_3). \end{aligned} \quad (4.1.12)$$

Thus $S_3 S_2 \mathbf{C}^3$ splits up as the direct sum of three irreducible Γ-modules. The first one (with highest weight t_1^6) has dimension 28 and is isomorphic to $S_6 \mathbf{C}^3$. The second one (with highest weight $t_1^4 t_2^2$) has dimension 27 and is isomorphic to the Weyl module in Example 4.1.15. The third and most interesting piece is the 1-dimensional representation $\rho : A \mapsto \det(A)^2$, or, equivalently,

$$W_{(2,2,2)} \mathbf{C}^3 \simeq \wedge_3 \mathbf{C}^3 \otimes \wedge_3 \mathbf{C}^3. \quad (4.1.13)$$

This submodule of $S_3 S_2 \mathbf{C}^3$ is spanned by the discriminant of the ternary quadric (4.1.10):

$$\Delta := a_{200} a_{020} a_{002} + 2 a_{011} a_{101} a_{110} - a_{200} a_{011}^2 - a_{020} a_{101}^2 - a_{002} a_{110}^2. \quad (4.1.14)$$

We have thus proved that an element $p \in S_3 S_2 \mathbf{C}^3$ satisfies $A \circ p = (\det A)^2 \cdot p$ for all $A \in \Gamma$ if and only if it is a multiple of the discriminant Δ. Using the same argument as in Lemma 3.2.3, we conclude that the multiples of Δ are the only $SL(\mathbf{C}^3)$-invariants in $S_3 S_2 \mathbf{C}^3$.

We have seen in the previous example that quite interesting Γ-modules can be obtained by repeated application of the operators S_d and $\wedge_d$. For instance,

- $S_m S_d \mathbf{C}^n$ is the space of homogeneous polynomials of degree d in the coefficients of a generic homogeneous polynomials of degree m in $x_1, \ldots, x_n$.
- $S_m \wedge_d \mathbf{C}^n$ is the space of homogeneous bracket polynomials of degree m in $\mathbf{C}[\Lambda(n,d)]$.

More generally, we can build new representations from old ones by applying the so-called *Schur functors* $W_\lambda(\cdot)$. Suppose that (ρ, V) is any Γ-module. Then we get a new Γ-module $(W_\lambda(\rho), W_\lambda(V))$ as follows. The underlying vector space is the Weyl module $W_\lambda(V)$ with respect to the representation ρ_λ of $GL(V)$. The new Γ-action on this space is defined by the composition $W_\lambda(\rho) := \rho_\lambda \circ \rho$.

By Corollary 4.1.9 there must exist corresponding functors in the theory of symmetric polynomials. These functors are called *plethysms*. We summarize this construction in the following proposition. Let $f_\rho(t_1, \ldots, t_n)$ be the formal character of the Γ-module (ρ, V), where $\Gamma = GL(\mathbf{C}^n)$, suppose that V has dimension m, and let $s_\lambda(z_1, z_2, \ldots, z_m)$ be the Schur polynomial which is the formal character of the $GL(V)$-module $W_\lambda(V)$. We write $f_\rho(t_1, \ldots, t_n)$ as the sum of m not necessarily distinct monomials of the form $t_1^{i_1} t_2^{i_2} \ldots t_n^{i_n}$.

Proposition 4.1.18. The formal character $f_{W_\lambda(\rho)}(t_1, t_2, \ldots, t_n)$ of the Γ-module $(W_\lambda(\rho), W_\lambda(V))$ is obtained by substituting the m monomials $t_1^{i_1} t_2^{i_2} \ldots t_n^{i_n}$ for the m variables z_i into the Schur polynomial $s_\lambda(z_1, z_2, \ldots, z_m)$ (in any order).

We illustrate this proposition by computing the formal character of the $GL(\mathbf{C}^4)$-module $\wedge_3 \wedge_2 \mathbf{C}^4$. Here $V = \wedge_2 \mathbf{C}^4$ has dimension $m = 6$ and formal character

$$f_\rho(t_1, t_2, t_3, t_4) = t_1 t_2 + t_1 t_3 + t_1 t_4 + t_2 t_3 + t_2 t_4 + t_3 t_4.$$

The $GL(V)$-module $\wedge_3 V = W_{(1,1,1)} V$ has the formal character

$$s_{(1,1,1)}(z_1, z_2, z_3, z_4, z_5, z_6) = \sum_{i=1}^{4} \sum_{j=i+1}^{5} \sum_{k=j+1}^{6} z_i z_j z_k.$$

By Proposition 4.1.18, the $GL(\mathbf{C}^4)$-module $W_{(1,1,1)}(V) = \wedge_3 \wedge_2 \mathbf{C}^4$ has the formal character

$$s_{(1,1,1)}(t_1t_2, t_1t_3, \ldots, t_3t_4) = t_1^3 t_2 t_3 t_4 + t_1^2 t_2^2 t_3^2 + 2t_1^2 t_2^2 t_3 t_4 + t_1^2 t_2^2 t_4^2$$
$$+ 2t_1^2 t_2 t_3^2 t_4 + 2t_1^2 t_2 t_3 t_4^2 + t_1^2 t_3^2 t_4^2 + t_1 t_2^3 t_3 t_4 + 2t_1 t_2^2 t_3^2 t_4 + 2t_1 t_2^2 t_3 t_4^2$$
$$+ t_1 t_2 t_3^3 t_4 + 2t_1 t_2 t_3^2 t_4^2 + t_1 t_2 t_3 t_4^3 + t_2^2 t_3^2 t_4^2.$$

Exercises

(1) Prove that every irreducible polynomial Γ-representation is homogeneous.
(2) Verify that the eight Young symmetrized vectors in Example 4.1.13 are a basis of $\otimes_3 \mathbf{C}^2$. Compute the determinant of the 8×8-transformation matrix with respect to the standard basis.
(3) * Describe the action of $GL(\mathbf{C}^n)$ on $S_m \wedge_d \mathbf{C}^n$.
(4) Determine the decomposition of $\wedge_3 \wedge_2 \mathbf{C}^4$ into irreducible $GL(\mathbf{C}^4)$-representations.

4.2. Binary forms revisited

We can now rephrase the main problem of invariant theory in the language of representation theory. The techniques developed in the previous section will then be applied to the rings of invariants and covariants of a binary form. In particular, we give an algorithm for computing the Hilbert (Molien) series of these invariant rings.

As before let $\Gamma = GL(\mathbf{C}^n)$. Let (V, ρ) be a Γ-representation of dimension m. We assume that (V, ρ) is homogeneous of degree d. For each integer $k \geq 0$ we get a representation $(S_k(V), S_k(\rho))$ of degree dk. The symmetric power $S_k(V)$ is a vector space of dimension $\binom{m+k-1}{k}$. We identify it with the space of homogeneous polynomial functions of degree k on V. The ring of polynomial functions on V is denoted $\mathbf{C}[V] = \bigoplus_{k=0}^{\infty} S_k(V)$.

A polynomial $f \in \mathbf{C}[V]$ is an *invariant* of index g provided $A \circ f = \det(A)^g \cdot f$ for all $A \in \Gamma$. Here $A \circ f$ is the polynomial function on V defined by $(A \circ f)(\mathbf{v}) = f(\rho(A) \cdot \mathbf{v})$ for $\mathbf{v} \in V$. This implies that f is a homogeneous polynomial of degree gn/d; in particular gn/d must be an integer. The invariant ring $\mathbf{C}[V]^\Gamma$ is the $\mathbf{C}$-linear span of all homogeneous invariants. Our main problem is to determine the invariant ring as explicitly as possible.

To this end we consider $\mathbf{C}[V]$ as an (infinite-dimensional) Γ-module. Its formal character is the generating function

$$f_{\mathbf{C}[V]}(t_1, \ldots, t_n) = \sum_{k=0}^{\infty} f_{S_k(V)}(t_1, \ldots, t_n), \tag{4.2.1}$$

where each summand $f_{S_k(V)}(t_1, \ldots, t_n)$ is a symmetric polynomial of degree dk. We can compute (4.2.1) using plethysms as in Proposition 4.1.18. Let $f_\rho(t_1, \ldots, t_n)$ denote the formal character of the m-dimensional representation (V, ρ). We write the symmetric polynomial f_ρ as the sum of m monomials (not necessarily distinct):

$$f_\rho(t_1, \ldots, t_n) = t_1^{i_{11}} t_2^{i_{12}} \cdots t_n^{i_{1n}} + t_1^{i_{21}} t_2^{i_{22}} \cdots t_n^{i_{2n}} + \ldots + t_1^{i_{m1}} t_2^{i_{m2}} \cdots t_n^{i_{mn}}. \quad (4.2.2)$$

Proposition 4.2.1. The formal character of the Γ-module $\mathbf{C}[V]$ is the generating function

$$f_{\mathbf{C}[V]}(t_1, \ldots, t_n) = \frac{1}{\prod_{\mu=1}^{m}(1 - t_1^{i_{\mu 1}} t_2^{i_{\mu 2}} \cdots t_n^{i_{\mu n}})}. \quad (4.2.3)$$

Proof. The formal character of $S_k(V)$ as a $GL(V)$-module equals

$$s_{(k)}(z_1, \ldots, z_m) = \sum_{\substack{i_1 + \ldots + i_m = k \\ i_1, \ldots, i_m \geq 0}} z_1^{i_1} z_2^{i_2} \cdots z_m^{i_m}.$$

Therefore the formal character of $\mathbf{C}[V]$ as a $GL(V)$-module is the generating function

$$\sum_{k=0}^{\infty} \sum_{\substack{i_1 + \ldots + i_m = k \\ i_1, \ldots, i_m \geq 0}} z_1^{i_1} z_2^{i_2} \cdots z_m^{i_m} = \frac{1}{\prod_{\mu=1}^{m}(1 - z_\mu)}. \quad (4.2.4)$$

Following Proposition 4.1.18, we now substitute $z_\mu \mapsto t_1^{i_{\mu 1}} t_2^{i_{\mu 2}} \cdots t_n^{i_{\mu n}}$ in (4.2.4) to get the formal character of $\mathbf{C}[V]$ as a Γ-module. ◁

Example 4.2.2. Let $n = 2$ and consider the space of binary quadrics $V = S_2 \mathbf{C}^2$. This is a three-dimensional Γ-module having formal character $f_\rho = t_1^2 + t_1 t_2 + t_2^2$. By Proposition 4.2.1, the polynomial ring $\mathbf{C}[V]$ is a Γ-module having the formal character

$$f_{\mathbf{C}[V]}(t_1, t_2) = \frac{1}{(1 - t_1^2)(1 - t_1 t_2)(1 - t_2^2)} = \quad (4.2.5)$$

$$1 + (t_1^2 + t_1 t_2 + t_2^2) + (t_1^4 + t_1^3 t_2 + t_1^2 t_2^2 + t_1 t_2^3 + t_2^4) + \underline{(t_1^2 t_2^2)}$$
$$+ (t_1^6 + t_1^5 t_2 + t_1^4 t_2^2 + t_1^3 t_2^3 + t_1^2 t_2^4 + t_1 t_2^5 + t_2^6) + (t_1^4 t_2^2 + t_1^3 t_2^3 + t_1^2 t_2^4)$$
$$+ (t_1^8 + t_2 t_1^7 + t_1^6 t_2^2 + t_1^5 t_2^3 + t_1^4 t_2^4 + t_1^3 t_2^5 + t_1^2 t_2^6 + t_1 t_2^7 + t_2^8)$$
$$+ (t_1^6 t_2^2 + t_1^5 t_2^3 + t_1^4 t_2^4 + t_1^3 t_2^5 + t_1^2 t_2^6) + \underline{(t_1^4 t_2^4)} + \ldots \ldots \text{ (higher terms)}.$$

Let (V, ρ) be any Γ-module of degree d. We can decompose the graded Γ-module $\mathbf{C}[V]$ into a unique direct sum of the irreducible Weyl modules:

$$\mathbf{C}[V] = \bigoplus_{k=0}^{\infty} S_k(V) = \bigoplus_{k=0}^{\infty} \bigoplus_{\lambda \vdash dk} c_\lambda W_\lambda \mathbf{C}^n \quad (4.2.6)$$

As in (4.1.19), the multiplicities c_λ can be read off from the corresponding

4.2. Binary forms revisited

decomposition of the formal character of $\mathbf{C}[V]$ into a positive linear combination of Schur polynomials

$$f_{\mathbf{C}[V]}(t_1, t_2 \ldots, t_n) = \sum_{k=0}^{\infty} \sum_{\lambda \vdash dk} c_\lambda \, s_\lambda(t_1, t_2, \ldots, t_n). \tag{4.2.7}$$

For instance, the identity (4.2.5) in Example 4.2.2 translates into the following irreducible decomposition of $GL(\mathbf{C}^2)$-modules:

$$\mathbf{C}[S_2\mathbf{C}^2] = W_{(0,0)}\mathbf{C}^2 \oplus W_{(2,0)}\mathbf{C}^2 \oplus W_{(4,0)}\mathbf{C}^2 \oplus W_{(2,2)}\mathbf{C}^2 \oplus W_{(6,0)}\mathbf{C}^2 \oplus W_{(4,2)}\mathbf{C}^2 \oplus \ldots$$

For each integer $g \geq 0$ there is a unique partition $\lambda = \underbrace{(g, g, \ldots, g)}_{n \text{ times}}$ of gn having a rectangular Ferrers diagram. The corresponding Weyl module $W_{(g,g,\ldots,g)}\mathbf{C}^n$ equals the one-dimensional Γ-module defined by $\rho = \det^g$. In precise terms, the underlying one-dimensional vector space should be written as $(\wedge_n \mathbf{C}^n)^{\otimes g}$. Its Schur polynomial equals $s_{(g,g,\ldots,g)} = t_1^g t_2^g \cdots t_n^g$.

A homogeneous polynomial in $f \in \mathbf{C}[V]$ is an invariant of index g if and only if it lies in a Γ-submodule isomorphic to $W_{(g,g,\ldots,g)}\mathbf{C}^n = ((\wedge_n \mathbf{C}^n)^{\otimes g}, \det^g)$. Using the notation in (4.2.6) we let $c_{(g,g,\ldots,g)}$ denote the multiplicity of this Weyl module in $\mathbf{C}[V]$. This number counts the linearly independent invariants of index g and hence of degree gn/d. The following theorem summarizes the representation-theoretic view on our main problem.

Theorem 4.2.3. Let V be any Γ-module. Then its invariant ring has the decomposition

$$\mathbf{C}[V]^\Gamma = \bigoplus_{g=0}^{\infty} c_{(g,g,\ldots,g)} \, W_{(g,g,\ldots,g)} \mathbf{C}^n$$

as a Γ-module. In particular, the Hilbert function of the invariant ring is given by

$$\dim \mathbf{C}[V]^\Gamma_k = \begin{cases} c_{(g,g,\ldots,g)} & \text{if } g = kd/n \text{ is an integer} \\ 0 & \text{otherwise}. \end{cases}$$

Example 4.2.2 (continued). Let $n = 2$, $d = 2$ and $V = S_2\mathbf{C}^2$, the space of binary quadrics. As can be seen from (4.2.5), the multiplicity of the Weyl module $W_{(g,g)}\mathbf{C}^2$ in $\mathbf{C}[V]$ equals $c_{(g,g)} = 1$ if g is even and $c_{(g,g)} = 0$ if g is odd. Therefore the invariant ring $\mathbf{C}[V]^\Gamma$ has the Hilbert function $1/(1-z^2)$. This proves that $\mathbf{C}[V]^\Gamma$ is generated as $\mathbf{C}$-algebra by one quadratic polynomial, which is the discriminant of the binary quadric.

For the remainder of this section we fix $n = 2$. We will develop the representation theory of $\Gamma = GL(\mathbf{C}^2)$ in detail and apply it to the study of binary forms. We begin with a description of the irreducible Γ-modules $W_\lambda \mathbf{C}^2$ and their Schur polynomials. Each Weyl module is indexed by a partition $\lambda = (i, j)$ having at most two distinct parts $i \geq j \geq 0$.

Lemma 4.2.4. The Weyl module $W_{(i,j)}\mathbf{C}^2$ is isomorphic to $S_{i-j}\mathbf{C}^2 \otimes (\wedge_2 \mathbf{C}^2)^{\otimes j}$.

Proof. For the partition $\lambda = (i, j)$ there are precisely $i - j + 1$ SSYTλ's with entries in $\{1, 2\}$. A typical such SSYTλ looks like

$$\begin{array}{ccccccccc} 1 & \cdots & 1 & 1 & \cdots & 1 & 1 & \ldots & 1 \\ 1 & \cdots & 1 & 2 & \cdots & 2 & & & \end{array}$$

By Corollary 4.1.14 (b), the formal character of $W_{(i,j)}\mathbf{C}^2$ equals the sum of the weights of all $i - j + 1$ SSYTλ's:

$$\begin{aligned} s_{(i,j)}(t_1, t_2) &= t_1^i t_2^j + t_1^{i-1} t_2^{j+1} + t_1^{i-2} t_2^{j+2} + \ldots + t_1^j t_2^i \\ &= (t_1^{i-j} + t_1^{i-j-1} t_2 + t_1^{i-j-2} t_2^2 + \ldots + t_2^{i-j}) \cdot (t_1^j t_2^j) \\ &= \frac{t_1^{i+1} t_2^j - t_1^j t_2^{i+1}}{t_1 - t_2} \end{aligned}$$

By Lemma 4.1.6, this Schur polynomial is the formal character of the tensor product $S_{i-j}\mathbf{C}^2 \otimes (\wedge_2 \mathbf{C}^2)^{\otimes j}$. Lemma 4.2.4 now follows directly from Theorem 4.1.7. ◁

Let $V = S_d \mathbf{C}^2$ denote the space of binary d-forms

$$a_0 x^d + \binom{d}{1} a_1 x^{d-1} y + \binom{d}{2} a_2 x^{d-2} y^2 + \binom{d}{3} a_3 x^{d-3} y^3 + \ldots + a_d y^d.$$

We identify the polynomial ring $\mathbf{C}[a_0, \ldots, a_d]$ with the Γ-module $\mathbf{C}[V] = \bigoplus_{k=0}^{\infty} S_k S_d \mathbf{C}^2$. More generally, the polynomial ring $\mathbf{C}[a_0, \ldots, a_d, x, y]$ is a rational Γ-module via the natural Γ-action defined in Sect. 3.6. For $A \in \Gamma$ this action is given by

$$\begin{aligned} (a_0, a_1, \ldots, a_d)^t &\mapsto \rho_{S_d \mathbf{C}^2}(A) \cdot (a_0, a_1, \ldots, a_d)^t \quad \text{and} \\ (x, y)^t &\mapsto A^{-1} \cdot (x, y)^t. \end{aligned} \tag{4.2.8}$$

Let $\mathbf{C}[a_0, \ldots, a_d, x, y]^\Gamma$ denote the ring of covariants, and let $\mathbf{C}[a_0, \ldots, a_d, x, y]^\Gamma_{(k,i)}$ denote the vector subspace of covariants of degree k, order i and index $g = \frac{dk-i}{2}$.

We now determine the decomposition of $\mathbf{C}[V]$ into irreducible Γ-modules. For $i = 0, 1, \ldots, \lfloor \frac{dk}{2} \rfloor$ let $m(d, k, i)$ denote the multiplicity of the Weyl module

$$W_{(\frac{dk+i}{2}, \frac{dk-i}{2})} \mathbf{C}^2 = S_i \mathbf{C}^2 \otimes (\wedge^2 \mathbf{C}^2)^{\otimes g}$$

4.2. Binary forms revisited

in $\mathbf{C}[V]$, or, equivalently, in $S_k S_d \mathbf{C}^2$. We write $\mathcal{W}_{(d,k,i)} \simeq m(d,k,i) \cdot \mathcal{W}_{(\frac{dk+i}{2}, \frac{dk-i}{2})} \mathbf{C}^2$ for the corresponding submodule of $\mathbf{C}[V]$. Let $f \in \mathbf{C}[V]$ and let g be the largest integer for which there exists $\tilde{f} \in \mathbf{C}[V, A]$ such that $A \circ f = \det(A)^g \cdot \tilde{f}$ for all $A \in \Gamma$. Then f lies in $\mathcal{W}_{(d,k,i)}$ if and only if f is homogeneous of degree k and the transformation $f \mapsto \tilde{f}$ is isomorphic to the action of Γ on $S_i \mathbf{C}^2$.

Given any polynomial $f \in \mathbf{C}[a_0, \ldots, a_d, x, y]$, we view f as a polynomial in x and y with coefficients in $\mathbf{C}[V]$. The *leading coefficient* of f is the polynomial $\text{lead}(f) := f(a_0, \ldots, a_d, 1, 0)$ in $\mathbf{C}[V]$.

Lemma 4.2.5. *The assignment $f \mapsto \text{lead}(f)$ defines a vector space monomorphism from $\mathbf{C}[a_0, \ldots, a_d, x, y]^\Gamma_{(k,i)}$ into $\mathcal{W}_{d,k,i}$.*

Proof. Let f be a covariant of degree k, order i and index $g = \frac{dk-i}{2}$. Then

$$f\left(\rho_{S_d \mathbf{C}^2}(A) \circ (a_0, \ldots, a_d), A^{-1} \circ (x, y)\right) = \det(A)^g \cdot f(a_0, \ldots, a_d, x, y).$$

Substituting $(x, y) \mapsto (1, 0)$ shows that $\text{lead}(f)$ is a polynomial of degree k in $\mathbf{C}[V]$, which satisfies the above defining condition of $\mathcal{W}_{(d,k,i)}$. The map $f \mapsto \text{lead}(f)$ is injective because every covariant f can be recovered from its leading coefficient $g = \text{lead}(f)$ as follows:

$$f(a_0, \ldots, a_d, x, y) = g\left(\rho_{S_d}\begin{pmatrix} x & 1 \\ y & 0 \end{pmatrix} \cdot (a_0, \ldots, a_d)\right). \triangleleft \qquad (4.2.9)$$

Theorem 4.2.6 (Robert's theorem). *The dimension of the space $\mathbf{C}[a_0, \ldots, a_d, x, y]^\Gamma_{(k,i)}$ of covariants of degree k and order i is equal to $m(d, k, i)$.*

Proof. Our argument follows Schur and Grunsky (1968: p. 28). Let $\tilde{m}(d, k, i)$ denote the dimension of the space of covariants of degree k and order i. By Lemma 4.2.5, the map $f \mapsto \text{lead}(f)$ defines a vector space monomorphism

$$\tilde{m}(d, k, i) \cdot \mathcal{W}_{(g,g)} \mathbf{C}^2 \simeq \mathbf{C}[a_0, \ldots, a_d, x, y]^\Gamma_{(k,i)}$$

$$\rightarrow \mathcal{W}_{(d,k,i)} \simeq m(d, k, i) \cdot \mathcal{W}_{(\frac{dk+i}{2}, \frac{dk-i}{2})} \mathbf{C}^2.$$

We need to show that $m(d, k, i) = \tilde{m}(d, k, i)$.

Consider the *unipotent subgroup* $U = \{\begin{pmatrix} 1 & \alpha \\ 0 & 1 \end{pmatrix} : \alpha \in \mathbf{C}\}$ in Γ. The action of U on the Γ-module $\mathcal{W}_{(\frac{dk+i}{2}, \frac{dk-i}{2})} \mathbf{C}^2$ has a unique fixed vector, up to scaling, say g. This follows, for instance, from the construction of the irreducible Γ-modules in Theorem 4.1.11. We now define $f \in \mathbf{C}[a_0, \ldots, a_d, x, y]$ by the formula (4.2.9). Then f is a covariant and $g = \text{lead}(f)$. This proves $m(d, k, i) \leq \tilde{m}(d, k, i)$.

Conversely, if f is any covariant, then $g = \operatorname{lead}(f)$ is a unipotent invariant. This implies $m(d,k,i) \geq \tilde{m}(d,k,i)$, and we are done. ◁

The multiplicities $m(d,k,i)$ can be expressed in terms of a certain explicit generating function. This technique is due to Cayley and Sylvester. The *q-binomial coefficient* is the expression

$$\begin{bmatrix} d+k \\ k \end{bmatrix}(q) := \frac{(1-q^{d+1})(1-q^{d+2})\cdots(1-q^{d+k})}{(1-q)(1-q^2)\cdots(1-q^k)}, \qquad (4.2.10)$$

where q is an indeterminate. The rational function (4.2.10) is a polynomial in q of degree kd. It is sometimes called the *Gaussian polynomial*. Its coefficients have the following combinatorial interpretation. If

$$\begin{bmatrix} d+k \\ k \end{bmatrix}(q) = \sum_{n=0}^{dk} p(d,k,n) \cdot q^n$$

then $p(d,k,n)$ equals the number of partitions of the integer n into at most k parts, with largest part $\leq d$. For the proof of this statement and related combinatorial interpretations we refer to Stanley (1986: chapter 1). We note that $\begin{bmatrix} d+k \\ k \end{bmatrix}(q)$ specializes to the usual binomial coefficient $\binom{d+k}{k}$ for $q=1$. Many of the familiar properties of binomial coefficients generalize to q-binomial coefficients: For instance, we have $\begin{bmatrix} d+k \\ k \end{bmatrix}(q) = \begin{bmatrix} d+k \\ d \end{bmatrix}(q)$ because $p(d,k,n) = p(k,d,n)$.

Theorem 4.2.7. *The dimension $m(d,k,i)$ of the space of covariants of degree k and order i of a binary d-form is given by the generating function*

$$(1-q) \cdot \begin{bmatrix} d+k \\ k \end{bmatrix}(q) = \sum_{g=0}^{\lfloor \frac{dk}{2} \rfloor} m(d,k,dk-2g) \cdot q^g + O(q^{\lfloor \frac{dk}{2} \rfloor + 1}). \qquad (4.2.11)$$

Proof. The formal character of the Γ-module $\mathbf{C}[V]$ equals

$$\frac{1}{(1-t_1^d)(1-t_1^{d-1}t_2)(1-t_1^{d-2}t_2^2)\cdots(1-t_2^d)} = \sum_{k=0}^{\infty} \begin{bmatrix} d+k \\ k \end{bmatrix}\left(\frac{t_1}{t_2}\right) \cdot t_2^{kd}. \qquad (4.2.12)$$

The identity (4.2.12) follows from known results on the partition function; see, e.g., formula (29) in Stanley (1986: p. 39). The summand on the right hand side of (4.2.12) equals the formal character of $S_k S_d \mathbf{C}^2$. By definition, $m(d,k,dk-2g)$

4.2. Binary forms revisited

is the multiplicity of $W_{(dk-g,g)}\mathbf{C}^2$ in $S_k S_d \mathbf{C}^2$. Therefore we have

$$\begin{bmatrix} d+k \\ k \end{bmatrix}\left(\frac{t_1}{t_2}\right) \cdot t_2^{kd} = \sum_{g=0}^{\lfloor \frac{kd}{2} \rfloor} m(d,k,dk-2g) \cdot \frac{t_2^{kd-g+1} t_1^g - t_2^g t_1^{kd-g+1}}{t_2 - t_1}. \qquad (4.2.13)$$

We substitute $t_1 = q$ and $t_2 = 1$ in (4.2.13), and we multiply both sides by $1 - q$. This gives the desired identity (4.2.11). ◁

Corollary 4.2.8. The number of linearly independent covariants of degree k and order i of a binary d-form equals

$$m(d,k,i) = p\left(d, k, \frac{dk-i}{2}\right) - p\left(d, k, \frac{dk-i-2}{2}\right), \qquad (4.2.14)$$

where $p(d,k,n)$ equals the number of partitions of n into $\leq k$ parts with largest part $\leq d$.

These results provide useful algorithmic tools for precomputing the Hilbert series (or parts thereof) for the rings of invariants and covariants of binary forms. For many examples of such calculations see Schur and Grunsky (1968); see also Springer (1977: section 3.4) for an asymptotic estimate of the number of fundamental invariants. A typical application of the enumerative method is the following duality result for invariants of binary forms.

Corollary 4.2.9 (Hermite reciprocity). The number $m(d,k,0)$ of degree k invariants of a binary d-form equals the number $m(k,d,0)$ of degree d invariants of a binary k-form.

Proof. The expression in terms of partition functions

$$m(d,k,0) = p\left(d, k, \frac{dk}{2}\right) - p\left(d, k, \frac{dk}{2} - 1\right)$$

shows that this function is symmetric in its two parameters d and k. ◁

Example 4.2.10. We illustrate the above techniques for the ring of covariants of a binary cubic ($d = 3$). The generating function (4.2.11) for the covariants of degree k equals

$$\frac{(1-q^{k+1})(1-q^{k+2})(1-q^{k+3})}{(1-q^2)(1-q^3)}. \qquad (4.2.15)$$

For small values of k the generating function (4.2.15) equals

$$k = 1 : \quad \underline{1} - q^4$$
$$k = 2 : \quad 1 + \underline{q^2} - q^5 - q^7$$
$$k = 3 : \quad 1 + q^2 + \underline{q^3} - q^7 - q^8 - q^{10} \qquad (4.2.16)$$
$$k = 4 : \quad 1 + q^2 + q^3 + q^4 + \underline{q^6} - O(q^7)$$
$$k = 5 : \quad 1 + q^2 + q^3 + q^4 + q^5 + q^6 - O(q^{10})$$
$$k = 6 : \quad 1 + q^2 + q^3 + q^4 + q^5 + 2q^6 + q^8 - O(q^{11})$$

Consider the four basic covariants of a binary cubic given in Sect. 3.7:

covariant	degree k	order i	index $g = (3k - i)/2$
f	1	3	0
H	2	2	2
T	3	3	3
D	4	0	6

These four covariants correspond to the four underlined terms in the generating functions in (4.2.16). We remark that the three covariants $f^2 D$, T^2 and H^3 all have degree 6, order 6 and index 6. Are these covariants linearly independent? Note that the coefficient of q^6 in the last line of (4.2.16) is only two, not three. This proves that $f^2 D$, T^2 and H^3 are linearly dependent. Indeed, we have the syzygy $f^2 D = T^2 + 4H^3$.

We will now sketch an alternative proof of Proposition 3.7.7. The algebra of covariants $\mathbf{C}[a_0, a_1, a_2, a_3, x, y]^\Gamma$ is a bigraded algebra via the degree k and the order i. Consider the subalgebra $\mathbf{C}[f, H, T, D]$. We need to show that both algebras have the same Hilbert function, and therefore are equal. Using Gröbner bases, we find that $\mathbf{C}[f, H, T, D]$ has the bigraded Hilbert function:

$$\sum_{i,k \geq 0} \dim \mathbf{C}[f, H, T, D]_{(i,k)} \, s^i \, t^k = \frac{1 - s^6 t^6}{(1 - s^1 t^3)(1 - s^4 t^0)(1 - s^2 t^2)(1 - s^3 t^3)}.$$

(4.2.17)

Here the term $s^6 t^6$ comes from the syzygy $f^2 D = T^2 + 4H^3$. It remains to be shown that the coefficient of $s^i t^k$ in (4.2.17) equals the coefficient $m(3, k, i)$ of $q^{(3k-i)/2}$ in (4.2.15). We leave this to the reader.

Exercises

(1) Show that the following identity holds for all integers $d \geq e \geq 0$:

$$\left(\frac{q^{d+1}-1}{q-1}\right) \cdot \left(\frac{q^{e+1}-1}{q-1}\right) = \sum_{i=0}^{e} q^i \cdot \left(\frac{q^{d+e+1-2i}-1}{q-1}\right).$$

(2) Deduce from (1) that for $d \geq e \geq 0$ there is an isomorphism of $GL(\mathbf{C}^2)$-modules:

$$S_d \mathbf{C}^2 \otimes S_e \mathbf{C}^2 \simeq S_{d+e} \mathbf{C}^2 \oplus W_{(d+e-1,1)} \mathbf{C}^2 \oplus W_{(d+e-2,2)} \mathbf{C}^2 \oplus \ldots \oplus W_{(d,e)} \mathbf{C}^2.$$

This identity is called the *Clebsch–Gordan formula*.

(3) * Compute the complete system of covariants for the binary quartic ($d = 4$). See Springer (1977: section 3.4.4).

(4) What is the number of linearly independent covariants of degree k and order i of the binary quintic ($d = 5$)?

4.3. Cayley's Ω-process and Hilbert finiteness theorem

The objective of this section is to prove Hilbert's famous finiteness theorem.

Theorem 4.3.1 (Hilbert's finiteness theorem). *Let (V, ρ) be any rational representation of the general linear group $\Gamma = GL(\mathbf{C}^n)$. Then the invariant ring $\mathbf{C}[V]^\Gamma$ is finitely generated as a $\mathbf{C}$-algebra.*

Like his contemporaries, Hilbert was mainly interested in invariants of homogeneous polynomials or forms, that is, relative Γ-invariants of the natural action on the symmetric power $V = S_d \mathbf{C}^n$. We will focus on invariants and covariants of forms in Sects. 4.4 and 4.5. Throughout Sect. 4.3 we are working with an arbitrary polynomial Γ-module V. The assumption of polynomiality is no restriction by Proposition 4.1.2.

The first proof of Theorem 4.3.1 appeared in Hilbert (1890). It was based on a radically new, non-constructive method, namely, the Hilbert basis theorem for polynomial ideals (Corollary 1.2.5). When Paul Gordan, "the king of invariants", first learned about this technique, he made his famous exclamation, *"Das ist Theologie und nicht Mathematik."*

Within three years Hilbert responded to Gordan's criticism by giving a constructive proof. This second proof, published in Hilbert (1893), is considerably deeper and more difficult than the first one. We will present this second proof and the resulting explicit algorithm for computing a finite algebra basis for $\mathbf{C}[V]^\Gamma$ in Sects. 4.6 and 4.7.

In this section we are concerned with Hilbert's non-constructive 1890 proof. One main ingredient of this proof is Cayley's Ω-process, a technical tool which was well-known in the 19th century. The Ω-process is a certain differential operator for the general linear group $\Gamma = GL(\mathbf{C}^n)$ which plays the part of the

Reynolds operator $*$ in the case of a finite matrix group Γ. Since Γ is a reductive group, there does exist also a full-fledged Reynolds operator, i.e., a $\mathbf{C}[V]^\Gamma$-linear map from $\mathbf{C}[V]$ to $\mathbf{C}[V]^\Gamma$ fixing $\mathbf{C}[V]^\Gamma$. But we will not need this Reynolds operator here, but instead we use the classical Ω-process.

Let $\mathbf{C}[\mathbf{t}]$ denote the polynomial ring generated by the n^2 entries in a generic $n \times n$-matrix $\mathbf{t} := (t_{ij})$. With each polynomial

$$f(t_{11}, t_{12}, \ldots, t_{nn}) = \sum_\nu a_\nu t_{11}^{\nu_{11}} t_{12}^{\nu_{12}} \cdots t_{nn}^{\nu_{nn}} \qquad (4.3.1)$$

we associate a corresponding differential operator

$$\mathcal{D}_f := f\left(\frac{\partial}{\partial t_{11}}, \frac{\partial}{\partial t_{12}}, \ldots, \frac{\partial}{\partial t_{nn}}\right) = \sum_\nu a_\nu \frac{\partial^{|\nu|}}{\partial t_{11}^{\nu_{11}} \partial t_{12}^{\nu_{12}} \cdots \partial t_{nn}^{\nu_{nn}}}. \qquad (4.3.2)$$

In (4.3.1) and (4.3.2) the sum is over a finite set of nonnegative integer matrices $\nu = (\nu_{ij})$ which serve as exponent matrices for monomials in $\mathbf{C}[\mathbf{t}]$. The norm $|\nu|$ of a nonnegative matrix ν is the sum of its entries.

Suppose that f is a homogeneous polynomial and hence $\mathcal{D}_f$ is a homogeneous differential operator. Now apply $\mathcal{D}_f$ to the polynomial f itself. When applying a differential monomial occurring in $\mathcal{D}_f$ to a monomial in f, then we get 0 unless the monomials correspond to the same exponent matrix ν. In this case the result equals the constant $\nu_{11}!\, \nu_{12}! \ldots \nu_{nn}!$. These observations prove the following.

Lemma 4.3.2. Each homogeneous polynomial $f = \sum_\nu a_\nu t_{11}^{\nu_{11}} \cdots t_{nn}^{\nu_{nn}}$ in $\mathbf{C}[\mathbf{t}]$ satisfies

$$\mathcal{D}_f(f) = \sum_\nu a_\nu^2\, \nu_{11}!\, \nu_{12}! \cdots \nu_{nn}!.$$

The next lemma states five basic rules for polynomial differential operators. The bilinearity rules (a), (b) and (c) follow immediately from the definition. Rules (b) and (c) reduce (d) to the case of differential monomials applied to monomials, which is easily checked. Property (e) is a consequence of (d).

Lemma 4.3.3.
(a) $\mathcal{D}_f(\phi_1 + \phi_2) = \mathcal{D}_f(\phi_1) + \mathcal{D}(\phi_2)$,
(b) $\mathcal{D}_{c \cdot f}(\phi) = \mathcal{D}_f(c \cdot \phi) = c \cdot \mathcal{D}_f(\phi)$ for constants $c \in \mathbf{C}$,
(c) $\mathcal{D}_{f+g}(\phi) = \mathcal{D}_f(\phi) + \mathcal{D}_g(\phi)$,
(d) $\mathcal{D}_{fg}(\phi) = \mathcal{D}_f \mathcal{D}_g(\phi)$, the composition of differential operators, and
(e) $\mathcal{D}_{f^p}(\phi) = \mathcal{D}_f^p(\phi)$.

We now choose a specific homogeneous polynomial $f(\mathbf{t})$, namely the determinant of $\mathbf{t}$. The resulting differential operator is called *Cayley's Ω-process* and is abbreviated $\Omega := \mathcal{D}_{\det} = \det\left(\frac{\partial}{\partial t_{ij}}\right)$. Using the familiar expansion of the

4.3. Cayley's Ω-process and Hilbert finiteness theorem

determinant, we get

$$\Omega(\phi) = \sum_{\pi \in S_n} \text{sign}(\pi) \frac{\partial^n \phi}{\partial t_{1,\pi_1} \partial t_{2,\pi_2} \cdots \partial t_{n,\pi_n}} \qquad (4.3.3)$$

for all $\phi \in \mathbf{C}[\mathbf{t}]$.

In what follows we consider three generic matrices $\mathbf{t} = (t_{ij})$, $\mathbf{s} = (s_{ij})$, $\mathbf{u} = (u_{ij})$. Their $3n^2$ entries are algebraically independent indeterminates over $\mathbf{C}$. All matrices have an associated Ω-process, which we denote with $\Omega_\mathbf{s}$, $\Omega_\mathbf{t}$ and $\Omega_\mathbf{u}$ respectively. Given a matrix-valued polynomial function ϕ, then the expression $\Omega_{\mathbf{st}}(\phi(\mathbf{st}))$ stands for the polynomial in $\mathbf{s}$ and $\mathbf{t}$ which is gotten by substituting the matrix product $\mathbf{u} = \mathbf{st}$ into the expression $\Omega_\mathbf{u}(\phi(\mathbf{u}))$. The expression $\Omega_\mathbf{t}(\phi(\mathbf{st}))$ denotes the result of applying the operator $\Omega_\mathbf{t}$ to $\phi(\mathbf{st})$, viewed as a polynomial function in $\mathbf{t}$ with parameters $\mathbf{s}$.

Theorem 4.3.4 (First main rule for the Ω-process). *In $\mathbf{C}[\mathbf{s}, \mathbf{t}]$ we have the identities*

$$\Omega_\mathbf{t}(\phi(\mathbf{st})) = \det(\mathbf{s}) \cdot \Omega_{\mathbf{st}}(\phi(\mathbf{st})) \quad \text{and} \quad \Omega_\mathbf{s}(\phi(\mathbf{st})) = \det(\mathbf{t}) \cdot \Omega_{\mathbf{st}}(\phi(\mathbf{st})). \qquad (4.3.4)$$

Proof. We prove the first identity by expanding the left hand side as in (4.3.3). For each term, corresponding to a permutation $\pi \in S_n$ we get

$$\frac{\partial^k \phi(\mathbf{st})}{\partial t_{1,\pi_1} \partial t_{2,\pi_2} \cdots \partial t_{n,\pi_n}}$$

$$= \sum_{\sigma_1, \sigma_2, \ldots, \sigma_n = 1}^{n} \frac{\partial^k \phi(\mathbf{u} = \mathbf{st})}{\partial u_{\sigma_1, \pi_1} \partial u_{\sigma_2, \pi_2} \cdots \partial u_{\sigma_n, \pi_n}} s_{\sigma_1, 1} s_{\sigma_2, 2} \cdots s_{\sigma_n, n}. \qquad (4.3.5)$$

Note that on the right hand side of (4.3.5) we have to sum over all index tuples σ and not only over permutations. Now if we antisymmetrize the expression (4.3.5) with respect to $\pi \in S_n$, then the left hand side becomes $\Omega_\mathbf{t}(\phi(\mathbf{st}))$. On the right hand side we get, after interchanging the two summations,

$$\sum_{\sigma_1, \sigma_2, \ldots, \sigma_n = 1}^{n} s_{\sigma_1, 1} s_{\sigma_2, 2} \cdots s_{\sigma_n, n} \left(\sum_{\pi \in S_n} \text{sign}(\pi) \frac{\partial^k \phi(\mathbf{u} = \mathbf{st})}{\partial u_{\sigma_1, \pi_1} \partial u_{\sigma_2, \pi_2} \cdots \partial u_{\sigma_n, \pi_n}} \right). \qquad (4.3.6)$$

The expression in the large bracket equals 0 by antisymmetry whenever the σ_i are not distinct. Otherwise $\sigma \in S_n$ is a permutation, and the bracket equals

$$\text{sign}(\sigma) \cdot \left(\sum_{\pi \in S_n} \text{sign}(\pi) \frac{\partial^k \phi(\mathbf{u} = \mathbf{st})}{\partial u_{1,\pi_1} \partial u_{2,\pi_2} \cdots \partial u_{n,\pi_n}} \right) = \text{sign}(\sigma) \cdot \Omega_{\mathbf{st}}(\phi(\mathbf{st})). \qquad (4.3.7)$$

Plugging (4.3.7) into (4.3.6), we get the desired result

$$\sum_{\sigma \in S_n} s_{\sigma_1,1} s_{\sigma_2,2} \ldots s_{\sigma_n,n} \cdot \text{sign}(\sigma) \cdot \Omega_{st}(\phi(st)) = \det(s) \cdot \Omega_{st}(\phi(st)).$$

The proof of the second identity in (4.3.4) is analogous. ◁

We can generalize the first main rule to the case of an iterated Ω-process. Application of the operator Ω_t to both sides of (4.3.4) yields

$$\Omega_t^2(\phi(st)) = \det(s) \cdot \Omega_t \Omega_{st}(\phi(st)) = \det(s)^2 \cdot \Omega_{st}^2(\phi(st)). \tag{4.3.8}$$

The second equation in (4.3.8) is gotten by applying the rule (4.3.4) to the function $\tilde{\phi}(\mathbf{u}) = \Omega_{\mathbf{u}} \phi(\mathbf{u})$, with the substitution $\mathbf{u} = st$. By iterating (4.3.8) we obtain the following result.

Corollary 4.3.5 (Generalized first main rule for the Ω-process). For each integer $p \geq 0$ we have the following two identities in $\mathbf{C}[\mathbf{s}, \mathbf{t}]$:

$$\begin{aligned}\Omega_t^p(\phi(st)) &= \det(s)^p \cdot \Omega_{st}^p(\phi(st)) \\ \text{and} \quad \Omega_s^p(\phi(st)) &= \det(t)^p \cdot \Omega_{st}^p(\phi(st)).\end{aligned} \tag{4.3.9}$$

Theorem 4.3.4 has the consequence that the Ω-process preserves the one-dimensional subalgebra which is generated by the determinant $\det(s)$ in $\mathbf{C}[\mathbf{s}]$.

Corollary 4.3.6. In $\mathbf{C}[\mathbf{s}]$ we have the identity

$$\Omega_s(\det(s)^p) = c_p \cdot \det(s)^{p-1}, \tag{4.3.10}$$

where c_p is a non-zero constant depending only on the integer p.

Proof. Applying (4.3.4) to the polynomial function $\phi(\mathbf{u}) := \det(\mathbf{u})^p$, we get

$$\Omega_t(\det(st)^p) = \det(s) \cdot \Omega_{st}(\det(st)^p). \tag{4.3.11}$$

On the other hand, Lemma 4.3.3 (b) implies

$$\Omega_t(\det(st)^p) = \Omega_t(\det(s)^p \det(t)^p) = \det(s)^p \cdot \Omega_t(\det(t)^p). \tag{4.3.12}$$

In the resulting identity

$$\Omega_{st}(\det(st)^p) = \det(s)^{p-1} \cdot \Omega_t(\det(t)^p) \tag{4.3.13}$$

4.3. Cayley's Ω-process and Hilbert finiteness theorem

we replace **t** by the unit matrix **1**, and we get the desired identity (4.3.10), where

$$c_p := \left\{ \Omega_t(\det(\mathbf{t})^p) \right\}_{\mathbf{t}:=1}.$$

It remains to be checked that the constant c_p is indeed non-zero. To this end we repeatedly apply Ω_s to (4.3.10). We obtain $\Omega_s^2(\det(\mathbf{s})^p) = c_p\, c_{p-1}\, \det(\mathbf{s})^{p-2}$, $\Omega_s^3(\det(\mathbf{s})^p) = c_p\, c_{p-1}\, c_{p-2}\, \det(\mathbf{s})^{p-3}$, and finally,

$$\Omega_s^p(\det(\mathbf{s})^p) = c_p\, c_{p-1}\, c_{p-2} \cdots c_2\, c_1. \qquad (4.3.14)$$

By Lemma 4.3.3 (e), Ω_s^p equals the differential operator $\mathcal{D}_{\det(\mathbf{s})^p}$ associated with the polynomial $\det(\mathbf{s})^p$. The expression (4.3.14) is a positive integer by Lemma 4.3.2. ◁

We will next derive the Second Main Rule for the Ω-Process. The ring $\mathbf{C}[V]$ of polynomial functions on the Γ-module V is written as $\mathbf{C}[\mathbf{v}]$, where $\mathbf{v}$ is a generic vector in V. For any polynomial function $f = f(\mathbf{v})$ we consider its image $\mathbf{t} \circ f = f(\mathbf{tv})$ under a generic linear transformation $\mathbf{t} \in \Gamma$. Thus $f(\mathbf{tv})$ is a polynomial in $n^2 + \dim(V)$ variables, namely, it is a polynomial both in $\mathbf{t}$ and in $\mathbf{v}$. For every nonnegative integer p the expression $\det(\mathbf{t})^q \cdot f(\mathbf{tv})$ is a polynomial function in $\mathbf{C}[\mathbf{v}, \mathbf{t}]$. The Cayley process Ω_t acts on these polynomials by regarding the coordinates of $\mathbf{v}$ as constants. After repeated application of Ω_t to $\det(\mathbf{t})^q \cdot f(\mathbf{tv})$, we can then replace $\mathbf{t}$ by the $n \times n$-zero matrix $\mathbf{0}$. This procedure always generates an element in the invariant ring $\mathbf{C}[\mathbf{v}]^\Gamma$:

Theorem 4.3.7 (Second main rule for the Ω-process). *Let $f \in \mathbf{C}[\mathbf{v}]$ be a homogeneous polynomial, and let $p, q \geq 0$ be arbitrary integers. Then*

$$I_{p,q}(f) := \left\{ \Omega_t^p \bigl(\det(\mathbf{t})^q \cdot f(\mathbf{tv}) \bigr) \right\}_{\mathbf{t}:=0} \qquad (4.3.15)$$

is a relative Γ-invariant.

Proof. We abbreviate

$$\phi(\mathbf{v}, \mathbf{t}) := \det(\mathbf{t})^q \cdot f(\mathbf{tv}). \qquad (4.3.16)$$

We will show that

$$[I_{p,q}(f)](\mathbf{v}) = \left\{ \Omega_t^p(\phi(\mathbf{v}, \mathbf{t})) \right\}_{\mathbf{t}:=0}$$

is either 0 or it is a relative Γ-invariant of index $p - q$. Let $\mathbf{s}$ be a second $n \times n$-matrix with generic entries. Then we find

$$\begin{aligned}\phi(\mathbf{v}, \mathbf{st}) &= \det(\mathbf{st})^q \cdot f(\mathbf{stv}) \\ &= \det(\mathbf{t})^q \cdot \det(\mathbf{s})^q \cdot f(\mathbf{s}(\mathbf{tv})) = \det(\mathbf{t})^q \cdot \phi(\mathbf{tv}, \mathbf{s}).\end{aligned} \qquad (4.3.17)$$

We now apply the differential operator Ω_s^p to both sides of (4.3.17), starting with the right hand side:

$$\det(\mathbf{t})^q \, \Omega_s^p\big(\phi(\mathbf{t}\mathbf{v},\,\mathbf{s})\big) = \Omega_s^p\big(\phi(\mathbf{v},\,\mathbf{s}\mathbf{t})\big) = \det(\mathbf{t})^p \, \Omega_{\mathbf{st}}^p\big(\phi(\mathbf{v},\,\mathbf{s}\mathbf{t})\big). \qquad (4.3.18)$$

Here the second equation is derived from Corollary 4.3.5. In the resulting identity

$$\Omega_s^p\big(\phi(\mathbf{t}\mathbf{v},\,\mathbf{s})\big) = \det(\mathbf{t})^{p-q} \cdot \Omega_{\mathbf{st}}^p\big(\phi(\mathbf{v},\,\mathbf{s}\mathbf{t})\big) \qquad (4.3.19)$$

we specialize $\mathbf{s}$ to the zero matrix $\mathbf{0}$. The left hand side then specializes to $[I_{p,q}(f)](\mathbf{t}\mathbf{v})$, while the right hand side specializes to $\det(\mathbf{t})^{p-q}[I_{p,q}(f)](\mathbf{v})$. This proves that $I_{p,q}(f)$, if non-zero, is a relative invariant with index $p-q$. ◁

Theorem 4.3.7 gives an explicit algorithm for generating an invariant $I_{p,q}(f)$ from an arbitrary polynomial function f in $\mathbf{C}[V]$. We will illustrate this algorithm in the next section. At this point we just note that $I_{p,q}(f)$ will often be simply zero. For instance, this is always the case when $p < q$.

We are now prepared to prove Hilbert's finiteness theorem.

Proof of Theorem 4.3.1. Let $\mathcal{I}_+^\Gamma \subset \mathbf{C}[\mathbf{v}]$ be the ideal generated by all homogeneous invariants of positive degree. By the Hilbert basis theorem, there exists a finite set $\{J_1, J_2, \ldots, J_r\}$ of homogeneous invariants such that

$$\mathcal{I}_+^\Gamma = \langle J_1, J_2, \ldots, J_r \rangle. \qquad (4.3.20)$$

We will show that the J_i form a fundamental system of invariants, i.e.,

$$\mathbf{C}[\mathbf{v}]^\Gamma = \mathbf{C}[J_1, J_2, \ldots, J_r]. \qquad (4.3.21)$$

Let $J \in \mathbf{C}[\mathbf{v}]^\Gamma$ be any homogeneous invariant of positive degree, and suppose that all homogeneous invariants of lower total degree lie in the subring $\mathbf{C}[J_1, J_2, \ldots, J_r]$. Write

$$J(\mathbf{v}) = \sum_{i=1}^r f_i(\mathbf{v}) \, J_i(\mathbf{v}) \qquad (4.3.22)$$

where $f_1, f_2, \ldots, f_r \in \mathbf{C}[\mathbf{v}]$ are homogeneous of degree $\deg(f_i) = \deg(J) - \deg(J_i)$. In (4.3.22) we replace $\mathbf{v}$ by $\mathbf{t}\mathbf{v}$ where $\mathbf{t}$ is a generic $n \times n$-matrix. If p and p_i are the indices of the invariants J and J_i respectively, then we get

$$\det(\mathbf{t})^p \, J(\mathbf{v}) = \sum_{i=1}^r \det(\mathbf{t})^{p_i} \, f_i(\mathbf{t}\mathbf{v}) \, J_i(\mathbf{v}). \qquad (4.3.23)$$

We now apply the differential operator Ω_t^p to the identity (4.3.23). On the left hand side we obtain $\Omega_t^p\big(\det(\mathbf{t})^p \, J(\mathbf{v})\big) = c \cdot J(\mathbf{v})$, where c is a positive integer,

4.4. Invariants and covariants of forms

by Lemma 4.3.2. By linearity on the right hand side,

$$c \cdot J(\mathbf{v}) = \sum_{i=1}^{r} J_i(\mathbf{v}) \cdot \Omega_{\mathbf{t}}^p \big(\det(\mathbf{t})^{p_i} f_i(\mathbf{t}\mathbf{v}) \big). \tag{4.3.24}$$

We finally replace the generic matrix $\mathbf{t}$ in (4.3.24) by the zero matrix $\mathbf{0}$, and we get

$$c \cdot J(\mathbf{v}) = \sum_{i=1}^{r} J_i(\mathbf{v}) \cdot [I_{p, p_i}(f_i)](\mathbf{v}). \tag{4.3.25}$$

By the second main rule (Theorem 4.3.7), all expressions on the right hand side lie in $\mathbf{C}[\mathbf{v}]^\Gamma$. Since the f_i have degree $< \deg(J)$, and the operator I_{p, p_i} is degree-preserving, the invariants $I_{p, p_i}(f_i)$ have lower total degree than J. Hence they are contained in $\mathbf{C}[J_1, J_2, \ldots, J_r]$. Since $c \ne 0$, the representation (4.3.25) implies that J lies in $\mathbf{C}[J_1, J_2, \ldots, J_r]$. ◁

Exercises

(1) Show that the invariant ring $\mathbf{C}[\mathbf{v}]^\Gamma$ is the $\mathbf{C}$-linear span of the expressions $I_{p,q}(m)$ where m runs over all monomials in $\mathbf{C}[\mathbf{v}]$.
(2) Consider the action of $GL(\mathbf{C}^n)$ by left multiplication on the space of $n \times s$-matrices and on the induced polynomial ring $\mathbf{C}[x_{ij}]$. Verify the First and Second Main Rule for the Ω-process for this representation.
(3) * Give an explicit formula for the operator $I_{p,q}(\cdot)$ in the case where $V = S_3\mathbf{C}^2$, the space of binary cubics. Find p, q and a monomial m such that $I_{p,q}(m)$ equals the discriminant of a binary cubic.

4.4. Invariants and covariants of forms

A homogeneous polynomial

$$f(x_1, x_2, \ldots, x_n) = \sum \binom{d}{i_1\, i_2\, \ldots\, i_n} \cdot a_{i_1 i_2 \ldots i_n} \cdot x_1^{i_1} x_2^{i_2} \ldots x_n^{i_n} \tag{4.4.1}$$

of total degree d in n variables $\mathbf{x} = (x_1, x_2, \ldots, x_n)$ is called an *n-ary form* of *degree d* (or short: *n-ary d-form*). These terms are traditionally in Latin; for instance, a *binary cubic* is a 2-ary 3-form and a *quaternary quintic* is a 4-ary 5-form. The sum in (4.4.1) is over the $\binom{n+d-1}{d}$-element set of nonnegative integer vectors $(i_1, i_2, \ldots, i_n)$ with $i_1 + i_2 + \ldots + i_n = d$. The *coefficients* $a_{i_1 i_2 \ldots i_n}$ are algebraically independent transcendentals over $\mathbf{C}$. It is customary (and essential for the symbolic representation in Sect. 4.5) to scale the $a_{i_1 i_2 \ldots i_n}$ by the multinomial coefficients $\binom{d}{i_1\, i_2\, \ldots\, i_n} = d!/(i_1! \cdot i_2! \cdots i_n!)$.

The set of *n*-ary *d*-forms is a $\binom{n+d-1}{d}$-dimensional complex vector space. We identify it with $S_d(\mathbf{C}^n)$, the *d-th symmetric power* of $\mathbf{C}^n$. This means that the form f is identified with the vector $\mathbf{a} := (\ldots, a_{i_1 i_2 \ldots i_n}, \ldots)$ of its coefficients.

Thus $f = f(\mathbf{x}) = f(\mathbf{a}, \mathbf{x})$ and $\mathbf{a}$ represent the same element of $S_d(\mathbf{C}^n)$. In some situations it is preferable to distinguish these two objects, in which case we refer to $\mathbf{a}$ as the *symmetric tensor of step d* associated with the d-form f.

Let $\mathbf{C}[\mathbf{a}, \mathbf{x}]$ denote the polynomial ring in the coefficients and the variables of f. This is the ring of polynomial functions on the vector space $S_d(\mathbf{C}^n) \oplus \mathbf{C}^n$. Its subring $\mathbf{C}[\mathbf{a}]$ is the ring of polynomial functions on $S_d(\mathbf{C}^n)$. The action of general linear group $\Gamma = GL(\mathbf{C}^n)$ on $\mathbf{C}^n$ induces a natural linear action on the space $S_d(\mathbf{C}^n) \oplus \mathbf{C}^n$. For each $T \in \Gamma$, the action $T : (\mathbf{a}, \mathbf{x}) \mapsto (\bar{\mathbf{a}}, \bar{\mathbf{x}})$ is defined by the equations

$$\mathbf{x} = T \cdot \bar{\mathbf{x}} \quad \text{and} \quad f(\mathbf{a}, \mathbf{x}) = f(\bar{\mathbf{a}}, \bar{\mathbf{x}}) \quad \text{for all } T = (t_{ij}) \in \Gamma. \tag{4.4.2}$$

It is crucial to note that this Γ-module is *not* the direct sum of the Γ-modules $S_d(\mathbf{C}^n)$ and $\mathbf{C}^n$. The Γ-module defined by (4.4.2) is the direct sum of $S_d(\mathbf{C}^n)$ and $\mathrm{Hom}(\mathbf{C}^n, \mathbf{C})$, the latter being the contragredient representation to $\mathbf{C}^n$. It is the rational (but not polynomial) Γ-module which has the formal character

$$\left(\frac{1}{t_1} + \frac{1}{t_2} + \ldots + \frac{1}{t_n}\right) \sum_{i_1 + \ldots + i_n = d} t_1^{i_1} t_2^{i_2} \ldots t_n^{i_n}.$$

A polynomial $I \in \mathbf{C}[\mathbf{a}, \mathbf{x}]$ is a *covariant* of f if it is a relative Γ-invariant, i.e.,

$$I(\bar{\mathbf{a}}, \bar{\mathbf{x}}) = \det(T)^g \cdot I(\mathbf{a}, \mathbf{x}) \tag{4.4.3}$$

for some non-negative integer g, which is called the *index* of the covariant I. The total degree of I with respect to the coefficient vector $\mathbf{a}$ is called the *degree* of the covariant I, and its total degree with respect to the old variables $\mathbf{x}$ is called the *order* of I. So, every d-form is a covariant of itself, having order d, degree 1 and index 0. An *invariant* of the n-ary d-form f is a covariant $I \in \mathbf{C}[\mathbf{a}]$ of order 0. The *covariant ring* of f is the graded $\mathbf{C}$-algebra $\mathbf{C}[\mathbf{a}, \mathbf{x}]^\Gamma$ generated by all covariants. The *invariant ring* of f is the subalgebra $\mathbf{C}[\mathbf{a}]^\Gamma$ generated by all invariants.

We also consider *joint covariants* and *joint invariants* of a collection of n-ary forms $f_1, f_2, \ldots, f_k$ of degrees $d_1, d_2, \ldots, d_k$. In that case the underlying Γ-module equals the direct sum $V = S_{d_1}\mathbf{C}^n \oplus S_{d_2}\mathbf{C}^n \oplus \ldots \oplus S_{d_k}\mathbf{C}^n \oplus \mathrm{Hom}(\mathbf{C}^n, \mathbf{C})$. This gives rise to the covariant ring $\mathbf{C}[V]^\Gamma = \mathbf{C}[\mathbf{a}_1, \mathbf{a}_2, \ldots, \mathbf{a}_k, \mathbf{x}]^\Gamma$ and the invariant ring $\mathbf{C}[\mathbf{a}_1, \ldots, \mathbf{a}_k]^\Gamma$. Both rings are subalgebras of the polynomial ring $\mathbf{C}[\mathbf{a}_1, \ldots, \mathbf{a}_k, \mathbf{x}]$. They are multigraded with respect to the "old variables" $\mathbf{x}$ and the coordinates of each symmetric tensor $\mathbf{a}_i$.

From Theorem 4.3.1 we infer that $\mathbf{C}[V]^\Gamma$ is finitely generated.

Corollary 4.4.1 (Hilbert's finiteness theorem). *The invariant ring or the covariant ring of one n-ary d-form or of several n-ary forms is finitely generated as a $\mathbf{C}$-algebra.*

This result had been proved in Theorem 3.7.1 for the case of binary forms ($n = 2$). But there is an essential difference between both finiteness proofs. In

4.4. Invariants and covariants of forms

Sect. 3.7 we gave an explicit algorithm for computing a finite set of fundamental set of invariants (or covariants). The general finiteness proof in Sect. 4.3 does not yield such an algorithm. The more difficult problem of giving an algorithm for Corollary 4.4.1 will be addressed in Sect. 4.6. The present section has two objectives. We discuss important examples of invariants and covariants, and we illustrate the practical use of the machinery developed in Sects. 4.1 and 4.3.

Our first example of a covariant of an n-ary d-form $f(\mathbf{a}, \mathbf{x})$ is the *Hessian*

$$H(\mathbf{a}, \mathbf{x}) = \det\left(\frac{\partial^2 f}{\partial x_i \partial x_j}\right). \tag{4.4.4}$$

Proposition 4.4.2. *The Hessian $H \in \mathbf{C}[\mathbf{a}, \mathbf{x}]$ of an n-ary form f of degree $d \geq 2$ is a covariant of index 2, degree n, and order $n(d-2)$.*

Proof. We consider $\mathbf{a}$ and $\mathbf{x}$ as polynomial functions in $\bar{\mathbf{a}}$, $\bar{\mathbf{x}}$ and $T = (t_{ij})$. Applying the differential operator $\frac{\partial^2}{\partial \bar{x}_i \partial \bar{x}_j}$ to the identity $f(\bar{\mathbf{a}}, \bar{\mathbf{x}}) = f(\mathbf{a}, \mathbf{x})$ in (4.4.2), we get

$$\frac{\partial^2 f}{\partial \bar{x}_i \partial \bar{x}_j}(\bar{\mathbf{a}}, \bar{\mathbf{x}}) = \sum_{k,l=1}^{n} \frac{\partial^2 f}{\partial x_i \partial x_j}(\mathbf{a}, \mathbf{x}) \cdot t_{ik} \cdot t_{jl}. \tag{4.4.5}$$

Forming the $n \times n$-determinant of these expressions for $1 \leq i, j \leq n$, we obtain

$$H(\bar{\mathbf{a}}, \bar{\mathbf{x}}) = \det\left(\frac{\partial^2 f}{\partial \bar{x}_i \partial \bar{x}_j}(\bar{\mathbf{a}}, \bar{\mathbf{x}})\right)$$
$$= \det(T)^2 \cdot \det\left(\frac{\partial^2 f}{\partial x_i \partial x_j}(\mathbf{a}, \mathbf{x})\right) = \det(T)^2 \cdot H(\mathbf{a}, \mathbf{x}). \tag{4.4.6}$$

This shows that the Hessian H is a covariant of index 2. Each expression $\partial^2 f/\partial x_i \partial x_j$ is homogeneous of degree 1 in $\mathbf{a}$ and of degree $(d-2)$ in $\mathbf{x}$, and hence H has degree n and order $n(d-2)$. ◁

In the case $d = 2$ the Hessian does not depend on $\mathbf{x}$. The resulting invariant $D(\mathbf{a}) := H(\mathbf{a}, \mathbf{x})$ is the *discriminant* of the n-ary quadratic form $f(\mathbf{a}, \mathbf{x}) = \sum_{i=1}^{n} \sum_{j=i}^{n} a_{ij} x_i x_j$. It is preferable to interpret the coefficient vector $\mathbf{a}$ as a symmetric matrix $\mathbf{a} = (a_{ij})$. The discriminant $D(\mathbf{a})$ equals the determinant of that matrix. Using matrix products, we write the given quadratic forms as $f(\mathbf{a}, \mathbf{x}) = \mathbf{x}^t \cdot \mathbf{a} \cdot \mathbf{x}$.

We illustrate this relabeling for the case $n = 3$. The ternary quadric

$$f(\mathbf{a}, \mathbf{x}) = a_{200} x_1^2 + a_{020} x_2^2 + a_{002} x_3^2 + 2a_{110} x_1 x_2 + 2a_{101} x_1 x_3 + 2a_{011} x_2 x_3$$
$$= \mathbf{x}^t \mathbf{a} \mathbf{x} = a_{11} x_1^2 + a_{22} x_2^2 + a_{33} x_3^2 + 2a_{12} x_1 x_2 + 2a_{13} x_1 x_3 + 2a_{23} x_2 x_3 \tag{4.4.7}$$

defines a general quadratic curve in the projective plane. Its discriminant

$$D(\mathbf{a}) = a_{200}\, a_{020}\, a_{002} + 2a_{110}\, a_{101}\, a_{011} - a_{110}^2\, a_{002} - a_{101}^2\, a_{020} - a_{011}^2\, a_{200} \quad (4.4.8)$$

vanishes if and only if the quadric form f factors into two linear factors, or, equivalently if the curve $\{f = 0\}$ is the union of two lines. Returning to the general quadratic form, we have the following solution to the problem of finding the invariants.

Theorem 4.4.3. *The discriminant D generates the invariant ring a quadratic n-ary form.*

Proof. In matrix notation Eq. (4.4.2) becomes

$$\overline{\mathbf{x}}^t\, \overline{\mathbf{a}}\overline{\mathbf{x}} = \mathbf{x}^t\, \mathbf{a}\mathbf{x} = (T\overline{\mathbf{x}})^t \mathbf{a}(T\overline{\mathbf{x}}) = \overline{\mathbf{x}}^t (T^t \mathbf{a} T)\overline{\mathbf{x}},$$

and the Γ-action on $S_2(\mathbf{C}^n)$ is expressed by the matrix equation $\overline{\mathbf{a}} = T^t \mathbf{a} T$.

In order to show $\mathbf{C}[\mathbf{a}]^\Gamma = \mathbf{C}[D]$, we let $I(\mathbf{a})$ be any homogeneous invariant of index g. Then

$$I(\overline{\mathbf{a}}) = I(T^t \mathbf{a} T) = \det(T)^g \cdot I(\mathbf{a}). \quad (4.4.9)$$

Writing **1** for the $n \times n$-unit matrix, we let $c := I(\mathbf{1}) \in \mathbf{C}$. This implies the equation

$$I(T^t T) = c \cdot \det(T)^g \quad \text{for all} \quad n \times n\text{-matrices } T \text{ over } \mathbf{C}. \quad (4.4.10)$$

Every symmetric matrix $\mathbf{a}$ admits a factorization $\mathbf{a} = T^t T$ over the complex numbers $\mathbf{C}$. Therefore (4.4.10) implies

$$I(\mathbf{a}) = c \cdot \det(\mathbf{a})^{g/2} \quad \text{in } \mathbf{C}[\mathbf{a}]. \quad (4.4.11)$$

But I was assumed to be a polynomial function, hence $g = 2p$ is even, and we conclude $I = cD^p \in \mathbf{C}[D]$. ◁

An important example of a joint covariant of several forms is the *Jacobian determinant J* of n forms

$$f_1(\mathbf{a}_1, \mathbf{x}),\ f_2(\mathbf{a}_2, \mathbf{x}) \ldots\ldots, f_n(\mathbf{a}_n, \mathbf{x})$$

of degrees $d_1, d_2, \ldots, d_n$ in n variables.

Example 4.4.4. The Jacobian determinant

$$J = J(\mathbf{a}_1, \ldots, \mathbf{a}_n, \mathbf{x}) = \det \begin{pmatrix} \frac{\partial f_1}{\partial x_1}(\mathbf{a}_1, \mathbf{x}) & \cdots & \frac{\partial f_1}{\partial x_n}(\mathbf{a}_1, \mathbf{x}) \\ \vdots & \ddots & \vdots \\ \frac{\partial f_n}{\partial x_1}(\mathbf{a}_n, \mathbf{x}) & \cdots & \frac{\partial f_n}{\partial x_n}(\mathbf{a}_n, \mathbf{x}) \end{pmatrix}$$

4.4. Invariants and covariants of forms

is a joint covariant of index 1, order $d_1+d_2+\ldots+d_n-n$, and degrees $1, 1, \ldots, 1$ in the coefficients.

We have the following general degree relation for joint covariants:

Proposition 4.4.5. Let $I(\mathbf{a}_1, \mathbf{a}_2, \ldots, \mathbf{a}_k, \mathbf{x})$ be a joint covariant of k n-ary forms

$$f_1(\mathbf{a}_1, \mathbf{x}), \ f_2(\mathbf{a}_2, \mathbf{x}), \ \ldots, \ f_k(\mathbf{a}_k, \mathbf{x})$$

of degrees $d_1, d_2, \ldots, d_k$. Suppose that I has index g, order m, and I is homogeneous of degrees $r_1, r_2, \ldots, r_k$ in the coefficient vectors $\mathbf{a}_1, \mathbf{a}_2, \ldots, \mathbf{a}_k$. Then

$$r_1 d_1 + r_2 d_2 + \ldots + r_k d_k = ng + m.$$

Proof. Consider the $n \times n$-diagonal matrix $T = \text{diag}(t, t, t, \ldots, t)$. By definition (4.4.2) of the induced transformation on the i-th symmetric tensor in question, we have $f_i(\mathbf{a}_i, \mathbf{x}) = f_i(\overline{\mathbf{a}}_i, \overline{\mathbf{x}}) = f_i(\overline{\mathbf{a}}_i, t^{-1}\mathbf{x})$. This implies $\overline{\mathbf{a}}_i = t^{d_i} \mathbf{a}_i$ for $i = 1, 2, \ldots, k$. The property that I is a covariant now implies

$$\begin{aligned} I(t^{d_1}\mathbf{a}_1, t^{d_2}\mathbf{a}_2, \ldots, t^{d_k}\mathbf{a}_k, t^{-1}\mathbf{x}) &= I(\overline{\mathbf{a}}_1, \overline{\mathbf{a}}_2, \ldots, \overline{\mathbf{a}}_k, \overline{\mathbf{x}}) \\ &= (\det T)^g \cdot I(\mathbf{a}_1, \ldots, \mathbf{a}_k, \mathbf{x}) = t^{gn} \cdot I(\mathbf{a}_1, \ldots, \mathbf{a}_k, \mathbf{x}). \end{aligned} \quad (4.4.12)$$

On the other hand, by the homogeneity assumption, the left hand side equals $t^{r_1 d_1} \ldots t^{r_k d_k} t^{-m} I(\mathbf{a}_1, \ldots, \mathbf{a}_k, \mathbf{x})$, and consequently $t^{r_1 d_1} \ldots t^{r_k d_k} t^{-m} = t^{gn}$. ◁

There are two invariants that are of great importance for elimination theory. These are the *(multivariate) discriminant* $D(\mathbf{a})$ of an n-ary d-form f, which vanishes if and only if the projective hypersurface $\{f = 0\}$ has a singularity, and the *(multivariate) resultant* $R(\mathbf{a}_1, \mathbf{a}_2, \ldots, \mathbf{a}_n)$ of n n-ary forms $f_1, f_2, \ldots, f_n$, which vanishes if and only if the polynomial system $\{f_1(\mathbf{x}) = f_2(\mathbf{x}) = \ldots = f_n(\mathbf{x}) = 0\}$ has a non-zero solution. For detailed introductions and recent results on resultants and discriminants we refer to Jouanalou (1991) and Gel'fand et al. (1993).

At this point we have reached a good understanding of the invariant theory of binary forms and of quadratic forms. Let us therefore proceed to the next case ($n = 3$, $d = 3$). The space $V = S_3\mathbf{C}^3$ of ternary cubics has dimension 10. A typical element in V is

$$\begin{aligned} f(x, y, x) = \ & a_{300}\, x^3 + 3a_{210}\, x^2 y + 3a_{201}\, x^2 z + 3a_{120}\, xy^2 \\ & + 6a_{111}\, xyz + 3a_{102}\, xz^2 + a_{030}\, y^3 + 3a_{021}\, y^2 z \\ & + 3a_{012}\, yz^2 + a_{003}\, z^3. \end{aligned} \quad (4.4.13)$$

We express the action $\mathbf{a} \mapsto T\mathbf{a}$ of a linear transformation for $T = (t_{ij}) \in \Gamma$ in coordinates:

$$a_{300} \mapsto t_{11}^3 a_{300} + 3t_{11}^2 t_{12} a_{210} + 3t_{11}^2 t_{13} a_{201} + 3t_{11} t_{12}^2 a_{120} + 6t_{11} t_{12} t_{13} a_{111}$$
$$+ 3t_{11} t_{13}^2 a_{102} + t_{12}^3 a_{030} + 3t_{12}^2 t_{13} a_{021} + 3t_{12} t_{13}^2 a_{012} + t_{13}^3 a_{003}$$

$$a_{210} \mapsto (t_{11}^2 t_{22} + 2t_{11} t_{12} t_{21}) a_{210} + (t_{11}^2 t_{23} + 2t_{11} t_{13} t_{21}) a_{201}$$
$$+ (2t_{11} t_{12} t_{22} + t_{12}^2 t_{21}) a_{120} + (2t_{11} t_{12} t_{23} + 2t_{11} t_{13} t_{22} + 2t_{12} t_{13} t_{21}) a_{111}$$
$$+ (2t_{11} t_{13} t_{23} + t_{13}^2 t_{21}) a_{102} + t_{12}^2 t_{22} a_{030} + (t_{12}^2 t_{23} + 2t_{12} t_{13} t_{22}) a_{021}$$
$$+ (2t_{12} t_{13} t_{23} + t_{13}^2 t_{22}) a_{012} + t_{13}^2 t_{23} a_{003} + t_{11}^2 t_{21} a_{300} \quad \text{etc., etc.}$$

In order to get some information about the invariants of a ternary cubic, we compute the formal character of the Γ-module $S_m S_3 \mathbf{C}^3$ for small values of m. By Proposition 4.1.18 we need to compute the plethysm $h_m \circ h_3$, where h_3 denotes the complete symmetric polynomial of degree 3 in t_1, t_2, t_3, and h_m denotes the complete symmetric polynomial of degree m in $10 = \dim(S_3 \mathbf{C}^3)$ variables. Using Algorithm 4.1.16, we determine the following decompositions into Schur polynomials $s_\lambda = s_\lambda(t_1, t_2, t_3)$.

$$h_2 \circ h_3 = s_{(6,0,0)} + s_{(4,2,0)},$$
$$h_3 \circ h_3 = s_{(9,0,0)} + s_{(7,2,0)} + s_{(6,3,0)} + s_{(5,2,2)} + s_{(4,4,1)}, \quad (4.4.13)$$
$$h_4 \circ h_3 = s_{(12,0,0)} + s_{(10,2,0)} + s_{(9,3,0)} + s_{(8,4,0)} + s_{(8,2,2)}$$
$$+ s_{(7,4,1)} + s_{(7,3,2)} + s_{(6,6,0)} + s_{(6,4,2)} + \underline{s_{(4,4,4)}}.$$

While there are no invariants in degrees 1, 2 and 3, the underlined term shows that there exists one invariant of degree 4 and index 4. Continuing this process, we find that there are no invariants in degrees 5, 7, 9 and 11, while the space of invariants is one-dimensional in the degrees 6, 8 and 10. In degree 12 there are two linearly independent invariants.

These enumerative results prove that there exists a unique (up to scaling) invariant S of degree 4 and a unique (up to scaling) invariant T of degree 6. All invariants of degree at most 12 lie in the subring $\mathbf{C}[S, T]$. The following result is to be proved in Exercise 4.7. (2), using the methods developed in Sect. 4.7.

Theorem 4.4.6. *The ring of invariants of a ternary cubic is generated by two invariants S and T of degrees 4 and 6 respectively.*

In the remainder of this section and in Sect. 4.5 we will address the question of how to generate and encode invariants such as S and T. Our first method for computing invariants is the Ω-process, which was crucial in the proof of finiteness theorem. We apply it to generate the degree 4 invariant S. The degree 6 invariant T will be generated in the next section; its monomial expansion in given in Example 4.5.3.

In order to find the invariant S we first choose a suitable degree 4 monomial

4.4. Invariants and covariants of forms

in $S_4S_3\mathbf{C}^3$, such as a_{111}^4. We apply to it the Ω-process an appropriate number of times. We compute

$$I_{4,0}(a_{111}^4) = \Omega_T^4(T \circ a_{111}^4),$$

where $\Omega_T = \dfrac{\partial^3}{\partial t_{11}\partial t_{22}\partial t_{33}} - \dfrac{\partial^3}{\partial t_{11}\partial t_{23}\partial t_{31}} - \dfrac{\partial^3}{\partial t_{12}\partial t_{21}\partial t_{33}}$

$+ \dfrac{\partial^3}{\partial t_{12}\partial t_{23}\partial t_{31}} + \dfrac{\partial^3}{\partial t_{13}\partial t_{21}\partial t_{32}} - \dfrac{\partial^3}{\partial t_{13}\partial t_{22}\partial t_{33}}.$

The expression $T \circ a_{111}^4$ is a polynomial in the two groups of variables $t_{11}, \ldots, t_{33}$ and $a_{300}, a_{210}, \ldots, a_{003}$. It is homogeneous of degree 4 in each group. The complete expansion of $T \circ a_{111}^4$ has 18,630 monomials.

Having computed this expansion, we successively apply the operator Ω_T. Each application decreases the degree in the t_{ij} by one while the degree in the a_{ijk} stays four. Thus $\Omega_T(T \circ a_{111}^4)$ has degree three in the t_{ij}. It has 7,824 monomials. The next polynomial $\Omega_T^2(T \circ a_{111}^4)$ is quadratic in the t_{ij}, and it has 3,639 monomials. The next polynomial $\Omega_T^3(T \circ a_{111}^4)$ is linear in the t_{ij}, and it has 150 monomials. Our final result $I_{4,0}(a_{111}^4) = \Omega_T^4(T \circ a_{111}^4)$ is a polynomial in the variables a_{ijk} alone. It has only 25 monomials. Each coefficient is an integer multiple of 18630, which we divide out for convenience.

Proposition 4.4.7. *The degree 4 invariant of a ternary cubic equals*

$S = \dfrac{1}{18630} \cdot I_{4,0}(a_{111}^4)$

$= a_{300}a_{120}a_{021}a_{003} - a_{300}a_{120}a_{012}^2 - a_{300}a_{111}a_{030}a_{003}$
$+ a_{300}a_{111}a_{021}a_{012} + a_{300}a_{102}a_{030}a_{012} - a_{300}a_{102}a_{021}^2 - a_{210}^2a_{021}a_{003}$
$+ a_{210}^2a_{012}^2 + a_{210}a_{201}a_{030}a_{003} - a_{210}a_{201}a_{021}a_{012} + a_{210}a_{120}a_{111}a_{003}$
$- a_{210}a_{120}a_{102}a_{012} - 2a_{210}a_{111}^2a_{012} + 3a_{210}a_{111}a_{102}a_{021} - a_{210}a_{102}^2a_{030}$
$- a_{201}^2a_{030}a_{012} + a_{201}^2a_{021}^2 - a_{201}a_{120}^2a_{003} + 3a_{201}a_{120}a_{111}a_{012}$
$- a_{201}a_{120}a_{102}a_{021} - 2a_{201}a_{111}^2a_{021} + a_{201}a_{111}a_{102}a_{030} + a_{120}^2a_{102}^2$
$- 2a_{120}a_{111}^2a_{102} + a_{111}^4.$

We now explain what it means that the monomial a_{111}^4 was "suitable". In what follows V can be any Γ-module. Let T denote the subgroup of diagonal matrices in Γ. This group is isomorphic to $(\mathbf{C}^*)^n$, and it is called the *maximal torus* of Γ. It can be assumed that a basis of V has been chosen such that T acts by scaling on the monomials in $\mathbf{C}[V]$.

We have the inclusion of graded invariant rings $\mathbf{C}[V]^\Gamma \subset \mathbf{C}[V]^T$. It is easy to see that the Ω-process preserves the ring of torus invariants $\mathbf{C}[V]^T$. Therefore the only *suitable* monomials m (with possibly $I_{p,q}(m) \neq 0$) are the monomials in $\mathbf{C}[V]^T$.

In order to compute the invariant ring $\mathbf{C}[V]^T$ we apply the methods presented in Sect. 1.4. Let us first suppose that $V = S_d \mathbf{C}^n$. A torus element $\mathrm{diag}(t_1, \ldots, t_n) \in T$ acts on a variable $a_{i_1 i_2 \ldots i_n}$ by multiplying it with $t_1^{i_1} t_2^{i_2} \cdots t_n^{i_n}$. Thus a monomial $\prod_{j=1}^m a_{v_{j1} v_{j2} \ldots v_{jn}}$ lies in $\mathbf{C}[V]^T$ if and only if

$$\prod_{j=1}^m t_1^{v_{j1}} t_2^{v_{j2}} \ldots t_n^{v_{jn}} = t_1^g t_2^g \cdots t_n^g \quad \text{where } g = md/n \tag{4.4.14}$$

Let r denote the least denominator of the rational number d/n. We abbreviate

$$\begin{aligned} \mathcal{A} = \{ r \cdot (i_1 - g, i_2 - g, \ldots i_n - g) : \\ i_1, \ldots, i_n \geq 0 \text{ integers with } i_1 + \ldots + i_n = d \}. \end{aligned} \tag{4.4.15}$$

We can reformulate (4.4.14) as follows.

Observation 4.4.8. The invariant ring $\mathbf{C}[V]^T$ equals the ring of invariants of the matrix group $\Gamma_{\mathcal{A}}$, as defined in Sect. 1.4.

This observation extends to an arbitrary Γ-module V. In (4.4.15) we need to take $\mathcal{A}$ to be the set of exponent vectors $(i_1, i_2, \ldots, i_n)$ appearing in the formal character of V. Algorithm 1.4.5 can be used to compute a Hilbert basis for $\mathbf{C}[V]^T$, and thus a vector space basis for the graded components $\mathbf{C}[V]_m^T = (S_m S_d \mathbf{C}^n)^T$. We summarize our first algorithm for computing all Γ-invariants in $S_m S_d \mathbf{C}^n$.

Algorithm 4.4.9.
Output: A spanning set for the $\mathbf{C}$-vector space of Γ-invariants in $S_m S_d \mathbf{C}^n$.
1. Using Algorithm 1.4.5, compute all T-invariant monomials in $S_m S_d \mathbf{C}^n$.
2. To each T-invariant monomial apply the m-fold Ω-process $I_{m,0}(\,\cdot\,)$.

Exercises

(1) Extend the list (4.4.13) by computing the plethysm $h_5 \circ h_3$ in terms of Schur polynomials.
(2) Apply Algorithm 4.4.9 to generate the fundamental covariants f, H, T, D of a binary cubic.
(3) Determine the ring of covariants of an n-ary quadratic form.
(4) * Determine the ring of joint invariants of two ternary quadratic forms.
(5) Compute the resultant R of three ternary quadratic forms.

4.5. Lie algebra action and the symbolic method

There are three different algorithms for computing a basis for the vector space of invariants of fixed degree.

(1) The Ω-process (Algorithm 4.4.9)
(2) Solving linear equations arising from the Lie algebra action (Theorem 4.5.2)
(3) Generating invariants in symbolic representation (Algorithm 4.5.8).

In this section we introduce the second and the third method. Both of these outperform the Ω-process in practical computations.

Let $\rho : \Gamma \to GL(V)$ be any rational representation of $\Gamma = GL(\mathbf{C}^n)$. The Lie algebra $\text{Lie}(\Gamma)$ can be identified with the vector space $\mathbf{C}^{n \times n}$ of $n \times n$-matrices. We choose the canonical basis $\{E_{ij}\}$ of matrix units for $\text{Lie}(\Gamma)$. The group homomorphism ρ induces a homomorphism of Lie algebras

$$\rho^* : \text{Lie}(\Gamma) \to \text{Lie}(GL(V)). \quad (4.5.1)$$

By **C**-linearity, it suffices to give the image of basis elements under ρ^*. We have

$$\rho^*(E_{ij}) = \left\{ \frac{\partial}{\partial t_{ij}} \rho(T) \right\}_{T=1} \quad (4.5.2)$$

where **1** denotes the $n \times n$-unit matrix.

Lemma 4.5.1. A vector $v \in V$ is a Γ-invariant of index g if and only if $\rho^*(E_{ij}) \cdot v = 0$ for all $i, j \in \{1, \ldots, n\}, i \neq j$ and $\rho^*(E_{ii}) \cdot v = g \cdot v$ for all $i \in \{1, \ldots, n\}$.

Proof. A vector $v \in V$ being a Γ-invariant of index g means that $\rho(T) \cdot v = \det(T)^g \cdot v$ for all $T = (t_{ij}) \in \Gamma$. Differentiating this identity with respect to the variables t_{ij} and substituting $T = \mathbf{1}$ thereafter, we obtain the only-if part of Lemma 4.5.1.

For the converse suppose that v is not a Γ-invariant. This means there exists an element in the Lie group $\rho(\Gamma) \cap SL(V)$ which moves v. Since this Lie group is connected, we can find such an element in any neighborhood of the identity. This implies that v is not fixed by the Lie algebra of $\rho(\Gamma) \cap SL(V)$. ◁

We work out an explicit description of the linear map (4.5.1) for the case $V = S_d \mathbf{C}^n$, $\rho = \rho_d$. Let $f = f(\mathbf{x})$ be any n-ary d-form. By (4.5.2), its image under $\rho_d^*(E_{ij})$ is the n-ary d-form

$$\left(\rho_d^*(E_{ij}) \cdot f\right)(\mathbf{x}) = \left\{\frac{\partial}{\partial t_{ij}} f(T\mathbf{x})\right\}_{T=1}$$

$$= \left\{\frac{\partial f}{\partial x_j}(T\mathbf{x}) \cdot x_i\right\}_{T=1} \quad (4.5.3)$$

$$= x_i \frac{\partial f}{\partial x_j}(\mathbf{x}).$$

On the monomial basis of $S_d\mathbf{C}^n$ this action is given by

$$\rho_d^*(E_{ij})(\mathbf{x}^\alpha) = \alpha_j \cdot \mathbf{x}^{\alpha+(e_i-e_j)}, \quad (4.5.4)$$

where $\mathbf{x}^\alpha = x_1^{\alpha_1} x_2^{\alpha_2} \cdots x_n^{\alpha_n}$ and e_i, e_j are unit vectors. This formula can be rewritten in terms of the canonical basis $\{E_{\alpha,\beta}\}$ of the Lie algebra Lie$(GL(V))$. We have

$$\rho_d^*(E_{ij}) = \sum_{\alpha:\alpha_j>0} \alpha_j \, E_{\alpha+(e_i-e_j),\alpha}. \quad (4.5.5)$$

We now iterate this construction and consider $W = S_m V$, first as a $GL(V)$-module and then as a Γ-module. These two representations are denoted ρ_m and $\rho_m \circ \rho_d$ respectively. In the following we represent forms $f = \sum_\alpha \binom{d}{\alpha} a_\alpha \mathbf{x}^\alpha$ in V by their symmetric tensors $\mathbf{a} = (a_\alpha)$. Elements in W are homogeneous polynomial functions $P = P(\mathbf{a})$. By (4.5.3) the Lie algebra Lie$(GL(V))$ acts on W via

$$\left(\rho_m^*(E_{\alpha,\beta})\right)(P) = a_\alpha \cdot \frac{\partial}{\partial a_\beta} P(\mathbf{a}).$$

The action of the smaller Lie algebra Lie(Γ) on W is described by the formula

$$\rho_m^*(\rho_d^*(E_{ij}))(P) = \sum_{\alpha:\alpha_j>0} \alpha_j \cdot a_{\alpha+(e_i-e_j)} \cdot \frac{\partial P}{\partial a_\alpha}. \quad (4.5.6)$$

Lemma 4.5.1 now implies the following theorem.

Theorem 4.5.2. Let $V = S_d\mathbf{C}^n$. An element $I = I(\mathbf{a}) \in \mathbf{C}[V]$ is a Γ-invariant of index g if and only if it satisfies the linear differential equations

$$\sum_\alpha \alpha_j \cdot a_\alpha \frac{\partial I}{\partial a_\alpha} = g \cdot I \quad \text{for } i = 1, 2, \ldots, n$$

$$\text{and} \quad \sum_\alpha \alpha_j \cdot a_{\alpha+(e_i-e_j)} \frac{\partial I}{\partial a_\alpha} = 0 \quad \text{for } i, j \in \{1, 2, \ldots, n\}, i \neq j. \quad (4.5.7)$$

The differential equations (4.5.7) translate into a system of linear equations

4.5. Lie algebra action and the symbolic method

on $W = S_m S_d \mathbf{C}^n$. The solution space to this system is precisely the vector space of degree m invariants of an n-ary d-form. The first group of equations in (4.5.7) states that I is invariant under the action of the maximal torus $T \simeq (\mathbf{C}^*)^n$ in Γ.

By the discussion at the end of Sect. 4.4, we can assume that a basis for the linear space of torus invariants W^T has been precomputed (e.g., using the methods in Sect. 1.4). We can restrict ourselves to the second group of $n^2 - n$ differential equations. These translate into a system of linear equations on W^T, whose solution space is W^Γ.

Example 4.5.3 (Ternary cubics). The degree 6 invariant T of the ternary cubic equals

$$a_{300}^2 a_{030}^2 a_{003}^2 - 6a_{300}^2 a_{030} a_{021} a_{012} a_{003} + 4a_{300}^2 a_{030} a_{012}^3 + 4a_{300}^2 a_{021}^3 a_{003}$$
$$- 3a_{300}^2 a_{021}^2 a_{012}^2 - 6a_{300} a_{210} a_{120} a_{030} a_{003}^2 + 18 a_{300} a_{210} a_{120} a_{021} a_{012} a_{003}$$
$$- 12 a_{300} a_{210} a_{120} a_{012}^3 + 12 a_{300} a_{210} a_{111} a_{030} a_{012} a_{003} - 24 a_{300} a_{210} a_{111} a_{021}^2 a_{003}$$
$$+ 12 a_{300} a_{210} a_{111} a_{021} a_{012}^2 + 6 a_{300} a_{210} a_{102} a_{030} a_{021} a_{003}$$
$$- 12 a_{300} a_{210} a_{102} a_{030} a_{012}^2 + 6 a_{300} a_{210} a_{102} a_{021}^2 a_{012} + 6 a_{300} a_{201} a_{120} a_{030} a_{012} a_{003}$$
$$- 12 a_{300} a_{201} a_{120} a_{021}^2 a_{003} + 6 a_{300} a_{201} a_{120} a_{021} a_{012}^2 + 12 a_{300} a_{201} a_{111} a_{030} a_{021} a_{003}$$
$$- 24 a_{300} a_{201} a_{111} a_{030} a_{012}^2 + 12 a_{300} a_{201} a_{111} a_{021}^2 a_{012} - 6 a_{300} a_{201} a_{102} a_{030}^2 a_{003}$$
$$+ 18 a_{300} a_{201} a_{102} a_{030} a_{021} a_{012} - 12 a_{300} a_{201} a_{102} a_{021}^3 + 4 a_{300} a_{120}^3 a_{003}^2$$
$$- 24 a_{300} a_{120}^2 a_{111} a_{012} a_{003} - 12 a_{300} a_{120}^2 a_{102} a_{021} a_{003} + 24 a_{300} a_{120}^2 a_{102} a_{012}^2$$
$$+ 36 a_{300} a_{120} a_{111}^2 a_{021} a_{003} + 12 a_{300} a_{120} a_{111}^2 a_{012}^2 + 12 a_{300} a_{120} a_{111} a_{102} a_{030} a_{003}$$
$$- 60 a_{300} a_{120} a_{111} a_{102} a_{021} a_{012} - 12 a_{300} a_{120} a_{102}^2 a_{030} a_{012} + 24 a_{300} a_{120} a_{102}^2 a_{021}^2$$
$$- 20 a_{300} a_{111}^3 a_{030} a_{003} - 12 a_{300} a_{111}^3 a_{021} a_{012} + 36 a_{300} a_{111}^2 a_{102} a_{030} a_{012}$$
$$+ 12 a_{300} a_{111}^2 a_{102} a_{021}^2 - 24 a_{300} a_{111} a_{102}^2 a_{030} a_{021} + 4 a_{300} a_{102}^3 a_{030}^2 + 4 a_{210}^3 a_{030} a_{003}^2$$
$$- 12 a_{210}^3 a_{021} a_{012} a_{003} + 8 a_{210}^3 a_{012}^3 - 12 a_{210}^2 a_{201} a_{030} a_{012} a_{003}$$
$$+ 24 a_{210}^2 a_{201} a_{021}^2 a_{003} - 12 a_{210}^2 a_{201} a_{021} a_{012}^2 - 3 a_{210}^2 a_{120}^2 a_{003}^2$$
$$+ 12 a_{210}^2 a_{120} a_{111} a_{012} a_{003} - 24 a_{120}^2 a_{111}^2 a_{102}^2 + 24 a_{120} a_{111}^4 a_{102}$$
$$+ 6 a_{210}^2 a_{120} a_{102} a_{021} a_{003} - 12 a_{210}^2 a_{120} a_{102} a_{012}^2 + 12 a_{210}^2 a_{111}^2 a_{021} a_{003}$$
$$- 24 a_{210}^2 a_{111}^2 a_{012}^2 - 24 a_{210}^2 a_{111} a_{102} a_{030} a_{003} - 27 a_{210}^2 a_{102}^2 a_{021}^2$$
$$+ 36 a_{210}^2 a_{111} a_{102} a_{021} a_{012} + 24 a_{210}^2 a_{102}^2 a_{030} a_{012} - 12 a_{210} a_{201}^2 a_{030} a_{021} a_{003}$$
$$+ 24 a_{210} a_{201}^2 a_{030} a_{012}^2 - 12 a_{210} a_{201}^2 a_{021}^2 a_{012} + 6 a_{210} a_{201} a_{120}^2 a_{012} a_{003}$$
$$- 60 a_{210} a_{201} a_{120} a_{111} a_{021} a_{003} + 36 a_{210} a_{201} a_{120} a_{111} a_{012}^2$$
$$+ 18 a_{210} a_{201} a_{120} a_{102} a_{030} a_{003} - 6 a_{210} a_{201} a_{120} a_{102} a_{021} a_{012}$$

$$+ 36a_{210}a_{201}a_{111}^2a_{030}a_{003} - 12a_{210}a_{201}a_{111}^2a_{021}a_{012}$$
$$- 60a_{210}a_{201}a_{111}a_{102}a_{030}a_{012} + 36a_{210}a_{201}a_{111}a_{102}a_{021}^2 + 6a_{210}a_{201}a_{102}^2a_{030}a_{021}$$
$$+ 12a_{210}a_{120}^2a_{111}a_{102}a_{003} - 12a_{210}a_{120}^2a_{102}^2a_{012} - 12a_{210}a_{120}a_{111}^3a_{003}$$
$$- 12a_{210}a_{120}a_{111}^2a_{102}a_{012} + 36a_{210}a_{120}a_{111}a_{102}^2a_{021} - 12a_{210}a_{120}a_{102}^3a_{030}$$
$$+ 24a_{210}a_{111}^4a_{012} - 36a_{210}a_{111}^3a_{102}a_{021} + 12a_{210}a_{111}^2a_{102}^2a_{030} + 4a_{201}^3a_{030}^2a_{003}$$
$$- 12a_{201}^3a_{030}a_{021}a_{012} + 8a_{201}^3a_{021}^3 + 24a_{201}^2a_{120}^2a_{021}a_{003} - 27a_{201}^2a_{120}^2a_{012}^2$$
$$- 24a_{201}^2a_{120}a_{111}a_{030}a_{003} + 36a_{201}^2a_{120}a_{111}a_{021}a_{012} + 6a_{201}^2a_{120}a_{102}a_{030}a_{012}$$
$$- 12a_{201}^2a_{120}a_{102}a_{021}^2 + 12a_{201}^2a_{111}^2a_{030}a_{012} - 24a_{201}^2a_{111}^2a_{021}^2$$
$$+ 12a_{201}^2a_{111}a_{102}a_{030}a_{021} - 3a_{201}^2a_{102}^2a_{030}^2 - 12a_{201}a_{120}^3a_{102}a_{003}$$
$$+ 12a_{201}a_{120}^2a_{111}^2a_{003} + 36a_{201}a_{120}^2a_{111}a_{102}a_{012} - 12a_{201}a_{120}^2a_{102}^2a_{021}$$
$$- 36a_{201}a_{120}a_{111}^3a_{012} - 12a_{201}a_{120}a_{111}^2a_{102}a_{021} + 12a_{201}a_{120}a_{111}a_{102}^2a_{030}$$
$$+ 24a_{201}a_{111}^4a_{021} - 12a_{201}a_{111}^3a_{102}a_{030} + 8a_{120}^3a_{102}^3 - 8a_{111}^6.$$

This invariant was generated as follows. We first computed all monomials of degree 6 which are invariant under the action of the maximal torus, using Observation 4.4.8. There are precisely 103 such monomials, namely, the monomials appearing in the above expansion of T. We then made an "ansatz" for T with 103 indeterminate coefficients. By Theorem 4.5.2 the invariant T is annihilated by the following six linear differential operators:

$$\mathcal{E}_{21} = 3a_{210}\frac{\partial}{\partial a_{300}} + 2a_{120}\frac{\partial}{\partial a_{210}} + 2a_{111}\frac{\partial}{\partial a_{201}} + a_{030}\frac{\partial}{\partial a_{120}} + a_{021}\frac{\partial}{\partial a_{111}}$$
$$+ a_{012}\frac{\partial}{\partial a_{102}}$$

$$\mathcal{E}_{31} = 3a_{201}\frac{\partial}{\partial a_{300}} + 2a_{111}\frac{\partial}{\partial a_{210}} + 2a_{102}\frac{\partial}{\partial a_{201}} + a_{021}\frac{\partial}{\partial a_{120}} + a_{012}\frac{\partial}{\partial a_{111}}$$
$$+ a_{003}\frac{\partial}{\partial a_{102}}$$

$$\mathcal{E}_{12} = a_{300}\frac{\partial}{\partial a_{210}} + 2a_{210}\frac{\partial}{\partial a_{120}} + a_{201}\frac{\partial}{\partial a_{111}} + 3a_{120}\frac{\partial}{\partial a_{030}} + 2a_{111}\frac{\partial}{\partial a_{021}}$$
$$+ a_{102}\frac{\partial}{\partial a_{012}}$$

$$\mathcal{E}_{32} = a_{201}\frac{\partial}{\partial a_{210}} + 2a_{111}\frac{\partial}{\partial a_{120}} + a_{102}\frac{\partial}{\partial a_{111}} + 3a_{021}\frac{\partial}{\partial a_{030}} + 2a_{012}\frac{\partial}{\partial a_{021}}$$
$$+ a_{003}\frac{\partial}{\partial a_{012}}$$

4.5. Lie algebra action and the symbolic method

$$\mathcal{E}_{13} = a_{300}\frac{\partial}{\partial a_{201}} + a_{210}\frac{\partial}{\partial a_{111}} + 2a_{201}\frac{\partial}{\partial a_{102}} + a_{120}\frac{\partial}{\partial a_{021}} + 2a_{111}\frac{\partial}{\partial a_{012}}$$
$$+ 3a_{102}\frac{\partial}{\partial a_{003}}$$

$$\mathcal{E}_{23} = a_{210}\frac{\partial}{\partial a_{201}} + a_{120}\frac{\partial}{\partial a_{111}} + 2a_{111}\frac{\partial}{\partial a_{102}} + a_{030}\frac{\partial}{\partial a_{021}} + 2a_{021}\frac{\partial}{\partial a_{012}}$$
$$+ 3a_{012}\frac{\partial}{\partial a_{003}}$$

Applying these operators to our ansatz for T, and equating the result with zero, we obtained a system of 540 linear equations in 103 variables. The solution space to this system is one-dimensional, in accordance with Theorem 4.4.6. The unique generator for this space, up to scaling, is the vector of coefficients for the above invariant.

We close Example 4.5.3 with a remark concerning the geometric significance of two invariants derived from S and T. The zero set defined by the ternary cubic (4.4.13) in the projective plane is a cubic curve. This curve is singular if and only if the *discriminant* Δ vanishes. Otherwise the curve is an *elliptic curve*. A classical invariant for distinguishing elliptic curves is the *j-invariant J*. This is a Γ-invariant rational function. We have the following formulas for the discriminant and the j-invariant in terms of the two basic invariants:

$$\Delta = T^2 - 64S^3 \quad \text{and} \quad J = \frac{S^3}{\Delta}. \tag{4.5.8}$$

Thus Theorem 4.4.6 implies the well known geometric fact that the moduli space of elliptic curves is birationally isomorphic to projective line P^1.

We now come to the *symbolic method* of classical invariant theory, which provides our third algorithm for computing invariants and covariants. One particularly nice feature of the symbolic method is a compact encoding for large invariants. Indeed, many of the 19th century tables of invariants are presented in symbolic notation. Familiarity with the subsequent material is thus a precondition for accessing many classical results and tables.

For simplicity of exposition we restrict ourselves to the Γ-module $V = S_d\mathbf{C}^n$, which is the case of invariants of a single form. The generalization to joint invariants and to covariants is straightforward and left as an exercise. Other generalizations, for instance to $V = \wedge_d\mathbf{C}^n$, are more difficult, as they involve the use of non-commutative algebras. For the state of the art regarding the symbolic method we refer to Grosshans et al. (1987).

It is our objective to construct the Γ-invariants in the vector space $S_m S_d \mathbf{C}^n$. Recall that the index g of these invariants satisfies the relation $g \cdot n = m \cdot d$. Consider the ring $\mathbf{C}[x_{ij}]$ of polynomial functions on a generic $m \times n$-matrix (x_{ij}), and let $\mathbf{C}[x_{ij}]_{(d,\ldots,d)}$ denote the subspace of polynomials that are homogeneous of degree d in each row $(x_{i1}, x_{i2}, \ldots, x_{in})$.

The natural monomial basis in $\mathbf{C}[x_{ij}]_{(d,\ldots,d)}$ is indexed by all non-negative $m \times n$-matrices (v_{ij}) having each row sum to d. We define the $\mathbf{C}$-linear map

$$\phi : \mathbf{C}[x_{ij}]_{(d,\ldots,d)} \to S_m S_d \mathbf{C}^n$$
$$\prod_{i=1}^{m} \prod_{j=1}^{n} x_{ij}^{v_{ij}} \mapsto \prod_{i=1}^{m} a_{v_{i1}, v_{i2}, \ldots, v_{in}}. \tag{4.5.9}$$

The map ϕ is sometimes called the *umbral operator*.

The symmetric group S_m acts on $\mathbf{C}[x_{ij}]$ by permuting rows. Images under ϕ are invariant under this action, that is, $\phi(\sigma P) = \phi(P)$ for all $P \in \mathbf{C}[x_{ij}]_{(d,\ldots,d)}$ and $\sigma \in S_m$. Let $*$ denote the Reynolds operator of the symmetric group S_m.

Lemma 4.5.4. The restriction of ϕ to the subspace of S_m-invariants defines a vector space isomorphism

$$\tilde{\phi} : \mathbf{C}[x_{ij}]_{(d,\ldots,d)}^{S_m} \simeq S_m S_d \mathbf{C}^n. \tag{4.5.10}$$

Proof. The inverse to $\tilde{\phi}$ is given by

$$(\tilde{\phi})^{-1} : \prod_{i=1}^{m} a_{v_{i1}, v_{i2}, \ldots, v_{in}} \mapsto (\prod_{i=1}^{m} \prod_{j=1}^{n} x_{ij}^{v_{ij}})^* = \frac{1}{m!} \sum_{\sigma \in S_m} \prod_{i=1}^{m} \prod_{j=1}^{n} x_{\sigma(i),j}^{v_{\sigma(i),j}}. \quad \triangleleft$$

We next show that $\tilde{\phi}$ preserves the invariants under the Γ-action on both spaces.

Theorem 4.5.5. An element $P \in \mathbf{C}[x_{ij}]_{(d,\ldots,d)}^{S_m}$ is a Γ-invariant of index g if and only if its image $\phi(P)$ in $S_m S_d \mathbf{C}^n$ is a Γ-invariant of index g.

Proof. We consider the action of the Lie algebra $\mathrm{Lie}(\Gamma)$ on both spaces. By Lemma 4.5.1 and Theorem 4.5.2, it suffices to show that these two actions commute.

For $S_m S_d \mathbf{C}^n$ the Lie algebra action had been determined above. We can rewrite (4.5.6) as follows:

$$E_{kl} : \prod_{i=1}^{m} a_{v_{i1}, v_{i2}, \ldots, v_{in}} \mapsto \sum_{j=1}^{m} v_{jl} \frac{a_{v_{j1}, \ldots, v_{jk}+1, \ldots, v_{jl}-1, \ldots, v_{jn}}}{a_{v_{j1}, v_{j2}, \ldots, v_{jn}}} \prod_{i=1}^{m} a_{v_{i1}, v_{i2}, \ldots, v_{in}}. \tag{4.5.11}$$

The action on $\mathbf{C}[x_{ij}]_{(d,\ldots,d)}$ can be described as follows. The image of a monomial $\mathbf{m} = \prod_{i=1}^{m} \prod_{j=1}^{n} x_{ij}^{v_{ij}}$ under the basis elements E_{kl} of $\mathrm{Lie}(\Gamma)$ is the coeffi-

4.5. Lie algebra action and the symbolic method

cient of the variable t in the expansion of

$$\prod_{i=1}^{m}(x_{il}+tx_{ik})^{v_{il}}\prod_{\substack{j=1\\j\neq l}}^{n}x_{ij}^{v_{ij}}.$$

This coefficient equals

$$\sum_{j=1}^{m}v_{jl}\cdot\frac{x_{jk}}{x_{jl}}\cdot\mathbf{m}=\sum_{j=1}^{m}v_{jl}\frac{x_{j1}^{v_{j1}}\cdots x_{jk}^{v_{jk}+1}\cdots x_{jl}^{v_{jl}-1}\cdots x_{jn}^{v_{jn}}}{x_{j1}^{v_{j1}}x_{j1}^{v_{j2}}\cdots x_{jn}^{v_{jn}}}\cdot\mathbf{m}. \quad (4.5.12)$$

The image of (4.5.12) under ϕ is equal to (4.5.11). This completes the proof. ◁

Recall from Sect. 3.2 that the Γ-invariants in $\mathbf{C}[x_{ij}]$ are precisely the rank n bracket polynomials on m letters. Using the First Fundamental Theorem 3.2.1, we may thus replace $\mathbf{C}[x_{ij}]^{\Gamma}$ by $\mathcal{B}_{m,n} = \mathbf{C}[\Lambda(m,n)]/I_{m,n}$, the bracket ring modulo the syzygy ideal. Let $\mathcal{B}_{m,n,g}$ denote the subspace of all rank n bracket polynomials of total degree g that are symmetric in the letters $1, 2, \ldots, m$.

Corallary 4.5.6. *The space $\mathcal{B}_{m,n,g}$ is isomorphic to the vector space of invariants $(S_m S_d \mathbf{C}^n)^{\Gamma}$.*

Example 4.5.7 (Cubic invariants of a ternary quadric, $n = 3$, $d = 2$, $m = 3$, $g = 2$). The space $\mathcal{B}_{3,3,2}$ consists of rank 3 bracket polynomials in $\{1, 2, 3\}$, homogeneous of bracket degree 2. This is a one-dimensional space, spanned by the bracket monomial $[1\,2\,3]^2$. Therefore the space $(S_3 S_2 \mathbf{C}^3)^{\Gamma}$ is spanned by the invariant $\phi([1\,2\,3]^2)$.

We evaluate this invariant in terms of the coefficients of the ternary quadric (4.1.10). In the following table the first column lists the monomial expansion of $[1\,2\,3]^2 = (x_{11}x_{22}x_{33} - x_{11}x_{23}x_{32} - x_{12}x_{21}x_{33} + x_{12}x_{23}x_{31} - x_{13}x_{21}x_{32} + x_{13}x_{22}x_{31})^2$. To each monomial we apply the operator ϕ. The results are listed in the second column:

$$
\begin{array}{ll}
x_{11}^2 x_{22}^2 x_{33}^2 & a_{200}a_{020}a_{002} \\
-2x_{11}^2 x_{22}x_{23}x_{32}x_{33} & -2a_{200}a_{011}^2 \\
+x_{11}^2 x_{23}^2 x_{32}^2 & +a_{200}a_{020}a_{002} \\
-2x_{11}x_{12}x_{21}x_{22}x_{33}^2 & -2a_{110}^2 a_{002} \\
+2x_{11}x_{12}x_{21}x_{23}x_{32}x_{33} & +2a_{110}a_{101}a_{011} \\
+2x_{11}x_{12}x_{22}x_{23}x_{31}x_{33} & +2a_{110}a_{101}a_{011} \\
-2x_{11}x_{12}x_{23}^2 x_{31}x_{32} & -2a_{110}^2 a_{002} \\
+2x_{11}x_{13}x_{21}x_{22}x_{32}x_{33} & +2a_{110}a_{101}a_{011}
\end{array}
$$

$$-2x_{11}x_{13}x_{21}x_{23}x_{32}^2 \qquad -2a_{101}^2 a_{020}$$
$$-2x_{11}x_{13}x_{22}^2 x_{31}x_{33} \qquad -2a_{101}^2 a_{020}$$
$$+2x_{11}x_{13}x_{22}x_{23}x_{31}x_{32} \qquad +2a_{110}a_{101}a_{011}$$
$$+x_{12}^2 x_{21}^2 x_{33}^2 \qquad +a_{200}a_{020}a_{002}$$
$$-2x_{12}^2 x_{21}x_{23}x_{31}x_{33} \qquad -2a_{101}^2 a_{020}$$
$$+x_{12}^2 x_{23}^2 x_{31}^2 \qquad +a_{200}a_{020}a_{002}$$
$$-2x_{12}x_{13}x_{21}^2 x_{32}x_{33} \qquad -2a_{200}a_{011}^2$$
$$+2x_{12}x_{13}x_{21}x_{22}x_{31}x_{33} \qquad +2a_{110}a_{101}a_{011}$$
$$+2x_{12}x_{13}x_{21}x_{23}x_{31}x_{32} \qquad +2a_{110}a_{101}a_{011}$$
$$-2x_{12}x_{13}x_{22}x_{23}x_{31}^2 \qquad -2a_{200}a_{011}^2$$
$$+x_{13}^2 x_{21}^2 x_{32}^2 \qquad +a_{200}a_{020}a_{002}$$
$$-2x_{13}^2 x_{21}x_{22}x_{31}x_{32} \qquad -2a_{110}^2 a_{002}$$
$$+x_{13}^2 x_{22}^2 x_{31}^2 \qquad +a_{200}a_{020}a_{002}$$

The sum over the second column in this table equals $\phi([1\,2\,3]^2) = 6 \cdot \Delta$, where Δ equals the discriminant (4.1.14).

From Theorem 4.5.5 and Corollary 4.5.6 we derive the following algorithm.

Algorithm 4.5.8.
Input: Integers m, d and n such that $g = \frac{m \cdot d}{n}$ is an integer.
Output: A basis $\mathcal{I}$ for the space of Γ-invariants in $S_m S_d \mathbf{C}^n$, in symbolic notation.

1. Let $\mathcal{T}$ be the set of rank n standard bracket monomials in the letters $\{1, 2, \ldots, m\}$ having degree g.
2. For each standard bracket monomial $t \in \mathcal{T}$ compute its S_m-symmetrization t^*.
3. Compute a basis $\mathcal{I}$ for the **C**-linear span of the bracket polynomials $\{t^* : t \in \mathcal{T}\}$.

Example 4.5.9 (Ternary cubics revisited, $m = 4$, $d = 3$, $n = 3$). The space of symmetrized rank 3 standard bracket polynomials on $\{1, 2, 3, 4\}$ having degree 4 is one-dimensional. It is spanned by the bracket monomial [123][124][134][234]. Therefore the space of degree 4 invariants is spanned by

$$\phi([123][124][134][234]) = 24 \cdot S. \qquad (4.5.13)$$

Here S denotes the familiar invariant given in Proposition 4.4.7.

Exercises

(1) Describe the action of the Lie algebra Lie(Γ) on the spaces $S_m \wedge_d \mathbf{C}^n$ and $\wedge_m S_d \mathbf{C}^n$.
(2) Give a symbolic representation for the degree 6 invariant of the ternary cubic.
(3) Compute a basis for the degree 8 invariants of the binary quintic.
(4) What is the smallest degree for an invariant of the ternary quartic?
(5) * Formulate the symbolic method for joint invariants and joint covariants of forms. State the map ϕ explicitly. Determine symbolic representations for the resultant and for the Jacobian of two binary cubics.
(6) * Determine the ring of joint invariants of k binary quadrics.

4.6. Hilbert's algorithm

We give an algorithm for computing a finite generating set for the invariant ring $\mathbf{C}[V]^\Gamma$ of an arbitrary polynomial Γ-module. Our discussion follows closely the original work of Hilbert (1893). One of the key concepts introduced in Hilbert (1893) is the *nullcone*. Computing the nullcone will be our theme in the first half of this section. Later on we need to pass from invariants defining the nullcone to the complete set of generators, which amounts to an integral closure computation. The complexity analysis, based on results of Hochster and Roberts (1974) and Popov (1981, 1982), will be presented in Sect. 4.7.

Let $\mathcal{I}_\Gamma$ denote the ideal in $\mathbf{C}[V]$ that is generated by all homogeneous Γ-invariants of positive degree. Let $\mathcal{N}_\Gamma$ denote the affine algebraic variety defined by $\mathcal{I}_\Gamma$. This subvariety of V is called the *nullcone* of the Γ-module V. The following result is crucial for our algorithm. Its proof is the very purpose for which Hilbert's Nullstellensatz was first invented.

Theorem 4.6.1 (Hilbert 1893). Let $I_1, \ldots, I_m$ be homogeneous invariants whose common zero set in V equals the nullcone $\mathcal{N}_\Gamma$. Then the invariant ring $\mathbf{C}[V]^\Gamma$ is finitely generated as a module over its subring $\mathbf{C}[I_1, \ldots, I_m]$.

Proof. By Theorem 4.3.1, the invariant ring is finitely generated as a $\mathbf{C}$-algebra: there exist invariants $J_1, \ldots, J_s$ such that

$$\mathbf{C}[V]^\Gamma = \mathbf{C}[J_1, J_2, \ldots, J_s]. \tag{4.6.1}$$

Let d denote the maximum of the degrees of $J_1, J_2, \ldots, J_s$. The nullcone $\mathcal{N}_\Gamma$, which is the variety defined by the fundamental invariants $J_1, \ldots, J_s$, coincides with the variety defined by $I_1, \ldots, I_m$. By Hilbert's Nullstellensatz, there exists an integer r such that

$$J_1^r, J_2^r, \ldots, J_s^r \in \langle I_1, I_2, \ldots, I_m \rangle. \tag{4.6.2}$$

Let $I \in \mathbf{C}[V]^\Gamma$ be any invariant of degree $\geq drs$. By (4.6.1) we can write I as a $\mathbf{C}$-linear combination of invariants of the form $J_1^{i_1} J_2^{i_2} \cdots J_s^{i_s}$, where $i_1 + i_2 + \ldots + i_s \geq sr$. Each such invariant lies in the ideal $\langle I_1, I_2, \ldots, I_m \rangle$, and so does I, by (4.6.2). We can write

$$I = f_1 I_1 + f_2 I_2 + \ldots + f_m I_m \quad \text{where } f_1, f_2, \ldots, f_s \in \mathbf{C}[V]. \tag{4.6.3}$$

Applying the Ω-process to the identity (4.6.3) as in (4.3.23), we see that the coefficients $f_1, f_2, \ldots, f_m$ may be chosen to be invariants. We now iterate this procedure for those invariants $f_i \in \mathbf{C}[V]^\Gamma$ whose degree is larger or equal to drs.

This proves that I is a linear combination of Γ-invariants of degree $< drs$, with coefficients in $\mathbf{C}[I_1, \ldots, I_m]$. Hence $\mathbf{C}[V]^\Gamma$ is finite over its subring $\mathbf{C}[I_1, \ldots, I_m]$. ◁

Corollary 4.6.2. Under the hypothesis of Theorem 4.6.1, the invariant ring $\mathbf{C}[V]^\Gamma$ equals the integral closure of $\mathbf{C}[I_1, \ldots, I_m]$ in the field $\mathbf{C}(V)$ of rational functions on V.

Proof. The ring $\mathbf{C}[V]^\Gamma$ is finite and therefore integral over $\mathbf{C}[I_1, \ldots, I_s]$. For the converse suppose that $f \in \mathbf{C}(V)$ is any rational function on V which is integral over $\mathbf{C}[I_1, \ldots, I_m]$. This means there exists an identity in $\mathbf{C}(V)$ of the form

$$f^n + p_{n-1}(I_1, \ldots, I_m) \cdot f^{n-1} + \ldots + p_1(I_1, \ldots, I_m) \cdot f + p_0(I_1, \ldots, I_m) = 0, \tag{4.6.4}$$

where the p_i are suitable polynomials. We write $f = g/h$, where g, h are relatively prime polynomials in $\mathbf{C}[V]$. The resulting identity

$$g^n/h = -p_{n-1}(I_1, \ldots, I_m) \cdot g^{n-1} - \ldots - p_1(I_1, \ldots, I_m) \cdot gh^{n-2}$$
$$- p_0(I_1, \ldots, I_m) h^{n-1} \tag{4.6.5}$$

shows that g^n/h is actually a polynomial in $\mathbf{C}[V]$. This means h divides g^n. Since g and h are assumed to be relatively prime, also h and g^n are relatively prime. This implies that h is a constant, and therefore $f = g/h$ lies in $\mathbf{C}[V]$.

It remains to be seen that f is a Γ-invariant. Since Γ is a connected group, it suffices to show that the orbit of f under Γ is a finite set. Consider the n roots of the polynomial in (4.6.4) in the algebraic closure of $\mathbf{C}(V)$. Since each coefficient $p_i(I_1, \ldots, I_m)$ is Γ-invariant, the set of roots is Γ-invariant. Therefore the orbit of f has at most n elements. ◁

The problem of computing fundamental invariants now splits up into two parts:
- Compute homogeneous invariants $I_1, \ldots, I_m$ whose variety equals the null-cone $\mathcal{N}_\Gamma$.
- Compute the integral closure of $\mathbf{C}[I_1, \ldots, I_m]$ in $\mathbf{C}(V)$.

4.6. Hilbert's algorithm

We first address the problem of computing the nullcone. Let $T \simeq (\mathbf{C}^*)^n$ denote the maximal torus in Γ, and consider its ring of invariants $\mathbf{C}[V]^T$. We know that $\mathbf{C}[V]^T$ is a monomial algebra, whose minimal generating set $\mathcal{H}$, the Hilbert basis, can be computed using Algorithm 1.4.5. Let $\mathcal{I}_T$ denote the monomial ideal in $\mathbf{C}[V]$ that is generated by $\mathcal{H}$ or, equivalently, by all homogeneous T-invariants of positive degree. The affine algebraic variety defined by $\mathcal{I}_I$ is denoted $\mathcal{C}_{T,\Gamma}$ and called the *canonical cone*. It is a union of linear coordinate subspaces of V.

For algebraic geometers we note that in geometric invariant theory (Mumford and Fogarty 1982) the points in $\mathcal{N}_\Gamma$ are called Γ-*semistable* and the points in $\mathcal{C}_{T,\Gamma}$ are called T-*semistable*. The following theorem relates the nullcone to the canonical cone.

Theorem 4.6.3. The nullcone equals the Γ-orbit of the canonical cone, i.e.,

$$\forall \mathbf{v}_0 \in V : \mathbf{v}_0 \in \mathcal{N}_\Gamma \iff \exists A_0 \in \Gamma : A_0 \circ \mathbf{v}_0 \in \mathcal{C}_{T,\Gamma}. \tag{4.6.6}$$

We will not present the proof of Theorem 4.6.3 here, but instead we refer to Hilbert (1893: § 15–16). Other proofs using modern algebraic geometry language can be found in Krafft (1985: § II.2.3) and in Mumford and Fogarty (1982).

The condition (4.6.6) can be rephrased in ideal-theoretic terms. Let $A = (a_{ij})$ be a generic $n \times n$-matrix, and let $\mathbf{C}[V, A]$ denote the polynomial ring generated by the variables a_{ij} and the coordinates of the generic point $\mathbf{v}$ of V. Let $A \circ \mathcal{I}_T$ denote the ideal in $\mathbf{C}[V, A]$ gotten by substituting $A \circ \mathbf{v}$ for $\mathbf{v}$ in $\mathcal{I}_T$. Then (4.6.6) is equivalent to

$$\text{Ideal}(\mathcal{N}_\Gamma) = \text{Rad}(\mathcal{I}_\Gamma) = \text{Rad}\big((A \circ \mathcal{I}_T) \cap \mathbf{C}[V]\big), \tag{4.6.6'}$$

where $\text{Ideal}(\cdot)$ stands for "the vanishing ideal of" and $\text{Rad}(\cdot)$ refers to the radical of ideals in $\mathbf{C}[V]$. As a consequence of Theorem 4.6.3 we get the following algorithm.

Algorithm 4.6.4 (Computing the nullcone).
Input: A polynomial Γ-module V.
Output: Homogeneous invariants whose affine variety in V equals the nullcone $\mathcal{N}_\Gamma$.

1. Compute the Hilbert basis $\mathcal{H}$ of the ring $\mathbf{C}[V]^T$ of torus invariants (Algorithm 1.4.5).
2. Using Gröbner bases, eliminate the variables $A = (a_{ij})$ in the ideal

$$A \circ \mathcal{I}_T = \langle A \circ m : m \in \mathcal{H} \rangle \subset \mathbf{C}[V, A].$$

Let $g_1, g_2, \ldots, g_s$ be the resulting (non-invariant!) generators of $(A \circ \mathcal{I}_T) \cap \mathbf{C}[V]$.

3. Generate invariants $I_1, I_2, I_3, \ldots$ degree by degree until

$$g_1, g_2, \ldots, g_s \in \mathrm{Rad}(\langle I_1, I_2, \ldots, I_m \rangle) \quad \text{for some } m \geq 0.$$

Step 3 is formulated rather vaguely, as is the specification of the input. One reasonable assumption is that the action of the Lie algebra $\mathrm{Lie}(\Gamma)$ on $\mathbf{C}[V]$ is given explicitly, that is, we know the system of differential equations (4.5.7). Equivalent to this is the knowledge of the system of linear equations $\mathrm{Lie}(V) \cdot S_m(V) = 0$ for each degree level m. The latter is the working assumption in Popov (1981). From the first group of equations in (4.5.7) we can read off the set of weights $\mathcal{A} \subset \mathbf{Z}^n$, which is the input for step 1.

Another possibility is that we might have an efficient subroutine for the symbolic method, which supplies a stream of invariants as in Algorithm 4.5.8. A third possibility is that we are given a subroutine for performing the Ω-process. This is closest in spirit to Algorithm 2.5.8 for finite groups. In all three cases we will naturally proceed one degree at a time in step 3, similar to Algorithm 2.5.8.

In order to improve Algorithm 4.6.4 and to analyze its complexity, we need to study the nullcone and the canonical cone more closely. Let $\mathbf{C}[A]$ denote the polynomial ring on the generic matrix $A = (a_{ij})$. The determinant $\det(A)$ is an element of degree n in $\mathbf{C}[A]$. For each fixed vector $\mathbf{v}_0 \in V$ we let $\mathbf{C}[A\mathbf{v}_0]$ denote the subring of $\mathbf{C}[A]$ generated by the coordinates of the transformed vector $A \circ \mathbf{v}_0$.

Lemma 4.6.5. A vector $\mathbf{v}_0 \in V$ does not lie in the nullcone $\mathcal{N}_\Gamma$ if and only if $\det(A)$ is integral over $\mathbf{C}[A\mathbf{v}_0]$.

Proof. Suppose that $\mathbf{v}_0$ does not lie in $\mathcal{N}_\Gamma$. Then there exists a homogeneous invariant $I \in \mathbf{C}[V]^\Gamma$ of positive degree such that $I(\mathbf{v}_0) \neq 0$. We have the identity

$$\det(A)^p = I(A\mathbf{v}_0)/I(\mathbf{v}_0) \quad \text{in } \mathbf{C}[A\mathbf{v}_0], \qquad (4.6.7)$$

where $p > 0$ is the index of I. This shows that $\det(A)$ is integral over $\mathbf{C}[A\mathbf{v}_0]$.

For the converse, suppose there exists an identity

$$\det(A)^p + \sum_{j=0}^{p-1} f_j(A\mathbf{v}_0) \det(A)^j = 0 \quad \text{in } \mathbf{C}[A\mathbf{v}_0]. \qquad (4.6.8)$$

We may assume that this identity is homogeneous, in particular each f_j is homogeneous. We replace $\mathbf{v}_0$ by the generic vector $\mathbf{v}$ and consider the homogeneous polynomial

$$\det(A)^p + \sum_{j=0}^{p-1} f_j(A\mathbf{v}) \det(A)^j \quad \text{in } \mathbf{C}[V, A]. \qquad (4.6.9)$$

We apply the p-fold Ω-process Ω_A^p to (4.6.9). By the results of Sect. 4.3, this

4.6. Hilbert's algorithm

transforms (4.6.9) into an expression

$$c + \sum_{j=0}^{p-1} I_j(\mathbf{v}) \quad \text{in } \mathbf{C}[V, A], \tag{4.6.10}$$

where c is a non-zero constant and $I_0, I_1, \ldots, I_{p-1}$ are homogeneous invariants of positive degree. Moreover, if in (4.6.10) the generic element $\mathbf{v}$ is replaced by the specific $\mathbf{v}_0$, then we get zero by (4.6.8). Since c is non-zero, there exists an index $j \in \{0, 1, \ldots, p-1\}$ such that $I_j(\mathbf{v}_0) \neq 0$. Therefore $\mathbf{v}_0 \notin \mathcal{N}_\Gamma$. ◁

We next give a criterion for $\mathbf{v}_0$ to lie in the canonical cone. Let $\mathbf{t} = \mathrm{diag}(t_1, t_2, \ldots, t_n)$ be a generic element of the maximal torus $T = (\mathbf{C}^*)^n$. For each fixed vector $\mathbf{v}_0 \in V$ we let $\mathbf{C}[\mathbf{t}\mathbf{v}_0]$ denote the subring of $\mathbf{C}[\mathbf{t}] = \mathbf{C}[t_1, t_2, \ldots, t_n]$ generated by the coordinates of the transformed vector $\mathbf{t} \circ \mathbf{v}_0$.

Lemma 4.6.6. For any vector $\mathbf{v}_0 \in V$ the following statements are equivalent:
(a) $\det(\mathbf{t}) = t_1 t_2 \ldots t_n$ is integral over $\mathbf{C}[\mathbf{t}\mathbf{v}_0]$;
(b) $(t_1 t_2 \ldots t_n)^p$ lies in $\mathbf{C}[\mathbf{t}\mathbf{v}_0]$ for some integer $p > 0$;
(c) $\mathbf{v}_0$ does not lie in the canonical cone $\mathcal{C}_{T,\Gamma}$.

Proof. The algebra $\mathbf{C}[\mathbf{t}\mathbf{v}_0]$ is generated by monomials in $t_1, \ldots, t_n$. Another monomial, such as $t_1 t_2 \ldots t_n$, lies in the integral closure of $\mathbf{C}[\mathbf{t}\mathbf{v}_0]$ if and only if one of its powers lies in $\mathbf{C}[\mathbf{t}\mathbf{v}_0]$. To see the equivalence of (b) and (c), we note that a non-constant monomial $m \in \mathbf{C}[V]$ is T-invariant if and only if $m(\mathbf{t}\mathbf{v}) = \det(\mathbf{t})^p \cdot m(\mathbf{v})$ for some $p > 0$. Now, (b) states that there exists a monomial $m \in \mathbf{C}[V]$ such that $\det(\mathbf{t})^p = m(\mathbf{t}\mathbf{v}_0) = t_1^{i_1} t_2^{i_2} \cdots t_n^{i_n} \cdot m(\mathbf{v}_0)$. This relation implies $i_1 = p, \ldots, i_n = p, m(\mathbf{v}_0) = 1$; so (b) is equivalent to the existence of a non-constant monomial $m \in \mathbf{C}[V]^T$ with $m(\mathbf{v}_0) = 1$, and hence to (c). ◁

The two previous lemmas suggest the following more refined approach to computing invariants. The correctness of Algorithm 4.6.7 follows from Theorem 4.6.3 and Lemma 4.6.5, and the elimination property of the chosen Gröbner basis monomial order.

Algorithm 4.6.7 ("The geometric invariant theory alternative").
Input: A Γ-module V of dimension N, and a point $\mathbf{v}_0 \in V$.
Output: Either a non-constant homogeneous invariant $I \in \mathbf{C}[V]^\Gamma$ such that $I(\mathbf{v}_0) \neq 0$, or a matrix $A_0 \in \Gamma$ such that $A_0 \mathbf{v}_0$ lies in the canonical cone $\mathcal{C}_{T,\Gamma}$.

1. Introduce new variables $D, y_1, \ldots, y_N$ and compute the reduced Gröbner basis $\mathcal{G}$ for the relations

$$\det(A) - D, \ A\mathbf{v}_0 - (y_1, \ldots, y_N)$$

with respect to an elimination monomial order $\{a_{ij}\} > D > \{y_i\}$ and let $\mathcal{G}' := \mathcal{G} \cap \mathbf{C}[D, y_1, \ldots, y_N]$.
2. Does there exist a relation of the form

$$D^p + \sum_{j=0}^{p-1} f_j(y_1, \ldots, y_N) D^j \quad \text{in } \mathcal{G}'?$$

If yes, then $\det(A)$ satisfies the integral relation (4.6.8). Proceed as in (4.6.9) and (4.6.10) to compute an invariant with $I(\mathbf{v}_0) \neq 0$.
3. If no, compute the minimal generating set $\mathcal{H}'$ of $\text{Rad}(\mathcal{I}_T)$, by taking the square-free part in each monomial in the Hilbert basis $\mathcal{H}$ of $\mathbf{C}[V]^T$.
4. Compute a common zero $(A_0, D_0, b_1, \ldots, b_N)$ of $\mathcal{G} \cup \mathcal{H}$ such that $D_0 = 1$. Then $A_0 \in \Gamma$, and $A_0 \mathbf{v}_0 = (b_1, \ldots, b_N)$ lies in $\mathcal{C}_{T,\Gamma}$.

In many examples of Γ-semi-stable points $\mathbf{v}_0 \in V \setminus \mathcal{N}_\Gamma$ it happens that the relation found in step 2 equals

$$D^p - I(y_1, y_2, \ldots, y_N),$$

where I is already an invariant. It clearly satisfies $I(\mathbf{v}_0) \neq 0$, so we do not need to invoke the second part of step 3 at all. We illustrate this nice behavior in the following example.

Example 4.6.8 (Two binary cubics). Let $V = S_3 \mathbf{C}^2$, the space of cubics in two variables x_1 and x_2. We consider the following two elements in V:

$$\mathbf{v}_0 = x_1^3 - 6x_1^2 x_2 + 11 x_1 x_2^2 - 6 x_2^3 = (x_1 - x_2)(x_1 - 2x_2)(x_1 - 3x_2)$$
$$\mathbf{v}_0' = x_1^3 - 8x_1^2 x_2 + 21 x_1 x_2^2 - 18 x_2^3 = (x_1 - 2x_2)(x_1 - 3x_2)^2.$$
(4.6.11)

Let $A = \begin{pmatrix} a_{11} & a_{12} \\ a_{21} & a_{22} \end{pmatrix}$ denote a generic matrix of Γ. The relations

$$\det(A) - D \quad \text{and} \quad A \circ \mathbf{v}_0 - (y_1, y_2, y_3, y_4), \qquad (4.6.12)$$

in the polynomial ring $\mathbf{C}[a_{11}, a_{12}, a_{21}, a_{22}, D, y_1, y_2, y_3, y_4]$ are explicitly given as:

$$a_{11} a_{22} - a_{12} a_{21} - D, \quad (a_{11}^3 - 6 a_{11}^2 a_{12} + 11 a_{11} a_{12}^2 - 6 a_{12}^3) - y_1,$$
$$(3 a_{21} a_{11}^2 - 12 a_{21} a_{11} a_{12} - 6 a_{11}^2 a_{22} + 11 a_{21} a_{12}^2 + 22 a_{11} a_{22} a_{12}$$
$$- 18 a_{22} a_{12}^2) - y_2,$$

4.6. Hilbert's algorithm

$$(3a_{21}^2 a_{11} - 6a_{21}^2 a_{12} - 12a_{21}a_{11}a_{22} + 22a_{21}a_{22}a_{12} + 11a_{11}a_{22}^2$$
$$- 18a_{22}^2 a_{12}) - y_3,$$
$$(a_{21}^3 - 6a_{21}^2 a_{22} + 11a_{21}a_{22}^2 - 6a_{22}^3) - y_4. \qquad (4.6.12')$$

The analogous relations for the second binary cubic v_0' are

$$a_{11}a_{22} - a_{12}a_{21} - D, \quad (a_{11}^3 - 8a_{11}^2 a_{12} + 21a_{11}a_{12}^2 - 18a_{12}^3) - y_1,$$
$$(3a_{21}a_{11}^2 - 16a_{21}a_{11}a_{12} - 8a_{11}^2 a_{22} + 21a_{21}a_{12}^2 + 42a_{11}a_{22}a_{12}$$
$$- 54a_{22}a_{12}^2) - y_2,$$
$$(3a_{21}^2 a_{11} - 8a_{21}^2 a_{12} - 16a_{21}a_{11}a_{22} + 42a_{21}a_{22}a_{12} + 21a_{11}a_{22}^2$$
$$- 54a_{22}^2 a_{12}) - y_3,$$
$$(a_{21}^3 - 8a_{21}^2 a_{22} + 21a_{21}a_{22}^2 - 18a_{22}^3) - y_4. \qquad (4.6.13)$$

Using the monomial order specified in step 1 of Algorithm 4.6.7, we now compute Gröbner bases $\mathcal{G}$ and $\mathcal{G}'$ for (4.6.12) and (4.6.13) respectively. In $\mathcal{G}$ we find the polynomial

$$D^6 - \frac{1}{4} y_2^2 y_3^2 + y_1 y_3^3 + y_2^3 y_4 - \frac{9}{2} y_1 y_2 y_3 y_4 + \frac{27}{4} y_1^2 y_4^2.$$

This trailing polynomial in y_1, y_2, y_3, y_4 is a multiple of the *discriminant* of the binary cubic. The discriminant is an invariant of index 6, which does not vanish at v_0.

The other Gröbner basis $\mathcal{G}'$ contains no such integral dependence, so we enter step 3 in Algorithm 4.6.7. The minimal defining set of the canonical cone equals

$$\mathcal{H}' = \{x_1 x_3, x_1 x_4, x_2 x_3, x_2 x_4\}.$$

From the Gröbner basis for $\mathcal{G}' \cup \mathcal{H}'$ we can determine the following common zero:

$$D = 1, \ a_{11} = 3, \ a_{12} = 1, \ a_{21} = 2, \ a_{22} = 1, \ y_1 = 0, \ y_2 = 0, \ y_3 = 1, \ y_4 = 0.$$

This tells us that the matrix $A_0 = \begin{pmatrix} 3 & 1 \\ 2 & 1 \end{pmatrix}$ transforms v_0' into the binary cubic

$$A_0 \circ v_0' = x_1 x_2^2 \in \mathcal{C}_{T,\Gamma}. \qquad \triangleleft$$

We now come to the problem of passing from the invariants $I_1, \ldots, I_m$

to the complete system of invariants. Equivalently, we need to compute the integral closure of $\mathbf{C}[I_1, \ldots, I_m]$ in $\mathbf{C}(V)$. The second task is related to the *normalization problem* of computing the integral closure of a given domain in its field of fractions. They are not quite the same problem because the field of fractions of $\mathbf{C}[I_1, \ldots, I_m]$ is much smaller than the ambient rational function field $\mathbf{C}(V)$. The normalization is a difficult computational problem in Gröbner basis theory, but there are known algorithms due to Traverso (1986) and Vasconcelos (1991). It is our objective to describe a reduction of our problem to normalization. It would be a worthwhile research problem to analyze the methods in Traverso (1986) and Vasconcelos (1991) in the context of invariant theory. In what follows we simply call "normalization" as a subroutine.

Algorithm 4.6.9 (Completing the system of fundamental invariants).
Input: Homogeneous invariants $I_1, \ldots, I_m$ whose affine variety equals the nullcone $\mathcal{N}_\Gamma$.
Output: A generating set $\{J_1, J_2, \ldots, J_s\}$ for the invariant ring $\mathbf{C}[V]^\Gamma$ as a $\mathbf{C}$-algebra.

1. Compute the integral closure R of the domain $\mathbf{C}\big[\det(A),\, A\mathbf{v},\, I_1(\mathbf{v}), \ldots, I_m(\mathbf{v})\big]$ in its field of fractions.
2. Among the generators of R choose those generators $J_1(\mathbf{v}), J_2(\mathbf{v}), \ldots, J_s(\mathbf{v})$ which do not depend on any of the variables $A = (a_{ij})$.

The correctness of Algorithm 4.6.9 is a consequence of Corollary 4.6.2 and the following result.

Proposition 4.6.10. The invariant ring equals the following intersection of a field with a polynomial ring:

$$\mathbf{C}[V]^\Gamma = \mathbf{C}\big(\det(A),\, A\mathbf{v},\, I_1(\mathbf{v}), \ldots, I_m(\mathbf{v})\big) \cap \mathbf{C}[V].$$

Proof. The inclusion "$\subseteq$" follows from the fact that every homogeneous invariant $J(\mathbf{v})$ satisfies an identity

$$J(\mathbf{v}) = \frac{J(A\mathbf{v})}{\det(A)^p}.$$

To prove the inclusion "$\supseteq$" we consider the Γ-action on $\mathbf{C}(A, V)$ given by

$$T : A \mapsto A \cdot T^{-1},\ \mathbf{v} \mapsto T \circ \mathbf{v}.$$

The field $\mathbf{C}\big(\det(A),\, A\mathbf{v},\, I_1(\mathbf{v}), \ldots, I_m(\mathbf{v})\big)$ is contained in the fixed field $\mathbf{C}(A, V)^\Gamma$. Therefore its intersection with $\mathbf{C}[V]$ is contained in $\mathbf{C}(A, V)^\Gamma \cap \mathbf{C}[V] = \mathbf{C}[V]^\Gamma$. ◁

4.7. Degree bounds

Exercises

(1) Compute the canonical cone for the following Γ-modules. In each case give the irreducible decomposition of $\mathcal{C}_{T,\Gamma}$ into linear coordinate subspaces:
   (a) $V = S_d \mathbf{C}^2$, the space of binary d-forms
   (b) $V = S_d \mathbf{C}^3$, the space of ternary d-forms
   (c) $V = S_3 \mathbf{C}^4$, the space of quaternary cubics (Hint: see Hilbert (1893: § 19).)
   (d) $V = \wedge_2 \mathbf{C}^4$

(2) This problem concerns the action of $\Gamma = GL(\mathbf{C}^2)$ on the space of $2 \times n$-matrices $\mathbf{C}^{2 \times n}$.
   (a) Compute the canonical cone.
   (b) Compute the nullcone, using Algorithm 4.6.4.
   (c) Choose one matrix in the nullcone and one matrix outside the nullcone, and apply Algorithm 4.6.7 to each of them.
   (d) Find a system of $2n - 3$ algebraically independent bracket polynomials, which define the nullcone set-theoretically. (Hint: see Hilbert (1893: § 11).)
   (e) Apply Algorithm 4.6.9 to your set of $2n - 3$ bracket polynomials in (d).

(3) * In general, is the invariant ring $\mathbf{C}[V]^\Gamma$ generated by the images of the Hilbert basis $\mathcal{H}$ of $\mathbf{C}[V]^T$ under the Ω-process? Give a proof or a counterexample.

(4) * Compute a fundamental set of invariants for the Γ-module $V = \wedge_2 \mathbf{C}^4$.

4.7. Degree bounds

We fix a homogeneous polynomial representation (V, ρ) of the general linear group $\Gamma = GL(\mathbf{C}^n)$ having degree d and dimension $N = \dim(V)$. It is our goal to give an upper bound in terms of n, d and N for the generators of the invariant ring $\mathbf{C}[V]^\Gamma$. From this we can get bounds on the computational complexity of the algorithms in the previous section. The results and methods to be presented are drawn from Hilbert (1893) and Popov (1981).

We proceed in two steps, just like in Sect. 4.6. First we determine the complexity of computing primary invariants as in Theorem 4.6.1.

Theorem 4.7.1. There exist homogeneous invariants $I_1, \ldots, I_m$ of degree less than $n^2 (dn+1)^{n^2}$ such that the variety defined by $I_1 = \ldots = I_m = 0$ equals the nullcone $\mathcal{N}_\Gamma$.

Note that this bound does not depend on N at all. For the proof of Theorem 4.7.1 we need the following lemma.

Lemma 4.7.2. Let $f_0, f_1, \ldots, f_s$ be homogeneous polynomials of degree t in s variables $y_1, \ldots, y_s$. Then there exists an algebraic dependency $P(f_0, f_1, \ldots, f_s) = 0$, where P is a homogeneous polynomial of degree $\leq s(t+1)^{s-1}$.

Proof. Let us compute an algebraic dependency P of $f_0, f_1, \ldots, f_s$ of minimum degree r, where r is to be determined. We make an "ansatz" for P with $\binom{r+s}{s}$ indeterminate coefficients. The expression $P(f_0, f_1, \ldots, f_s)$ is a homogeneous polynomial of degree rt in s variables. Equating it to zero and collecting terms with respect to $y_1, \ldots, y_s$, we get a system of $\binom{rt+s-1}{s-1}$ linear equations for the coefficients of P. In order for this system to have a non-trivial solution, it suffices to choose r large enough so that

$$\binom{r+s}{s} = \frac{(r+1)\cdots(r+s)}{1 \cdot 2 \cdots s} \geq \frac{(rt+1)\cdots(rt+s-1)}{1 \cdot 2 \cdots (s-1)} = \binom{rt+s-1}{s-1}. \tag{4.7.1}$$

If we set $r = s(t+1)^{s-1}$, then $(r+1)^s > s(rt+s-1)^{s-1}$ and (4.7.1) is satisfied. ◁

Proof of Theorem 4.7.1. We need to show the following statement: For any $\mathbf{v}_0 \in V \setminus \mathcal{N}_\Gamma$ there exists an invariant I of degree $< n^2(dn+1)^{n^2}$ such that $I(\mathbf{v}_0) \neq 0$. We apply step 1 of Algorithm 4.6.7 and identify $y_1, \ldots, y_N$ with the coordinates of $\mathbf{A}\mathbf{v}_0$.

Let s denote the Krull dimension of $\mathbf{C}[\mathbf{A}\mathbf{v}_0] = \mathbf{C}[y_1, \ldots, y_N]$. Clearly, $s \leq n^2$. By the Noether normalization lemma, there exist s algebraically independent linear combinations $z_i = \sum_{j=1}^N \lambda_{ij} y_j$, such that $\mathbf{C}[\mathbf{A}\mathbf{v}_0]$ is integral over $\mathbf{C}[z_1, \ldots, z_s]$. By Lemma 4.6.5, $D = \det(A)$ is integral over $\mathbf{C}[\mathbf{A}\mathbf{v}_0]$, and hence it is integral over $\mathbf{C}[z_1, \ldots, z_s]$.

Each of the polynomials $D^d, z_1^n, z_2^n, \ldots, z_s^n$ is homogeneous of degree nd in the variables $A = (a_{ij})$. Since D^d is integrally dependent upon the algebraically independent polynomials $z_1^n, z_2^n, \ldots, z_s^n$, there exists a unique homogeneous affine dependency of minimum degree of the form

$$P(D^d, z_1^n, z_2^n, \ldots, z_s^n) = D^{dp} - \sum_{i=0}^{p-1} P_i(z_1^n, z_2^n, \ldots, z_s^n) D^{ip} = 0. \tag{4.7.2}$$

By Lemma 4.7.2, the degree of this relation and hence the degree of each P_i is bounded above by $s(nd+1)^{s-1} < n^2(nd+1)^{n^2}$. Applying the Ω-process as in the proof of Lemma 4.6.5, we obtain a homogeneous invariant of degree $< n^2(nd+1)^{n^2}$ which does not vanish at $\mathbf{v}_0$. ◁

In order to derive degree bounds for the fundamental invariants from Theorem 4.7.1, we first need to state a very important structural property of the invariant ring $\mathbf{C}[V]^\Gamma$, for $\Gamma = SL(\mathbf{C}^n)$.

Theorem 4.7.3 (Hochster and Roberts 1974). *The invariant ring $\mathbf{C}[V]^\Gamma$ is a Cohen–Macaulay and Gorenstein domain.*

4.7. Degree bounds

The Cohen–Macaulay property for invariants of finite groups was proved in Sect. 2.3. We refer to Hochster and Roberts (1974) or Kempf (1979) for the general proof in the case of a reductive group, such as $\Gamma = SL(\mathbb{C}^n)$. The fact that $\mathbb{C}[V]^\Gamma$ is an integral domain follows from the connectedness of Γ, similarly to the proof of Corollary 4.6.2.

What we need here is the fact that $\mathbb{C}[V]^\Gamma$ is Gorenstein. For a Cohen–Macaulay ring the Gorenstein property is equivalent to an elementary symmetry property of the Hilbert series. Recall that the Hilbert series of any finitely generated graded $\mathbb{C}$-algebra is a rational function (Atiyah and Macdonald 1969). The following theorem combines results of Stanley (1978) and Kempf (1979). The Hilbert series $H(\mathbb{C}[V]^\Gamma, z)$ is also called the *Molien series* of the Γ-module V.

Theorem 4.7.4. *The Molien series satisfies the following identity of rational functions:*

$$H\left(\mathbb{C}[V]^\Gamma, \frac{1}{z}\right) = \pm z^q \cdot H(\mathbb{C}[V]^\Gamma, z), \tag{4.7.3}$$

where q is a non-negative integer.

The fact that q is non-negative is due to Kempf (1979). Stanley (1979a) has shown that for most representations we have in fact $q \geq \dim(V)$.

Just like in the case of finite groups, one would like to precompute the Molien series $H(\mathbb{C}[V]^\Gamma, z)$ before running the algorithms in Sect. 4.6. In practise the following method works surprisingly well. As in (4.2.2) let

$$f_\rho = t_1^{i_{11}} t_2^{i_{12}} \cdots t_n^{i_{1n}} + t_1^{i_{21}} t_2^{i_{22}} \cdots t_n^{i_{2n}} + \ldots + t_1^{i_{m1}} t_2^{i_{m2}} \cdots t_n^{i_{mn}} \tag{4.7.4}$$

be the formal character of the given representation. Consider the following generating function in $t_1, \ldots, t_n$ and one new variable z.

$$\frac{\prod_{1 \leq i < j \leq n}(t_i - t_j) \prod_{i=1}^n t_i^{i-1}}{\prod_{\mu=1}^m (1 - z \cdot t_1^{i_{\mu 1}} t_2^{i_{\mu 2}} \cdots t_n^{i_{\mu n}})}. \tag{4.7.5}$$

Algorithm 4.7.5 (Precomputing the Molien series).
Input: The formal character (4.7.4) of a polynomial Γ-module V, and an integer $M \geq 0$.
Output: The truncated Molien series $H_{\leq M}(\mathbb{C}[V]^\Gamma, z) = \sum_{m=0}^M \dim_{\mathbb{C}}(\mathbb{C}[V]_m^\Gamma) z^m$.

1. Compute the Taylor series expansion of (4.7.5) with respect to z up to order M.
2. Let $P(z, t_1, \ldots, t_n)$ denote its normal form with respect to $\{t_1 t_2 \cdots t_n \to 1\}$. (This singleton is a Gröbner basis.)
3. The constant term $P(z, 0, 0, \ldots, 0)$ equals the desired truncated Molien series.

By studying the truncated Molien series for increasing M it is sometimes possible to guess (and then prove) a formula for the rational function $H(\mathbf{C}[V]^\Gamma, z)$. We present two alternative proofs of correctness for Algorithm 4.7.5.

Proof I. We multiply the both sides of the identity (4.2.7) by the Vandermonde determinant $\prod_{1 \leq i < j \leq n}(t_i - t_j)$. The truncation of the resulting formal power series equals

$$\sum_{k=0}^{\lfloor \frac{Mn}{d} \rfloor} \sum_{\lambda \vdash dk} c_\lambda \, s_\lambda(t_1, t_2, \ldots, t_n) \prod_{1 \leq i < j \leq n}(t_i - t_j) = \sum_{k=0}^{\lfloor \frac{Mn}{d} \rfloor} \sum_{\lambda \vdash dk} c_\lambda \, a_\lambda(t_1, t_2, \ldots, t_n),$$

where a_λ denotes the antisymmetrization of the monomial $t_1^{\lambda_1+n-1} t_2^{\lambda_2+n-2} \cdots t_n^{\lambda_n}$, as in Sect. 1.1. Multiply the right hand side by $\prod_{i=1}^n t_i^{i-1}$. After the reduction $t_1 t_2 \cdots t_n \to 1$ in step 1, the only terms free from $t_1, \ldots, t_n$ are those arising from partitions of the form $\lambda = (g, g, \ldots, g)$. Theorem 4.2.3 now implies the correctness of Algorithm 4.7.6. ◁

Proof II. The continuous generalization of Molien's Theorem 2.2.1 states that

$$H(\mathbf{C}[V]^\Gamma, z) = \int_{g \in \Gamma} \frac{dg}{\det(1 - zg)}, \qquad (4.7.6)$$

where dg denotes the Haar probability measure concentrated on the maximal compact subgroup of Γ. Using *Weyl's character formula* (Weyl 1926), the integral (4.7.6) can be expressed as an integral over the maximal compact torus in $T' = T \cap SL(\mathbf{C}^n) = \{\text{diag}(t_1, t_2, \ldots, t_n) : t_1 t_2 \cdots t_n = 1\}$. Denoting elements of T' by $\mathbf{t} = \text{diag}(t_1, \ldots, t_n)$, we have

$$H(\mathbf{C}[V]^\Gamma, z) = \int_{\mathbf{t} \in T'} \frac{\Delta(\mathbf{t}) \, d\mathbf{t}}{\det(1 - z\mathbf{t})}, \qquad (4.7.7)$$

where $\Delta(\mathbf{t}) = \prod_{1 \leq i < j \leq n}(1 - \frac{t_j}{t_i})$. This integral is with respect to the Haar probability measure concentrated on the maximal compact subgroup of T'. Such an integral can be evaluated formally using the following rule: the integral of any non-constant character $t_1^{i_1} t_2^{i_2} \cdots t_n^{i_n}$ over T' is zero. Noting that $\det(1 - z\mathbf{t})$ equals the denominator in (4.7.5), we find that evaluating the integral in (4.7.7) is equivalent to Algorithm 4.7.5. ◁

Example 4.7.6 (The number of invariants of a ternary quartic). We compute the Molien series of $V = S_4 \mathbf{C}^3$. The generating function (4.7.5) equals

$$\frac{(t_1 - t_2)(t_1 - t_3)(t_2 - t_3) t_2 t_3^2}{(1 - zt_3^4)(1 - zt_2 t_3^3)(1 - zt_2^2 t_3^2)(1 - zt_2^3 t_3)(1 - zt_2^4)(1 - zt_1 t_3^3) \cdots (1 - zt_1^4)}. \qquad (4.7.8)$$

4.7. Degree bounds

Using Algorithm 4.7.5 we easily compute the Molien series up to degree $M = 21$:

$$H(\mathbf{C}[V]^\Gamma, z) = 1 + z^3 + 2z^6 + 4z^9 + 7z^{12} + 11z^{15} + 19z^{18} + 29z^{21} + \ldots$$

For analyzing the ternary quartic this suggests that we start by looking for one fundamental invariant in degree 3 and another fundamental invariant in degree 6.

Let us return to our discussion of general degree bounds. The bound given in the subsequent Theorem 4.7.7 is doubly-exponential, and it is certainly not best possible. At present, it is unknown whether there exists a degree bound for Hilbert's finiteness theorem which is single-exponential in n^2, the dimension of the group Γ.

Theorem 4.7.7 (Popov 1981). *The invariant ring $\mathbf{C}[V]^\Gamma$ is generated as a $\mathbf{C}$-algebra by homogeneous invariants of degree at most $N\bigl(n^2(dn+1)^{n^2}\bigr)!$.*

Proof.
Let $I_1, \ldots, I_m$ be homogeneous polynomials of degree at most $n^2(dn+1)^{n^2}$ such that $\mathbf{C}[V]^\Gamma$ is integral over $\mathbf{C}[I_1, \ldots, I_m]$. Both algebras have the Krull dimension, say r. Clearly, $r \leq N$.

We perform a homogeneous Noether normalization: we replace $I_1, \ldots, I_m$ by pure powers $I_1^{d_1}, \ldots, I_m^{d_m}$ which are homogeneous of the same degree, say S. The integer S can be generously bounded above by the factorial $\bigl(n^2(dn+1)^{n^2}\bigr)!$. Let $\theta_1, \ldots, \theta_r$ be generic $\mathbf{C}$-linear combinations of $I_1^{d_1}, \ldots, I_m^{d_m}$. Then $\theta_1, \ldots, \theta_r$ are algebraically independent polynomials of degree S such that $\mathbf{C}[I_1, \ldots, I_m]$ is integral over $\mathbf{C}[\theta_1, \ldots, \theta_r]$.

In summary, we have found a homogeneous system of parameters $\theta_1, \ldots, \theta_r$ of the same degree $S \leq \bigl(n^2(dn+1)^{n^2}\bigr)!$ for the invariant ring $\mathbf{C}[V]^\Gamma$. By Theorem 4.7.3, the invariant ring is Cohen–Macaulay, which means there exists a Hironaka decomposition as in (2.3.1). Let $\eta_1, \eta_2, \ldots, \eta_t$ be a basis for $\mathbf{C}[V]^\Gamma$ as free $\mathbf{C}[\theta_1, \ldots, \theta_r]$-module, and let $e_1 \leq e_2 \leq \ldots \leq e_t$ denote the degrees of $\eta_1, \eta_2, \ldots, \eta_t$. It suffices to prove that $e_t \leq N \cdot S$.

By Corollary 2.3.4, the Molien series equals

$$H(\mathbf{C}[V]^\Gamma, z) = \frac{z^{e_1} + z^{e_2} + \ldots + z^{e_t}}{(1 - z^S)^r}. \tag{4.7.9}$$

From Theorem 4.7.4 we get the following identity of rational functions:

$$\frac{z^{q+e_1} + \ldots + z^{q+e_t}}{(1 - z^S)^r} = \pm \frac{z^{-e_1} + \ldots + z^{-e_t}}{(1 - z^{-S})^r} = \pm \frac{z^{rS-e_1} + \ldots + z^{rS-e_t}}{(1 - z^S)^r}.$$

Clearing denominators results in an identity of polynomials. Equating highest terms on both sides, we see that $q + e_t = rS - e_1$ and consequently

$$e_t = rS - e_1 - q \leq rS \leq N \cdot S.$$

This completes the proof of Theorem 4.7.7. ◁

Exercises

(1) Compute the unique degree three invariant of the ternary quartic. What does its vanishing mean geometrically?
(2) Prove Theorem 4.4.6, using the following steps:
    (a) Show that the invariants S and T are algebraically independent.
    (b) Show that the invariants S and T define the nullcone set-theoretically.
    (c) Using Theorem 4.7.4, give an upper bound M for the degrees in a minimal fundamental set of invariants.
    (d) Using Algorithm 4.7.5, compute the Molien series up to order M.
    (e) Conclude that $H(\mathbb{C}[V]^\Gamma, z) = \frac{1}{(1-z^4)(1-z^6)}$.

References

Atiyah, M. F., Macdonald, I. G. (1969): Introduction to commutative algebra. Addison-Wesley, Reading, Massachusetts.

Barnabei, M., Brini, A., Rota, G.-C. (1985): On the exterior calculus of invariant theory. J. Algebra 96: 120–160.

Bayer, D., Stillman, M. (1992): Computation of Hilbert functions. J. Symb. Comp. 14: 31–50.

Becker, T., Kredel, H., Weispfenning, V. (1993): Gröbner bases. Springer, Berlin Heidelberg New York Tokyo.

Bigatti, A., Caboara, M., Robbiano, L. (1991): On the computation of Hilbert–Poincaré series. Appl. Algebra Eng. Comm. Comput. 2: 21–33.

Bokowski, J., Sturmfels, B. (1989): Computational synthetic geometry. Springer, Berlin Heidelberg New York Tokyo (Lecture notes in mathematics, vol. 1355).

Buchberger, B. (1985): Gröbner bases – an algorithmic method in polynomial ideal theory. In: Bose, N. K. (ed.): Multidimensional systems theory. D. Reidel, Dordrecht, pp. 184–232.

Buchberger, B. (1988): Applications of Gröbner bases in non-linear computational geometry. In: Rice, J. R. (ed.): Mathematical aspects of scientific software. Springer, Berlin Heidelberg New York Tokyo, pp. 59–87.

Chevalley, C. (1955): Invariants of finite groups generated by reflections. Am. J. Math. 67: 778–782.

Conti, P., Traverso, C. (1991): Buchberger algorithm and integer programming. In: Mattson, H. F., Mora, T., Rao, T. R. N. (eds.): Applied algebra, algebraic algorithms and error-correcting codes. AAECC-9. Springer, Berlin Heidelberg New York Tokyo, pp. 130–139 (Lecture notes in computer science, vol. 539).

Cox, D., Little, J., O'Shea, D. (1992): Ideals, varieties and algorithms. Springer, Berlin Heidelberg New York Tokyo (Undergraduate texts in mathematics).

Dalbec, J. (1992): Straightening Euclidean invariants. Ann. Math. Artif. Intell. (to appear).

DeConcini, C., Procesi, C. (1976): A characteristic free approach to invariant theory. Adv. Math. 21: 330–354.

Désarmenien, J., Kung, J. P. S., Rota, G.-C. (1978): Invariant theory, Young bitableaux and combinatorics. Adv. Math. 27: 63–92.

Dickson, L. E. (1914): Algebraic invariants. John Wiley, New York.

Dieudonné, J., Carell, J. B. (1971): Invariant theory – old and new. Academic Press, New York.

Dixmier, J., Lazard, D. (1988): Mimimum number of fundamental invariants for the binary form of degree 7. J. Symb. Comput. 6: 113–115.

Doubilet, P., Rota, G.-C., Stein, J. (1974): On the foundations of combinatorial theory: IX. Combinatorial methods in invariant theory. Stud. Appl. Math. 53: 185–215.

Eisenbud, D., Sturmfels, B. (1992): Finding sparse regular sequences. J. Pure Appl. Algebra (to appear).

Fogarty, J. (1969): Invariant theory. W. A. Benjamin, New York.

Fulton, W., Harris, J. (1991): Representation theory. Springer, Berlin Heidelberg New York Tokyo (Graduate texts in mathematics, vol. 129).

Gaal, L. (1988): Classical Galois theory, 4th edn. Chelsea, New York.

Garsia, A., Stanton, D. (1984): Group actions on Stanley–Reisner rings and invariants of permutation groups. Adv. Math. 51: 107–201.

Gatermann, K. (1990): Symbolic solution of polynomial equation systems with symmetry. In: Proceedings of ISSAC '90 (Tokyo, Japan, August 20–24, 1990). ACM Press, New York, pp. 112–119.

Gel'fand, I. M., Kapranov, M., Zelevinsky, A. (1993): Discriminants and resultants. Birkhäuser, Boston (to appear).

Glenn, O. (1915): A treatise on the theory of invariants. Ginn and Company, Boston.

Gordan, P. (1868): Beweis, dass jede Covariante und Invariante einer binären Form eine ganze Funktion mit numerischen Coefficienten einer endlichen Anzahl solcher Formen ist. J. Reine Angew. Math. 69: 323–354.

Gordan, P. (1900): Les invariants des formes binaires. J. Math. Pures Appl. 6: 141–156.

Grace, J. H., Young, A. (1903): The algebra of invariants. Cambridge University Press, Cambridge.

Greub, W. H. (1967): Multilinear algebra. Springer, Berlin Heidelberg New York (Grundlehren der mathematischen Wissenschaften, vol. 136).

Grosshans, F. D., Rota, G.-C., Stein, J. (1987): Invariant theory and superalgebras. American Mathematical Society, Regional Conference Series 69.

Grove, L. C., Benson, C. T. (1985): Finite reflection groups, 2nd edn. Springer, Berlin Heidelberg New York Tokyo (Graduate texts in mathematics, vol. 99).

Gurevich, G. B. (1964): Foundations of the theory of algebraic invariants. Noordhoff, Groningen.

Havel, T. (1991): Some examples for the use of distances as coordinates in Euclidean geometry. J. Symb. Comput. 11: 595–618.

Hilbert, D. (1890): Über die Theorie der algebraischen Formen. Math. Ann. 36: 473–531 (also: Gesammelte Abhandlungen, vol. II, pp. 199–257, Springer, Berlin Heidelberg New York, 1970).

Hilbert, D. (1893): Über die vollen Invariantensysteme. Math. Ann. 42: 313–370 (also: Gesammelte Abhandlungen, vol. II, pp. 287–344, Springer, Berlin Heidelberg New York, 1970).

Hochster, M. (1972): Rings of invariants of tori, Cohen–Macaulay rings generated by monomials, and polytopes. Ann. Math. 96: 318–337.

Hochster, M., Eagon, J. A. (1971): Cohen–Macaulay rings, invariant theory, and the generic perfection of determinantal loci. Am. J. Math. 93: 1020–1058.

Hochster, M., Roberts, J. (1974): Rings of invariants of reductive groups acting on regular rings are Cohen–Macauley. Adv. Math. 13: 115–175.

Hodge, W. V. D., Pedoe, D. (1947): Methods of algebraic geometry. Cambridge University Press, Cambridge.

References

Huffman, W. C., Sloane, N. J. A. (1979): Most primitive groups have messy invariants. Adv. Math. 32: 118–127.

Jarić, M. V., Birman, J. L. (1977): New algorithms for the Molien function. J. Math. Phys. 18: 1456–1458.

Jouanolou, J. P. (1991): Le formalisme du résultant. Adv. Math. 90: 117–263.

Kapur, D., Madlener, K. (1989): A completion procedure for computing a canonical basis of a k-subalgebra. In: Kaltofen, E., Watt, S. (eds.): Proceedings of Computers and Mathematics 89. MIT, Cambridge, June 1989, pp. 1–11.

Kempf, G. (1979): The Hochster–Roberts theorem of invariant theory. Michigan Math. J. 26: 19–32.

Kempf, G. (1987): Computing invariants. In: Koh, S. S. (ed.): Invariant theory. Springer, Berlin Heidelberg New York Tokyo (Lecture notes in mathematics, vol. 1278).

Klein, F. (1872): Vergleichende Betrachtungen über neue geometrische Forschungen. Programm zum Eintritt in die philosophische Fakultät und den Senat der Universität zu Erlangen. A. Deichert, Erlangen.

Klein, F. (1914): Elementarmathematik vom höheren Standpunkt aus, Teil II: Geometrie. Teubner, Leipzig.

Knuth, D. E. (1970): Permutations, matrices, and generalized Young tableaux. Pacific J. Math. 34: 709–727.

Krafft, H. (1985): Geometrische Methoden in der Invariantentheorie. Vieweg, Braunschweig.

Kung, J. P. S., Rota, G.-C. (1984): The invariant theory of binary forms. Bull. Am. Math. Soc. 10: 27–85.

Kunz, E. (1985): Introduction to commutative algebra and algebraic geometry. Birkhäuser, Boston.

Landau, S. (1985): Computing Galois groups in polynomial time. J. Comput. Syst. Sci. 30: 179–208.

Lee, C. (1989): The associahedron and triangulations of the n-gon. Eur. J. Combin. 10: 551–560.

Logar, A. (1988): A computational proof of the Noether normalization lemma. In: Mora, T. (ed.): Applied algebra, algebraic algorithms, and error-correcting codes. AAECC-6. Springer, Berlin Heidelberg New York Tokyo, pp. 259–273 (Lecture notes in computer science, vol. 357).

Loos, R. (1982): Generalized polynomial remainder sequences. In: Buchberger, B., Collins, G. E., Loos, R. (eds.): Computer algebra. Symbolic and algebraic computation. Springer, Wien New York, pp. 115–137.

Macdonald, I. (1979): Symmetric functions and Hall polynomials. Oxford University Press, Oxford.

McMillen, T., White, N. (1991): The dotted straightening algorithm. J. Symb. Comput. 11: 471–482.

Meyer, W. Fr. (1892): Bericht über den gegenwärtigen Stand der Invariantentheorie. Jahresber. Dtsch. Math. Verein. 1892: 79–292.

Molien, T. (1897): Über die Invarianten der linearen Substitutionsgruppe. Sitzungsber. Königl. Preuss. Akad. Wiss.: 1152–1156.

Morgenstern, J. (1991): Invariant and geometric aspects of algebraic complexity theory I. J. Symb. Comput. 11: 455–469.

Mourrain, B. (1991): Approche effective de la théorie des invariants des groupes classiques. Ph. D. dissertation, École Polytechnique Palaiseau.

Mumford, D., Fogarty, F. (1982): Geometric invariant theory, 2nd edn. Springer, New York Berlin Heidelberg.

Nagata, M. (1959): On the 14th problem of Hilbert. Am. J. Math. 81: 766–772.

Noether, E. (1916): Der Endlichkeitssatz der Invarianten endlicher Gruppen. Math. Ann. 77: 89–92.

Pottier, L. (1992): Sub-groups of Z^n, standard bases, and linear diophantine systems. Research report 1510, I. N. R. I. A. Sophia Antipolis.

Procesi, C. (1982): A primer in invariant theory. Brandeis lecture notes 1. Brandeis University, Waltham, Massachusetts.

Popov, V. (1981): Constructive invariant theory. Astérique 87–88: 303–334.

Popov, V. (1982): The constructive theory of invariants. Math. USSR Izvest. 19: 359–376.

Reeves, A., Sturmfels, B. (1992): A note on polynomial reduction. Manuscript, Cornell University.

Reiner, V. S. (1992): Quotients of Coxeter complexes and P-partitions. Mem. AMS, vol. 95, no. 460.

Robbiano, L. (1988): Introduction to the theory of Gröbner bases. In: The Curves seminar at Queen's, vol. V. Kingston, Ontario, pp. B1–B29 (Queen's papers in pure and applied mathematics).

Robbiano, L., Sweedler, M. (1990): Subalgebra bases. In: Bruns, W., Simis, A. (eds.): Commutative algebra. Springer, Berlin Heidelberg New York Tokyo, pp. 61–87 (Lecture notes in mathematics, vol. 1430).

Rota, G.-C., Stein, J. (1976): Applications of Cayley algebras. In: Colloquio Internazionale sulle Teorie Combinatorie (1973). Accademia Nazionale dei Lincei, Roma, pp. 71–97.

Rota, G.-C., Sturmfels, B. (1990): Introduction to invariant theory in superalgebras. In: Stanton, D. (ed.): Invariant theory and tableaux. Springer, Berlin Heidelberg New York Tokyo, pp. 1–35 (The IMA volumes in mathematics and its applications, vol. 19).

Schrijver, A. (1986): Theory of linear and integer programming. Wiley-Interscience, Chichester.

Schur, I., Grunsky, H. (1968): Vorlesungen über Invariantentheorie. Springer, Berlin Heidelberg New York.

Shephard, G. C., Todd, J. A. (1954): Finite unitary reflection groups. Canad. J. Math. 6: 274–304.

Sloane, N. J. A. (1977): Error-correcting codes and invariant theory: new applications of a nineteenth-century technique. Am. Math. Monthly 84: 82–107.

Springer, T. A. (1977): Invariant theory. Springer, Berlin Heidelberg New York (Lecture notes in mathematics, vol. 585).

Stanley, R. P. (1978): Hilbert functions of graded algebras. Adv. Math. 28: 57–83.

Stanley, R. P. (1979a): Combinatorics and invariant theory. Am. Math. Soc. Proc. Symp. Pure Math. 34: 345–355.

Stanley, R. P. (1979b): Invariants of finite groups and their applications to combinatorics. Bull. Am. Math. Soc. 1: 475–511.

Stanley, R. P. (1983): Combinatorics and commutative algebra. Birkhäuser, Boston.

Stanley, R. P. (1986): Enumerative combinatorics, vol. I. Wadsworth & Brooks/Cole, Monterey, California.

Study, E. (1982): Methoden zur Theorie der ternären Formen. (Reprint of the original 1889 edition.) Springer, Berlin Heidelberg New York.

Sturmfels, B. (1991): Computational algebraic geometry of projective configurations. J. Symb. Comput. 11: 595–618.

Sturmfels, B., White, N. (1989): Gröbner bases and invariant theory. Adv. Math. 76: 245–259.

Sturmfels, B., White, N. (1990): All 11_3 and 12_3-configurations are rational. Aequ. Math. 39: 254–260.

Sturmfels, B., White, N. (1991): Computing combinatorial decompositions of rings. Combinatorica 11: 275–293.

Sturmfels, B., Whiteley, W. (1991): On the synthetic factorization of projectively invariant polynomials. J. Symb. Comput. 11: 439–454.

Sturmfels, B., Zelevinsky, A. (1993): Maximal minors and their leading terms. Adv. Math. 98: 65–112.

Traverso, C. (1986): A study of algebraic algorithms: the normalization. Rendi. Sem. Mat. Torino [Fasc. Spec. "Algebraic varieties of small dimension"]: 111–130.

Turnbull, H. W., Young, A. (1926): The linear invariants of ten quaternary quadrics. Trans. Cambridge Philos. Soc. 23: 265–301.

van der Waerden, B. L. (1971): Moderne Algebra I, 8th edn. Springer, Berlin Heidelberg New York.

Vasconcelos, W. (1991): Computing the integral closure of an affine domain. Proc. Am. Math. Soc. 113: 633–638.

Wehlau, D. L. (1991): Constructive invariant theory for tori. Manuscript, University of Toronto.

Weyl, H. (1926): Zur Darstellungstheorie und Invariantenabzählung der projektiven, der Komplex- und der Drehungsgruppe. In: Weyl, H.: Gesammelte Abhandlungen, vol. III. Springer, Berlin Heidelberg New York, pp. 1–25 (1968).

Weyl, H. (1939): The classical groups – their invariants and representations. Princeton University Press, Princeton.

White, N. (1990): Implementation of the straightening algorithm of classical invariant theory, In: Stanton, D. (ed.): Invariant theory and tableaux. Springer, Berlin Heidelberg New York Tokyo, pp. 36–45 (The IMA volumes in applied mathematics and its applications, vol. 19).

White, N. (1991): Multilinear Cayley factorization. J. Symb. Comput. 11: 421–438.

White, N., Whiteley, W. (1983): The algebraic geometry of stresses in frameworks. SIAM J. Algeb. Discrete Methods 4: 481–511.

Whiteley, W. (1991): Invariant computations for analytic projective geometry. J. Symb. Comput. 11: 549–578.

Young, A. (1928): On quantitative substitutional analysis (3rd paper). Proc. London Math. Soc. [2] 28: 255–292.

Subject index

Abelian group 66
Associahedron 132
Atomic extensor 113

Basic set 72
Binary form 118, 147, 182
Bracket 18, 77, 127
Bracket ring 78, 125, 130
Bundle condition 102

Canonical cone 179, 181
Cardano's formula 69
Catalecticant 121, 128
Cayley factorization 99, 104, 111, 115
Cayley's Ω-process 156, 167, 180
Circular straightening 131
Coding theory 34
Cohen–Macaulay 39, 51, 84
Common transversal 105, 107
Complete symmetric polynomial 12
Covariant 119, 134, 162
Cross ratio 128
Cyclic group 57, 64

Dade's algorithm 56
Desargues' theorem 99
Descent monomial 75
Diagonal monomial order 81, 90
Dihedral group 33, 36, 62, 73
Discriminant 106, 120, 161, 163, 165, 173, 183

Elemental bracket monomial 132
Elimination ideal 13
Extensor 95
Exterior algebra 37, 94
Exterior power 138
Euclidean invariants 17

Ferrers diagram 140
Final polynomial 102
First fundamental theorem 18, 85, 125
Formal character 139, 148
Fundamental system of covariants 133
Fundamental system of invariants 4, 14, 32, 68, 184

Galois theory 69
Gaussian polynomial 152
Gordan's finiteness theorem 129
Gorenstein 187
Grassmann–Cayley algebra 94, 96
Grassmann–Plücker syzygy 10, 80
Grassmann variety 79, 95
Gröbner bases 8, 23, 50, 58, 71
Grothendieck ring 140

Hermite reciprocity 153
Heron's formula 17
Hessian 121, 163
Hilbert basis theorem 11, 160
Hilbert basis (for monoids) 19, 22, 68, 168, 179
Hilbert's finiteness theorem 26, 155, 162
Hilbert's Nullstellensatz 45, 53, 61, 177
Hilbert series 29, 32, 40, 135, 154
Hironaka decomposition 39, 51, 54, 57, 69
Homogeneous system of parameters 37, 55

Index 147
Initial ideal 8
Initial monomial 3
Integer programming 19, 21
Invariant subring 14
Invariant variety 62

Subject index

Irreducible representation 140

Lexicographic order 3, 8, 50
Lie algebra 169, 180

Meet operation 96
Molien series 29, 47, 187
Multilinear Cayley expression 111

Noether normalization 37, 189
Noether's degree bound 27
Normal form 8
Normalization 184
Nullcone 53, 177

Orbit variety 16, 58

Partition 4, 153
Pascal's theorem 103
Permutation group 64
Plethysm 166
Plücker coordinates 107
Primary invariants 41, 54
Popov's degree bound 189
Power sum 4, 17

Quadrilateral set 91

Reductive group 21
Reflection group 44
Relative invariant 37, 85
Representation of $GL(n, \mathbf{C})$ 137
Resultant 121, 165
Reynolds operator 25, 43, 52, 57, 156

Robert's theorem 151

SAGBI basis 91
Schur polynomial 6, 143, 149, 166
Secondary invariants 41, 54, 65, 73
Second fundamental theorem 84
Semi-stable 179
Simple Grassmann–Cayley expression 98
Simplicial complex 84, 132
Shephard–Todd–Chevalley theorem 44
Standard monomial/tableau 8, 81, 90, 140, 142, 176
Straightening law 82, 142
Subduction algorithm 91
Symbolic method 173
Symmetric polynomial 2, 9, 15, 72, 143, 145, 166
Symmetric power 138, 161
Syzygy 14, 18, 58, 79, 134, 154

Tensor product 138
Ternary cubic 166, 171, 178
Torus 20, 171

Umbral operator 174

Van der Waerden syzygy 79

Weight enumerator 35
Weyl module 141, 148
Weyl's character formula 188

Young symmetrizer 141
Young tableau 18, 81, 140

Texts and Monographs in Symbolic Computation

Editors: B. Buchberger, G.E. Collins

This series publishes research monographs and textbooks by researchers and visiting researchers of the Research Institute for Symbolic Computation at the University of Linz, Austria. According to the basic philosophy of the institute, the emphasis of the book series lays on computer sciences based on mathematics (logic, formal methods) and on computer-assisted mathematics. An important part of the scope of the series is symbolic computation. Occasionally, also the proceedings of conferences organized by the Institute will be included in the series.

In preparation:

Wen-Tsün Wu

Mechanical Theorem Proving in Geometries

Basic Principles

Translated from Chinese by Xiaofan Jin and Dongming Wang
1993. Approx. 270 pages.
Soft cover approx. öS 560,–, DM 80,–
ISBN 3-211-82506-1

This book is a translation of Professor Wu's seminal Chinese book of 1984 on Automated Geometric Theorem Proving. The translation is done by his former student Dongming Wang jointly with Xiaofan Jin so that authenticity is guaranteed. Meanwhile, automated geometric theorem proving based on Wu's method of characteristic sets has become one of the fundamental, practically successful methods in this area that has drastically enhanced the scope of what is computationally tractable in automated theorem proving.
This book is a source book for students and researchers who want to study both the intuitive first ideas behind the method and the formal details together with many examples.

Springer-Verlag Wien New York

Sachsenplatz 4-6, P.O. Box 89, A-1201 Wien · Heidelberger Platz 3, D-14197 Berlin
175 Fifth Avenue, New York, NY 10010, USA · 37-3, Hongo 3-chome, Bunkyo-ku, Tokyo, 113, Japan